INSTRUCTOR'S MAN

UNIVERSE

FOURTH EDITION

by William J. Kaufmann III

George A. Carlson
CITRUS COLLEGE

W. H. FREEMAN AND COMPANY
NEW YORK

ISBN 0-7167-2532-0

Copyright © 1994 by W. H. Freeman and Company

No part of this book may be reproduced by any mechanical, photographic, or electronic process, or in the form of a phonographic recording, nor may it be stored in a retrieval system, transmitted, or otherwise copied for public or private use, without written permission from the publisher.

Printed in the United States of America

2 3 4 5 6 7 8 9 0   VB   9 9 8 7 6 5 4 3

CONTENTS

    Foreword   v
    A Note on the *Universe* Package   vii

Instructor's Manual

1. Astronomy and the Universe   1
    Essay—Sandra M. Faber: "Why Astronomy?"   3
2. Knowing the Heavens   10
3. Eclipses and Ancient Astronomy   15
    Essay—Owen Gingerich: "Astrology and Astronomy"   16
4. Gravitation and the Motions of Planets   20
5. The Nature of Light and Matter   34
6. Optics and Telescopes   45
7. Our Solar System   51
8. Our Living Earth   64
9. Our Barren Moon   70
10. Sun-Scorched Mercury   76
11. Cloud-Covered Venus   82
12. The Martian Invasions   87
    Essay—Marcia Neugebauer: "Exploring the Planets"   89
13. Jupiter: Lord of the Planets   93
14. The Galilean Satellites of Jupiter   98
15. The Spectacular Saturnian System   104
16. The Outer World   112
17. Interplanetary Vagabonds   120
18. Our Star   128
    Essay—Arthur B. C. Walker, Jr.: "The Solar Corona"   130
19. The Nature of Stars   139
20. The Birth of Stars   157
21. Stellar Maturity and Old Age   163
22. The Deaths of Stars   171
23. Neutron Stars   179
24. Black Holes   185
25. Our Galaxy   196
26. Galaxies   205

	Essay—Alan Dressler: "The Great Attractor"	207
27	Quasars, Blazars, and Active Galaxies	213
28	Cosmology: The Creation and Fate of the Universe	223
	Essay—Stephen W. Hawking: "The Edge of Spacetime"	225
29	Exploring the Early Universe	232
	Afterword: The Search for Extraterrestrial Life	240

RESOURCE GUIDE 242

APPENDICES

1	Addresses of Organizations and Audio-Visual Material Suppliers	381
2	References on Teaching Astronomy	382
3	Astronomy Lab Manuals	383
4	A Selected Bibliography for Sky Observing	384
5	References on Computer Software for Teaching or Learning Astronomy	387

FOREWORD

For the Instructor's Manual and Resource Guide to the fourth edition of **Universe** by William J. Kaufmann III, very much is owed to the two authors of the Instructor's Guide to the previous edition—Thomas Robertson and Andrew Fraknoi. Both were involved with many editions of Dr. Kaufmann's textbooks. Their hard work and dedication to the projects on which they worked was and is greatly appreciated. The structure and a large amount of their work remains in this edition. It seems appropriate then to quote a section from Professor Robertson's foreword to the third edition of **Universe** in which he explains some of the motivations behind the work and its structure.

"For those instructors who are teaching an astronomy course for the first time, an enormous amount of time is required in becoming familiar with the text, planning lectures, reviewing material that may not be fresh in one's mind, and learning many new areas. Often it is not obvious what difficulties the students will have with the material. In such cases it is tempting to follow the text as closely as possible. After several passes at a course, one develops a better feel for the material and the students. Different approaches to the presentation of the material are eventually developed to provide variety for the students. One feels much more comfortable in such a situation, and that feeling is usually transferred to the students. The Chapter Summary and Teaching Hints and Strategies sections of this guide were designed to make the transformation from new instructor to experienced lecturer as rapid and as efficient as possible."

Professsor Robertson's Instructor's Manual sections—the Chapter Summaries, Teaching Hints and Strategies, and Answers to Computational Questions—have been ably revised and updated by George A. Carlson. Professor Fraknoi's contribution—the Resource Guide section—was updated by Peter Scott Campbell, a reference librarian at the Queens Borough Public Library. To a large extent he relied upon Professor Fraknoi's contribution to the Research Guide for **Discovering the Universe**, which was published in 1993. The publisher and Mr. Campbell cordially acknowledge Professor Fraknoi's tremendous contributions both to this work as it stands and to astronomy reference listings in general. The structure of this work remains the same: chapter listings are broken down into Reading Materials for Instructors, Reading Materials for Students, Recent Audio-Visual Materials, and Topics for Discussions and Papers. Five useful appendices of more general sources and addresses of astronomy-related organizations and suppliers follow the chapter listings.

From George A. Carlson

Usually it is not easy to edit or revise another person's work. When I began revision of the present Instructor's Guide, I found the previous version by

Professor Thomas Robertson to be so clear and logical that I was surprised at how easily I could make the required changes. I am greatly indebted to him for this excellent baseline from which I could work. I have retained the form and style of the guide, keeping Professor Robertson's words where possible and adding my own for the new subjects covered in the textbook.

Each chapter in this guide corresponds to a chapter in the textbook. A concise overall Chapter Summary is followed by a brief section-by-section summary. Next, the Teaching Hints and Strategies section helps in identifying areas where students have difficulty and suggesting ways of making the concepts more clear. These include demonstrations, analogies, projects, and various other methods that have proven useful in my own teaching. I have checked all the computational question solutions and worked out the new ones. Since I have either written or carefully read these parts of the guide, I assume responsibility for any errors, and would appreciate having them pointed out. I would also be grateful for any suggestions or techniques that instructors have found helpful in their classes. My address is given below.

I have found working with Dr. Kaufmann and the publishers to be an enjoyable experience. Patrick Shriner and Jeff Sands in New York were extremely helpful. I have been singularly fortunate to have Dr. Allan Sandage as a mentor and friend. Thanks, Uncle Allan. I have also benefitted enormously from my long association with Dr. Jerome Kristian. Thanks, Jerry.

Dr. George A. Carlson
Citrus College
1000 W. Foothill Blvd.
Glendora, California 91741

A NOTE ON THE UNIVERSE PACKAGE

In addition to this work, a notable set of supplements has been assembled to accompany **UNIVERSE,** Fourth Edition. They offer a variety of ways to facilitate your job of teaching, and will at the same time make the course more stimulating and enjoyable for your students. As such, they should appeal to first-time instructors and veterans alike. The supplements are:

OVERHEAD TRANSPARENCIES (ISBN 0-7167-2535-5)
SLIDE SET (ISBN 0-7167-2536-3)
Both contain 100 full-color drawings from the spectacular art program in the textbook, which should greatly enhance your lectures.

TEST BANK by William J. F. Wilson and T. Alan Clark of the University of Calgary (printed version: ISBN 0-7167-2533-9), (computerized Macintosh version: 0-7167-2534-7), (computerized IBM version: 0-7167-2540-1)
This collection of 2331 multiple-choice questions are revised and supplemented from the test bank to the third edition. Each question contains a reference to the text page or pages from which it was drawn. The questions have been designed to meet the needs of a first-year course, and to provide goals and incentives for learning. The authors suggest how each question can be used by assigning either an "A" (for assignment) or a "T" (for test). The computerized versions utilize new, uncomplicated testing software that make generating tests and quizzes quite easy. The computerized versions also contain a separate section that lists the outlines for each chapter. These can be printed out to give to your students, or used as the basis for lecture notes.

ACTIVE SUN VIDEOS produced by Lockheed Research Laboratory (Video #1: 0-7167-2550-9), (Video #2: 0-7167-2551-7), (Video #3: 0-7167-2552-5)
These self-contained 10-12 minute videos highlight recent advances in high-resolution solar astronomy. Video #1 focuses on the photosphere, #2 discusses sunspots, and #3 is about the Sun's corona. They are highly compatible with the textbook's discussions of solar astronomy and are intended to be shown in class to supplement your lectures on these topics. Lockheed and W. H. Freeman are interested in hearing instructors' thoughts on possible future videos, and you are invited to write the publisher at the address below with any ideas about what topics would be useful to you.

W. H. Freeman also has two 45-minute video lectures delivered by Dr. Kaufmann. **Black Holes and Warped Spacetime** (ISBN 0-7167-2471-5) covers most of the material in Chapter 24 and could be used as a "guest lecture" for that chapter. **Cosmology and the Creation of the Universe** (ISBN: 0-7167-2486-3) can be used as an introduction to the material in Chapter 28.

A Note on the *Universe* Package

Gems of Hubble (ISBN: 0-7167-2568-1) is an Electronic PictureBook offering produced by the Space Telescope Science Institute. It contains 29 exceptional Macintosh HyperCard images recently obtained by the Hubble Space Telescope. All the necessary software accompanies the package.

W. H. Freeman is interested in any suggestions, comments, or questions you have about any of the supplements in the Kaufmann package. All comments and suggestions will be taken into consideration during the planning of future editions. Please address them to:

> W. H. Freeman and Company
> Supplements Editor
> 41 Madison Avenue
> New York, New York 10010

If you would like information on ordering titles from W. H. Freeman, please contact us at this address:

> W. H. Freeman and Company
> Sales Support
> 41 Madison Avenue
> New York, New York 10010

CHAPTER 1

Astronomy and the Universe

Chapter Summary

This chapter contains a preview of the major classes of objects to be discussed in subsequent chapters. A brief introduction to the tools, such as angular sizes and angular separations and scientific notation, used by astronomers to describe and investigate astronomical objects, is also included. Careers in astronomy are identified. The last section presents astronomy as an adventure of the human mind.

1-1 Astronomers use the laws of physics and construct testable theories to understand the universe

The use of physical laws in astronomy to construct optical instruments, to understand the physical nature of celestial objects and to explore space is discussed as justification for mastery of the material in Chapters 4 through 6. Some elements of scientific methodology are reviewed.

1-2 By exploring the planets, astronomers uncover clues about the formation of the solar system

The value of planetary exploration efforts, which are described in Chapters 7 through 17, is outlined. Some of the benefits of these efforts are a better understanding of geology, geophysics, weather and climate, as well as a more comprehensive understanding of our planet as one of many such objects.

1-3 By studying stars and nebulae, astronomers discover how stars are born, grow old, and eventually die

The properties of stars and the concept of stellar evolution and exotic examples of stages in stellar evolution, to be covered in Chapters 18 and 19 and in Chapters 20 through 24, respectively, are noted.

1-4 By observing galaxies, astronomers learn about the creation and fate of the universe

The types and distributions of stellar systems and related extragalactic objects are enumerated. A brief note is made about the ultimate value of basic research.

1-5 Astronomers use angles to denote the apparent sizes and positions of objects in the sky

The basic units of angular measure are reviewed in the context of determining the angular sizes of and angular separations between celestial objects.

1-6 Powers-of-ten notation is a useful shorthand system of writing numbers

The basic elements of expressing numbers in scientific or powers-of-ten notation are presented with several examples of decimal equivalents. The arithmetic processing of such numbers is addressed in Box 1-2.

Box 1-1 The small-angle formula

The conversion of angular sizes and angular separations into linear distances is described and an example is solved using data for the planet Jupiter.

1-7 Astronomical distances are often measured in AUs, parsecs or light-years

The traditional units used by astronomers to express distances in the solar system and in stellar systems, astronomical units, light-years and parsecs, are defined.

Box 1-2 Arithmetic with exponents

The rules for multiplying and dividing two numbers expressed as powers of ten are explained and examples are solved.

1-8 Astronomy is an adventure of the human mind

The final section of this chapter encourages readers to keep in mind the contributions of astronomy and of modern science as examples of the power of the human mind to advance our understanding of nature and to control and enrich our lives.

Box 1-3 Units of length, time, and mass

A review is provided of the basic units of measure used by scientists to express mass, time, speed and length. Metric and English equivalents are provided. The

SI system is introduced. Density is defined and the density of water is noted in both cgs and SI units.

Box 1-4   Astronomy as a profession

Some of the academic requirements for a career in astronomy are discussed. A brief overview is given of the major employment possibilities for professional astronomers.

Essay by Sandra Faber
Why Astronomy?

Essay Summary

This essay focuses on four main points of interest which might enhance student appreciation of astronomy. 1) The study of astronomy is described as an expansion of the standard general educational goal of developing an historical perspective of the place of man in history to the cosmic time scale. 2) The study of astronomy is viewed as a challenge to many of the belief systems commonly held by college students. 3) The cosmic timescale is noted in the context of evolution and the reality of perpetual change in the universe. 4) Astronomy is viewed as providing a new perspective on the nature of human existence in philosophical and religious terms.

This is an excellent opportunity to provide some incentive for students to develop an additional sense of value for the subject matter of the course. Students who take astronomy because it fulfills a general education science requirement often fail to see why they should make more than a minimal effort in a course which is far removed from practical applications. Surveys at our university show that, while faculty consider issues like those delineated in this essay to be central to general education, students rate them far below selection of a major and mastery of basic skills in relative importance.

TEACHING HINTS AND STRATEGIES

The foundation of modern astronomy is based on our understanding of the *laws of physical science* (section 1-1). Since astronomers generally can't experiment, we must assume that physical laws derived in our vicinity must also be valid elsewhere in the universe. If we did not assume that the same laws of nature apply in the far reaches of the universe, then we could not hope to make any progress in our investigations. Our assumptions about the universal applicability of these basic laws have been carefully examined and shown to be consistent with our observations.

The introduction of terminology associated with *scientific methods* (section 1-1) provides an excellent opportunity to call attention to some confusion about the use of the words hypothesis, theory and law in scientific contexts. I like to point out that the words theory and hypothesis can be used interchangeably in common usage, but that Einstein's theory of relativity is much more than just one man's guess. It can be helpful to ask students to think in terms of theories with a small "t" and with a capital "T". The difference is that a Theory has been successfully tested and found to be accurate in its ability to predict additional applications and has been verified. This distinction is helpful in combating the public impression that the Theory of Relativity and the Theory of Evolution are of diminished value because these are only "theories." One should also try to emphasize the similarities of "laws" and "Theories." It can be misleading for students to try to comprehend the relative merits of Newton's laws of motion and Einstein's theories of relativity.

While an understanding of the formation of the *solar system* (section 1-2) might be important for the astronomers involved, it has little direct appeal for more practical students. The need to better understand the changes currently taking place in our atmosphere, such as the increases in carbon dioxide levels and resulting climatic temperature changes, will certainly have a direct impact on all inhabitants of our planet in our lifetimes. Our attempts to understand the current nature and evolution of planetary atmospheres is a logical step in learning to manage our own environment.

An attempt should be made to demonstrate that modern astronomers are trying to determine the physical nature of celestial objects and the relationships between those various objects in the context of *stellar evolution* (section 1-3). Since our Sun is a star, it is in our direct interest to know how it will change with time and when it will change. The fact that it must change is very difficult for some students to accept. If a student accepts that the Sun is constantly losing energy in the form of light and heat and that the finite Sun cannot contain an infinite store of energy, then the ultimate death of the Sun is the only logical conclusion.

The investigation of *stellar systems and the universe* (section 1-4) is difficult to present in a practical context. It does provide an opportunity to discuss the differences between applied and basic science. Much of the modern technology we enjoy can be traced to advances in our understanding of the basic laws of nature. Our understanding of such fundamental laws would likely not be rapidly advanced if all scientists engaged in applied science.

The review of *angles* (section 1-5) is an excellent opportunity to emphasize that astronomy is an observational rather than an experimental science, principally because of the vast distances between celestial objects. While the space program is very important in the advance of modern astronomy, man has only visited one other celestial object (the Moon) and has landed spacecraft on only two others (Mars and Venus). Much of the space program has been devoted to using remote observatories which are still passive observers and which do not conduct active experiments beyond the solar system. Due to the vast distances between astronomical objects, their true sizes and distances are very frequently unknown.

The introduction of *scientific notation* (section 1-6) can be justified by noting that astronomy is the science of not only the big, but also the small. Call attention to the left end of Figure 1-12. It is helpful to remind the students that the observational (passive) nature of astronomical investigations means that all of the information about the physical conditions existing on celestial objects, and about their past and future conditions, must be extracted from an understanding of the nature of atoms and the constituent parts of atoms, the smallest entities of the universe.

Application of the *small angle formula* (Box 1-1) to objects having the same angular size but vastly different distances will reinforce the limitations of angular size and angular separation data.

Try to encourage your students to apply *arithmetic with exponents* (Box 1-2) to the determination of the relative orders of magnitudes of numbers.

The use of different systems of *distance units* in astronomy (section 1-7) can serve to emphasize the role of units in general in the physical sciences and to illustrate the justification of establishing different systems of units for a simple quantity such as length. Many students fail to associate units with numbers. They have dealt with numbers in mathematics courses, but have had little experience in the measuring activities which lie at the heart of the physical sciences. A brief discussion of how units give meaning to numbers can be helpful in making students aware of the units. The value of using different units of length in astronomy is that it permits the use of simple fractions and whole numbers in comparing sizes and distances which are of comparable magnitude. While any system of units can be used with powers-of-ten notation, it is not as easy to visualize the relative magnitudes of two numbers having very large or very small values. Comparisons of the sizes of and distances between objects of vastly different dimensions are more readily accomplished in powers-of-ten notation. It is useful to encourage students to view the differences between exponents in the context of order-of-magnitude comparisons. It is always a good idea to emphasize that light-years are units of length and not units of time.

It is a good idea to review what is meant by the word "size." Usually it is used in the context of length measurement. Many students confuse the basic concepts of size, area, volume, mass, weight and density. A clear presentation at this point might avoid much confusion in the future. Be sure to identify radius and diameter as length measurements related to circles and spheres which represent their sizes.

Units (Box 1-3) can be emphasized by having students identify the type of quantity expressed by dimensional analysis. The units used in the fourth edition of *Universe* are generally on the SI system rather than cgs units typically used in astrophysics and frequently used in earlier editions.

Information about *careers in astronomy* (Box 1-4) should be of interest to all students as future taxpayers and citizens. Point out that even though there are only a few professional astronomers, the observational contributions by the legions of amateurs is extensive, more so than any other science.

The concept of the *adventure of the human mind* (section 1-8) is clearly appropriate in modern astronomy. If modern astronomers are less immediately connected with applied science than were astronomers of the past, we certainly are engaged in one of the most intriguing intellectual exercises of all time. It is part of the challenge of modern astronomy education to convey to students the fact that the science of modern astronomy is involved in understanding and explaining objects as interesting as the most innovative science fiction writers can imagine. The exploration of the universe with the mind as well as with manned or unmanned spacecraft is the next logical step in the journeys of exploration which have always driven mankind to expand beyond his boundaries since before recorded history.

Answers to Chapter 1 Computational Questions

3. There are 60″ per arc minute and 60′ per degree, so
$$1° = (60 \times 60)″ = 3600″$$

9. (a) 10^7

   (b) 4×10^5

   (c) 6×10^{-2}

   (d) 1.7×10^{10}

11. Use the small-angle formula given in Box 1-1 on page 8 with
$$d = 18.687 \text{ AU}$$
$$D = 51,118 \text{ km}$$
$$D = \frac{\alpha d}{206,265}$$
$$\alpha = \frac{D(206,265)}{d} = \frac{(51,118 \text{ km})(206,265)}{(18.687 \text{ AU})(1.496 \times 10^8 \text{ km/AU})}$$
$$= \frac{(5.112)(2.06265)}{(1.862)(1.496)} = 3.8 \text{ arc sec}$$

12. (a) $2.65 \text{ pc} \times (3.09 \times 10^{13} \text{ km/pc}) = 8.19 \times 10^{13} \text{ km}$

    (b) Convert Sirius' distance from parsecs to light-years as follows:
$$2.65 \text{ pc} \times (3.26 \text{ ly/pc}) = 8.64 \text{ ly}$$

Since a light-year is the distance traveled by light in one year, the light from Sirius takes about 8.64 years to reach us.

13. $\frac{3}{4}(1.99 \times 10^{30}$ kg$) = 1.49 \times 10^{30}$ kg = the mass of hydrogen in the Sun. The Sun therefore contains

$$\frac{1.49 \times 10^{30} \text{ kg}}{1.67 \times 10^{-27} \text{ kg/atom}} = 0.892 \times 10^{30+27} \text{ atoms.}$$

$$= 8.92 \times 10^{56} \text{ hydrogen atoms.}$$

14. You must first express the Moon's diameter in seconds of arc:

Moon's diameter = 1/2 ° = 30 ′ = (30 × 60) ″ = 1800 ″

Now you can use the small-angle formula in Box 1-1 on page 8:

$$D = \frac{\alpha d}{206{,}265} = \frac{(1800)(384{,}000 \text{ km})}{206{,}265}$$

$$= \frac{(1.8)(3.84)}{2.06265} \times 10^{3+5-5} \text{ km}$$

$$= 3350 \text{ km}$$

NOTE: If you had used the more accurate observation that the Moon's angular diameter is 31′ 05″, you would have calculated the lunar diameter to be 3472 km, which is quite close to the value given in Appendix 3 in the textbook.

15. These three questions require that you rewrite the small-angle formula as follows:

$$d = 206{,}265 \left(\frac{D}{\alpha}\right)$$

(a) $\alpha = 1$ degree = 3600 arc sec and $D = 2.5$ cm, so

$$d = 206{,}265 \left(\frac{2.5}{3600}\right) = 143 \text{ cm} = 1.43 \text{ m}$$

(b) $\alpha = 1$ arc min = 60 arc sec and $D = 2.5$ cm, so

$$d = 206{,}265 \left(\frac{2.5}{60}\right) = 8594 \text{ cm} = 85.9 \text{ m}$$

(c) $\alpha = 1$ arc sec and $D = 2.5$ cm, so

$$d = 206{,}265 \left(\frac{2.5}{1}\right) = 515{,}660 \text{ cm} = 5.16 \text{ km}$$

16. The distance from the Sun to the Earth is 1 AU = 1.496×10^8 km. If

you express the Sun–Earth distance in kilometers, you must express the speed of light in km/s. The light-travel-time is 1 AU divided by the speed of light:

$$d = vt$$

$$t = \frac{d}{v} = \frac{1.496 \times 10^8 \text{ km}}{3 \times 10^5 \text{ km/s}} = 500 \text{ s} = 8.3 \text{ minutes}$$

17. You must first calculate the Earth's volume in cubic meters:

$$V = \frac{4}{3}\pi\left(\frac{1}{2} \times 1.2756 \times 10^7 \text{ m}\right)^3$$

$$= 1.087 \times 10^{21} \text{ m}^3$$

The Earth's average density is therefore:

$$\frac{5.98 \times 10^{24} \text{ kg}}{1.087 \times 10^{21} \text{ m}^3} = 5500 \text{ kg/m}^3$$

To compare this with the density of water, you must convert 1 g/cm^3 into units of kilograms per cubic meter:

$$\frac{1 \text{ g}}{\text{cm}^3}\left(\frac{1 \text{ kg}}{10^3 \text{ g}}\right)\left(\frac{10^2 \text{ cm}}{1 \text{ m}}\right)^3 = 10^{6-3} \text{ kg/m}^3 = 1000 \text{ kg/m}^3$$

Thus the average density of the Earth is 5.5 times greater than that of water.

18. You must first express the radius of the observable universe in the same units of length as the radius of the hydrogen atom:

$$(2 \times 10^{10} \text{ ly})(9.46 \times 10^{12} \text{ km/ly})(10^5 \text{ cm/km})$$

$$= 1.9 \times 10^{28} \text{ cm}$$

Dividing the radius of the observable universe by the radius of a hydrogen atom, we get:

$$\frac{19 \times 10^{27}}{5 \times 10^{-9}} = 3.8 \times 10^{36}$$

So the universe is about 4×10^{36} times bigger than a hydrogen atom.

20. Set up a proportion, but be sure that you express all the distances in the same units (e.g., centimeters). The diameter of the Sun is to the size of a basketball as the distance to Proxima Centauri is to the unknown distance (x), so

$$\frac{1.4 \times 10^{11} \text{ cm}}{30 \text{ cm}} = \frac{(4.3 \text{ ly})(9.46 \times 10^{17} \text{ cm/ly})}{x}$$

Rearranging terms we get

$$x = \frac{(4.3)(9.46 \times 10^{17})(30 \text{ cm})}{1.4 \times 10^{11}}$$

$$= \frac{(4.3)(9.46)(3.0)}{1.4} \times 10^{17+1-11} \text{ cm}$$

$$= 8.7 \times 10^{8} \text{ cm} = 8700 \text{ km}$$

In other words, if the Sun were the size of a basketball, the nearest star would be 8700 km away, which is roughly the distance from Los Angeles to Tokyo.

21. Using the information given in the previous question, you first convert the distance to Proxima Centauri to the same unit of length as the Sun's diameter. You then form a ratio by dividing the distance to Proxima Centauri by the Sun's diameter:

$$\frac{(4.3 \text{ ly})(9.46 \times 10^{17} \text{ cm/ly})}{1.4 \times 10^{11} \text{ cm}}$$

$$= \frac{(4.3)(9.46)}{1.4} \times 10^{17-11} = 2.9 \times 10^{7}$$

Thus, 29 million Suns placed side by side would reach to the nearest star.

22. Calculate how far light travels in 4 hours as follows:

$$(3 \times 10^{5} \text{ km/s})(4 \text{ hr})(3600 \text{ s/hr})$$

$$= 4.3 \times 10^{9} \text{ km}$$

23. The distance to the Moon is 384,000 km (see Question 11 or Appendix 3). Using the small-angle formula with $\alpha = 2''$, you obtain

$$D = \frac{2 \times 384,000}{206,265} \text{ km} = 3.7 \text{ km}$$

So the smallest lunar crater you can see through your telescope is about 4 km in diameter.

24. Convert 20 billion years into seconds as follows:

$$(2 \times 10^{10} \text{ yr})(3.16 \times 10^{7} \text{ s/yr}) = 6 \times 10^{17} \text{ s}$$

CHAPTER 2

KNOWING THE HEAVENS

Chapter Summary

This chapter discusses the nature of constellations and the appearance of the daytime and nighttime skies. The changes in the appearance of the sky which accompany changes in observer location on the Earth, as well as changes which occur with the passage of time are described. The physical explanations of these changes are provided in the context of the spherical shape of the Earth and its motions in space. The origins of naked-eye astronomy are noted.

2-1 Eighty-eight constellations cover the entire sky

Constellations are described as regions of the sky similar to geographic regions on the Earth. The changing patterns of stars visible in the early evening sky from one month to the next is reviewed briefly. Diurnal motion is described.

Box 2-1 Star names and catalogues

The use of Bayer designations for stars is discussed. The major catalogues used by astronomers to refer to the brightest stars and non-stellar objects are identified.

2-2 It is often convenient to imagine that the stars are located on the celestial sphere

The concept of the celestial sphere is developed. The celestial equator and celestial poles are defined. Right ascension and declination are introduced as celestial analogs of terrestrial longitude and latitude.

2-3 The seasons are caused by the tilt of the Earth's axis of rotation

The principal cause of seasons is discussed. The definitions of the vernal and autumnal equinoxes and the summer and winter solstices as epochs, or points in time, in the calendric cycle are presented. The variations of the sunrise and sunset azimuths and of the altitude of the Sun at noon with the seasons are explained for an observer in the northern hemisphere. The position of the vernal equinox as a point is defined.

Box 2-2 Celestial coordinates

The coordinates of the equatorial coordinate system, right ascension and declination are defined. The vernal equinox is defined here once again and is noted as the origin of the right ascension coordinate.

2-4 Precession is a slow, conical motion of the Earth's axis of rotation

Precession is described and the roles of the small inclination of the lunar orbit, gravitational attractions of the Sun and Moon, and the oblate shape of the Earth in producing precession are presented. The impact of precession on the apparent positions of stars over long periods of time is reviewed and the changes which it produces on the equinoxes and on the seasonal cycle are noted.

Box 2-3 Tropics and circles

The connection between the Arctic Circle, Antarctic Circle, Tropic of Cancer and Tropic of Capricorn and the orientation of and motions of the Earth in space are described.

2-5 Keeping track of time is traditionally a responsibility of astronomers

The basic elements of defining a time system, such as definitions of the meridian and transits, the differences between mean and apparent time and distinctions between solar and sidereal time are introduced. The concept of time zones is explained.

Box 2-4 Sidereal time

A more detailed treatment of the relationship between solar and sidereal time intervals is provided with explanations of the causes of these differences. The relationships between time units and angular measure are listed.

2-6 Astronomical observations led to the development of the modern calendar

A brief history of the development of our modern calendar is presented which includes an explanation of the primary function of a calendar. Descriptions are provided for some of the early calendars and their problems in accurately predicting the precise beginning of the seasonal cycle. The use of leap years to adjust the calendar periodically is described. The identification of the sidereal year as the true orbital period of the Earth is made.

12   CHAPTER 2

Teaching Hints and Strategies

This chapter covers most of that which students associate with astronomy. It provides opportunities for a number of both short and long range observing projects which might be completed on one night, or may extend over a few or several weeks or even the entire course. As many students have very poor spatial reasoning skills, such projects can be very useful and are generally well received by the students.

The nature of *constellations* is discussed (section 2-1). Students invariably confuse constellations and star clusters, so be sure to stress the differences. Stars in a cluster have common distances, ages and chemical composition while stars in a constellation are not necessarily related. An introduction to astronomical nomenclature can transform some of the mystery experienced by students when they read popular astronomy articles into a feeling of familiarity. It is helpful to relate the constellations to geographic regions such as countries and states to develop an appreciation for their value in locating celestial objects. Patterns made by bright stars are generally called "asterisms." These patterns can be only a part of a constellation (e.g., the Big Dipper is part of Ursa Major), the entire constellation (e.g., the Northern Cross contains all the bright stars of Cygnus), or involve stars from more than one constellation (e.g., the Summer Triangle).

Rotation, revolution and precession of the Earth should be defined and attributed to the observations described in these sections. A planetarium presentation can be very helpful here, if one is available. Students should be encouraged to go out at night and observe *diurnal motion* (section 2-1). The rotation of the Earth on its axis should be identified as the basis for the geographic coordinate system. A brief discussion of that system provides a natural lead-in to the celestial sphere and celestial coordinates. Rotation can be defined as motion of a body around an axis passing through the body while revolution is the motion around an axis not passing through the body.

Point out that the location of every point on the Earth's surface can be specified by only two numbers–the longitude and latitude. Since the celestial sphere (section 2-2) is an apparent spherical surface, the same system can be used there. Students are amused by the fact that all directions are due south if one is standing at the north pole.

The *causes of seasons* (section 2-3) are misunderstood by a very large number of students. The effects of the varying altitude of the Sun at noon can be demonstrated by using a flashlight held at different angles to a wall. The variation in the length of day is well known to everyone. However, many students look at elliptical orbits and conclude that seasons are caused by the varying distance of the Earth from the Sun. Be sure to point out that the Earth is closest to the Sun in January and farthest away in July. Also point out that the northern and southern hemispheres have opposite seasons. Note that the names of the equinoxes and solstices are for northern hemisphere observers.

Precession (section 2-4) can be illustrated using a simple toy top in the classroom. When the top is spinning upright, the tendency of the axis to point in a fixed direction can be seen. When the top begins to fall over due to the external gravitational force of the Earth, the resulting precession can be attributed to the compromise the top makes between falling over and not falling over. The necessity of rotation for precession is easy to show and the precise nature of the motion is obvious. Point out that the rate of spin is greater than the rate of precession for both the top and the Earth. It can be instructive to discuss what summer and winter constellations are, how they are defined and how they will appear in 13,000 years.

The discussion of *celestial coordinates* (Box 2-2) notes that right ascension is measured in hours. You might want to explain that this convention was established to make the changes in the nighttime sky more meaningful to observers. The diurnal motion of heavenly objects is due to the rotation of the Earth. Just as a watch tells us the position of the Sun in the sky, astronomers think of the celestial sphere as a clock which is keeping sidereal time. Remind students that celestial coordinates are slowly changing due to precession and thus are exactly true for only one point in time. You might provide examples of the epochs indicated on star charts or in star catalogues to reinforce this point.

A description of the astronomical origins of *tropics and circles* (Box 2-3) can be valuable in the context of causes of seasons. Point out the impact that the "land of the midnight sun" has on warm summer weather and contrast this with cold polar air masses in winter. You might also discuss the cultural importance of the date of zenith transit of the Sun for civilizations like the pre-Columbian civilizations in the western hemisphere at latitudes between the Tropics of Cancer and Capricorn.

The role of astronomers in the development of our *time systems* (section 2-5) and their connection with the Earth's rotation should be noted here as an example of an early application of astronomy.

The practical applications of *sidereal time* (Box 2-4) by astronomers should be stressed. You might discuss the problems which are encountered by the telescope operators of the IUE who work shifts determined by sidereal time rather than solar time.

Stress that the calendar modifications were done in an attempt to preserve the synchronization of the calendar to the seasons (section 2-6). It is said that when Pope Gregory XIII eliminated 10 days from the calendar, there were riots because the people felt that their lives were being shortened by those 10 days.

Answers to Chapter 2 Computational Questions

18. Procyon rises 4 minutes earlier each evening, so in two weeks it rises 14 × 4 = 56 minutes earlier, at 9:04 PM.

19. (a) They would become shorter and the solar day would become more nearly equal to the sidereal day in length.

(b) They would both become longer and they would become less similar in length.

(c) If the Earth's rotation were retrograde the solar day would be shorter than the sidereal day.

20. Sidereal time is the right ascension of any object on the meridian. The vernal equinox has a right ascension of 0^{hr}, because it is the starting place from which right ascension is measured. The meridian is 90° (= 6 hours of time) from the east point, where the vernal equinox is located. Remembering the direction in which right ascension is measured, we see that the sidereal time is 18:00.

21. Sidereal time is the right ascension of any object on the meridian. The Sun is on the meridian at noon, which is 12:00 solar time. The autumnal equinox has a right ascension of 12^{hr}, because it is directly opposite the vernal equinox. Therefore, sidereal time is nearly equal to solar time on the date of the autumnal equinox. They are exactly equal at high noon on that date (September 22).

27. Midnight corresponds to 0:00 solar time. See answer to question 21 to understand that, on the date of the autumnal equinox, sidereal time is nearly equal to solar time. The right ascension of a star on the meridian, which equals the sidereal time, is almost exactly $0^{hr}\ 00^{min}\ 00^{sec}$.

28. At noon on the date of the winter solstice, the Sun is $23\frac{1}{2}°$ south of the celestial equator. Your latitude (40°) is the angular elevation of the north celestial pole above the northern horizon. Your latitude is also the angular distance, along the meridian, from the zenith to the celestial equator. So the Sun is $40° + 23\frac{1}{2}° = 63\frac{1}{2}°$ from the zenith, which is $90° - 63\frac{1}{2}° = 26°$ above the southern horizon.

CHAPTER

3

ECLIPSES AND ANCIENT ASTRONOMY

Chapter Summary

This chapter covers the orbital motion of the Moon, lunar phases, lunar synodic and sidereal periods as well as the geometries of lunar and solar eclipses. Early attempts to determine the size of the Earth by Eratosthenes and the relative sizes and distances of the Sun and Moon by Aristarchus are described. Eclipse classification, eclipse cycles and the historical origins of the magnitude system are presented.

3-1 Lunar phases are caused by the Moon's orbital motion

The geometry of lunar phases is detailed and the differences between sidereal and synodic months are explained. The variable lengths of lunar periods are noted.

3-2 Ancient astronomers measured the size of the Earth and attempted to determine distances to the Sun and Moon

The descriptions of the methods of Eratosthenes to estimate the size of the Earth and of Aristarchus to determine the relative sizes and distances of the Sun and Moon provide an introduction to the discussion of eclipse geometry in the next section.

3-3 Eclipses occur only when the Sun and Moon are both on the line of nodes

The line of nodes is defined in terms of the orbital planes of Earth and Moon. The changing orientation of the line of nodes is mentioned briefly in this section and is explained in greater detail in Box 3-1. The concept of eclipse seasons is not mentioned explicitly, however, Figure 3-9 is well suited to an explanation of this concept and Tables 3-2 and 3-3 clearly reflect this semi-annual pattern.

Box 3-1 Some details of the Moon's orbit

The concepts of perigee and apogee are reviewed and the line of apsides defined. The regression of the line of nodes is described and its cause is identified.

3-4 Lunar eclipses can be either partial, total, or penumbral depending on the alignment of the Sun, Earth, and Moon

Eclipse classification is covered with definitions of the umbra and penumbra of the shadow of the Earth and the geometries which result in the different classes of eclipses. A table is provided which lists the most conspicuous lunar eclipses between 1993 and 1999.

3-5 Solar eclipses can be either partial, total, or annular depending on the positions of the Sun, Earth, and Moon

Solar eclipse classification is covered along with the geometries involved. The eclipse path is defined and annular eclipses are explained. Table 3-3 lists the solar eclipses from 1993 to 1999.

3-6 Ancient astronomers achieved a limited ability to predict eclipses

Historical attempts to establish some patterns to eclipse cycles are briefly reviewed. The concepts of an eclipse year and a saros are introduced and discussed.

3-7 Ancient astronomers devised star catalogues and established a brightness scale that is still used

The development of the magnitude system by Hipparchus is described and the peculiarities of the system are noted. The extensions of the magnitude system to represent the brightnesses of the brightest and faintest celestial objects are explained.

Essay by Owen Gingerich
Astrology and Astronomy

Essay Summary

The common origins of astronomy and astrology are reviewed. Some basic terms used in astrology are noted. The general thesis of astrology is reviewed and Kepler's connections to astrology are noted. The modern divergence of astronomy and astrology is detailed. The statistical research of Gauquelin is reviewed.

Many students tend to associate *astronomy* with astrology and with learning the names of constellations. Few students understand that modern astronomy is a branch of physics. It is relatively easy to involve students in a test of the predicting powers of astrology. This might involve very general horoscopes published in the paper or more detailed predictions generated by computer programs. Such an approach might achieve a more lasting impression than a lecture citing results of specific studies.

Teaching Hints and Strategies

The changing *phases of the Moon* (section 3-1) are probably among the most familiar celestial variations for students, even though their cause is not well understood. The lunar phases can be demonstrated by illuminating a sphere with a flood lamp at the front of the room. Move the students to the center of the classroom to represent observers on the Earth and move the sphere to various positions around the students to represent the positions of the Moon. The concept of time can also be discussed having the students rotate once for each passing day during this demonstration. The location of the Moon in the sky relative to the Sun during its various phases should be noted.

The demonstration of phases of the Moon can be greatly simplified if one can use a video camera to represent an observer on the Earth. The view as seen by an observer on the Earth (the camera) can be easily displayed on a monitor for the entire class to see. This same technique can be used to demonstrate the phases of planets as seen from Earth.

It is useful to ask students to keep an observation journal during the term in which they describe any observations of astronomical interest made by them. A small spiral-bound notebook is ideal for this purpose. Entries should include date, time, phenomenon observed, atmospheric condition, location in the sky, etc. When writing is required, the students force themselves to think about what they are saying in order to produce a coherent description. Also, writing helps one's memory. Moon phases are ideal subjects for the journals.

When discussing the determination of the *size of the Earth* (section 3-2) by Eratosthenes, remember that many students are surprised to learn that the ancients were aware that the Earth was spherical and not flat. The legends of Christopher Columbus sailing off the end of the Earth are hard to forget.

Colored tape can be used on the classroom walls to indicate the apparent paths of the Sun and Moon on the sky. The inclination of the lunar orbit is thus easily demonstrated. A piece of string or ribbon can be used to trace the *line of nodes* (section 3-3). Move a light source to demonstrate the apparent motion of the Sun each day (1 degree per day) and move a sphere representing the Moon (13 degrees per day). Two slide projectors at the center of the room containing slides of the Sun and Moon can add some realism to the changing positions of these celestial objects. Eclipse seasons occur only when both the Sun and Moon are lined up near the line of nodes. This occurs twice each month for the Moon but only twice each year for the Sun. Thus the Sun's apparent motion (due to the revolution of the Earth) controls the frequency of eclipse seasons.

Eclipse classification (sections 3-4 and 3-5) is often confusing for students. Remind them that eclipses are classified by naming the object which is supposed to be visible but disappears. In solar eclipses the Sun disappears. In lunar eclipses the Moon disappears. You might point out that due to the angular size of the Sun and its apparent rate of motion along the ecliptic, at least one eclipse of each type must occur during each eclipse season. If only one eclipse occurs, it will likely be a central eclipse (total or annular solar or total lunar). If more than one eclipse occurs during an eclipse season, it will probably be non-central (partial solar or partial or penumbral lunar). Be sure to emphasize that not all lunar eclipses are listed in Table 3-2. This makes the identification of the eclipse season cycle more difficult.

The *varying distance of the Moon* (Box 3-1) from the Earth is a key element in distinguishing between annular and total solar eclipses. It also produces periodic variations in the intensities of tides and in the apparent rate of motion of the Moon among the stars. You might also point out here that the strongest tides on Earth usually occur when the Moon is at perigee when it is new or full and the Earth is at perihelion (in late December and early January).

One must be careful when discussing *prehistoric astronomy* (section 3-6) not to confuse calendric applications of archaeo-astronomical sites with eclipse predicting applications. While evidence for the former is very strong, evidence of the latter is far from universally accepted. While solar and lunar cycles are easily established from relatively simple observations over periods of a few years for the Sun, and of a few decades in the case of the Moon, eclipse cycles are much more complicated and require extensive records over several decades and even centuries. Such records seem inconsistent with what we know about the constructors of many of these prehistoric sites.

Any description of the *apparent magnitude system* (section 3-7) should discuss how the seemingly logical approach of the initial system has resulted in the backwards numbering system which is so confusing to the novice. It is also instructive to point out that the planets known to the ancients were those with apparent magnitudes comparable to the twenty or so brightest stars. Additional comments and suggestions are provided in the Teaching Hints and Strategies section of this *Instructor's Guide* for Chapter 19.

Answers to Chapter 3 Computational Questions

2. Figure 3-3 is helpful in answering this question. (a) waning crescent, (b) full moon, (c) waxing crescent, provided the Sun sets before 9 PM, (d) at sunrise, (e) at noon, (f) at sunrise.

3. Remember that the Sun is on the vernal equinox on the first day of spring. You might find it helpful to consult Figure 2-9 while answering this

question. Also keep in mind that, as seen from Earth, the Moon appears to move eastward among the stars as it orbits the Earth. (a) new moon, (b) first quarter, (c) full moon, (d) third quarter.

7. There is one more sidereal month in a year than synodic months. As the Earth orbits the Sun, the Moon must travel a little more than 360° to complete a synodic month (examine Figure 3-5). In one full year, these extra increments add up to a full rotation with respect to the stars, which is equivalent to one sidereal month.

20. (a) If you just added 18 years and 11.3 days to July 11, 1991, you got the right answer: July 22, 2009. The extra third of a day means that the eclipse will occur about 120° east of Hawaii, in southeast Asia. But you should have realized that there are five leap years between 1991 and 2009, so you should have added only 18 years and 10.3 days. In crossing the international dateline to get from Hawaii to southeast Asia, however, you must advance the date by one day, again arriving at July 22, 2009.

(b) The Hawaiians will see the next eclipse of that series after three saroses. Two of these saroses have five leap years and one has four. So you should add 2(18 years 10.3 days)+(18 years 11.3 days) = 54 years 32 days to July 11, 1991. The answer therefore is August 12, 2045. This eclipse will be seen by many people in North America, because the eclipse path cuts across the United States from California to Florida.

Incidentally, the total solar eclipse of July 11, 1991 belongs to "saros cycle 136," which began on June 14, 1360, with a partial eclipse near the south pole and will end on September 11, 2694, with a similar eclipse at the north pole. This saros cycle is noteworthy because it yielded three totalities exceeding 7 minutes between 1937 and 1973. The longest occurred on June 20, 1955; it lasted 7 minutes and 8 seconds, just 23 seconds short of the greatest duration possible. The 1991 eclipse, with its 6-minute and 58-second duration, is a cousin of the 6-minute and 50-second eclipse of May 29, 1919. Both are separated from the 1955 eclipse by two saroses.

21. Basically, you need to figure out the angle that Jupiter subtended during the occultation. If you knew how fast the Moon was moving, you could multiply the Moon's angular speed by the time it took from beginning to full occultation and thereby find the angular distance the Moon had covered. This angular distance would be equal to the angular size of Jupiter.

You know it took 90 s for full occultation, but you need to compute the angular speed of the Moon along its orbit. The Moon covers 360° in one sidereal month (27^d 7^h 43^m = 2,360,591 seconds $\approx 2.3 \times 10^6$ s). So the Moon's angular speed is 360° divided by 2.36×10^6 s. The angular distance the Moon covered in 90 seconds, and therefore the angular diameter of Jupiter during the occultation is 90 s ($360° / 2.36 \times 10^6$ s) = 1.37×10^{-2} = 0.0137° $\approx$ 49 arc sec.

CHAPTER

4

GRAVITATION AND THE MOTIONS OF PLANETS

Chapter Summary

The history of the development of modern theories of motion is traced and major contributors are identified. Cosmologic theories of Ptolemy and Copernicus are described and compared. The achievements of Tycho Brahe are reviewed. Contributions of Kepler and Galileo to Newtonian mechanics are noted. Newton's laws of motion and universal gravitation are discussed and the theories of relativity are introduced.

4-1 Ancient astronomers invented geocentric cosmologies to explain planetary motions

The contributions of the ancient Greeks to early models of the universe are described and the observational basis for such theories is identified. Concepts of direct and retrograde planetary motions along the zodiac are introduced. The use of epicycles and deferents to explain the observations is reviewed.

4-2 Nicolaus Copernicus devised the first comprehensive heliocentric cosmology

The heliocentric cosmology is introduced. Characteristics of inferior and superior planets are identified and planetary configurations are defined. Planetary synodic and sidereal periods are described. The method of plotting planetary orbits developed by Copernicus, described in detail in Box 4-2, is noted briefly.

Box 4-1 Synodic and sidereal periods

The relationships between synodic and sidereal periods of inferior and superior planets are derived and sample applications are provided.

Box 4-2 Copernicus' method of determining the sizes of orbits

A geometric description is provided for the method employed by Copernicus to plot the orbits of the planets using observations made from a moving Earth.

4-3 Tycho Brahe made astronomical observations that disproved ancient ideas about the heavens

Tycho's observations of the supernova of 1572 are noted and his contributions to observational astronomy in general, and to the cosmological debate in particular, are briefly noted.

4-4 Johannes Kepler proposed elliptical paths for the planets about the Sun

Kepler's three laws of planetary motion are stated and described in a historical context. Basic elements of ellipses are presented.

Kepler's first law—The orbit of a planet about the Sun is an ellipse with the Sun at one focus.

Kepler's second law (law of equal areas)—A line joining a planet and the Sun sweeps out equal areas in equal intervals of time.

Kepler's third law (harmonic law)—The squares of the sidereal periods of the planets are proportional to the cubes of their semimajor axes. Some of the applications of Kepler's laws to orbital motions of objects other than planets are noted.

4-5 Galileo's discoveries with a telescope strongly supported a heliocentric cosmology

The importance of the telescopic observations and insightful analyses of Galileo and their relevance to the debate regarding heliocentric vs. geocentric cosmologies are stressed.

4-6 Isaac Newton formulated three laws that describe fundamental properties of physical reality

Newton's laws of motion are stated and explained in detail.

Newton's first law (law of inertia)—A body remains at rest, or moves in a straight line at a constant speed, unless acted upon by an outside force.

Newton's second law—$F = ma$

Newton's third law—Whenever one body exerts a force on a second body, the second body exerts an equal and opposite force on the first body.

4-7 Newton's description of gravity accounts for Kepler's laws and explains the motions of the planets

Newton's law of universal gravitation is stated and explained in detail.

Universal law of gravitation—Two bodies attract each other with a force that is directly proportional to the product of the masses and inversely proportional to the square of the distance between them.

$$F = \frac{Gm_1m_2}{r^2}$$

The derivation of Kepler's laws of planetary motion from Newtonian mechanics is used to demonstrate the predictive merit of his system and demonstrate the transformation from descriptive, empirical laws to more general explanations. The power and limitations of Newtonian mechanics are discussed.

Box 4-3   Ellipses and orbits

The mathematical characteristics of ellipses are presented and the concept of the center of mass is introduced. The inclusion of a mass term in Newton's derived form of Kepler's third law is discussed.

Box 4-4   The conservation laws

The concept of a conservation law is described in the context of energy, momentum and angular momentum conservation. Kinetic and potential energy are defined.

Box 4-5   Supercomputing—A new mode of science

The applications of supercomputers to problems which require very extensive calculations is summarized. Numerical solutions to the laws of physics which describe such diverse phenomena as black hole accretion discs, galactic encounters and orbit calculations in celestial mechanics are noted.

4-8   Albert Einstein's theory states that gravity affects the shape of space and the flow of time

The introduction of the theories of relativity are reviewed in the context of electromagnetic theory. The basic assumptions upon which Newtonian mechanics and relativity are based are contrasted. The fundamental predictions of special and general relativity are briefly presented.

Teaching Hints and Strategies

This chapter deals with the development of modern concepts for motion and our revised perceptions of space and time. It provides an opportunity to discuss the evolution of the nature of scientific investigation and to introduce the concept of the use of scientific models as descriptive tools. The constant change in man's understanding of the universe is very disturbing for students who just want to know what is "right." An understanding of this constant change should be a central issue in any general education science course.

As an alternative to the traditional emphasis on the scientific method, one might approach the history of science in the context of the "critical thinking" models of educational psychology presented by William G. Perry in *Forms of Intellectual and Ethical Development in the College Years* (New York; Holt, Reinhart and Winston; 1970). While many general education students consider scientific methods to be irrelevant to their lives, few would argue that the development of critical thinking skills is important for all educated persons.

The *comparison of the Ptolemaic and Copernican systems* (sections 4-1 and 4-2) provides an excellent opportunity to discuss the verification process in science. Divide the class in half and have each group adopt one of these theories and present explanations for diurnal motion, annual motions of the Sun, Moon and planets relative to the stars, and retrograde motion.

It can be instructive to point out to the students that *sidereal periods* (Box 4-1) increase with increasing distance from the Sun. *Synodic periods* are highest for planets near the Earth and decrease as distance from the Earth increases as is shown in Appendix 1 in the textbook.

To illustrate the difference in meaning between the two periods you can use the analogy of two long distance runners in a race. The faster runner (Earth) completes one lap when he/she again reaches the start line (a fixed position) having gone around the track once. This is the "sidereal period" of this runner. However, since the other slower runner (e.g., Mars) has moved in the meantime, it takes some additional time for the first runner to catch up to and gain a lap on the second. This is the "synodic period" that it takes for the runners to arrive at the same configuration (side by side) as at the start.

You might have your students apply the *method of Copernicus for determining orbit sizes* (Box 4-2) to information provided from the Astronomical Almanac to compute the sizes of the orbits of Uranus, Neptune and Pluto.

Tycho Brahe (section 4-3) is one of the most interesting characters in the history of astronomy. His personal life is in stark contrast to that of Kepler, one of his employees, and has been documented in a variety of works. It is interesting to note the vast improvement in the accuracy of celestial observations which resulted from the efforts of Tycho. One might question how modern astronomy would have fared without the meticulous detail in his observational data which made possible the discoveries of Kepler and, ultimately, the revision of mechanics by Newton.

Kepler's laws (section 4-4) require some elaboration because students are unfamiliar with ellipses. The discussion of the ellipse can be done in parallel to that of a circle (with which students are familiar). Point out that it takes two numbers to completely specify an elipse but only one for the circle. This makes the ellipse only twice as complicated as a circle. You can attach three pegs to the chalkboard around which you drape a loop of string. Use one peg as the center of a circle and the other two as the foci of the ellipse. The definitions of the circle and the ellipse involve the constancy of the loop length.

For Kepler's second law, a good analogy to use is that of a roller coaster. The top of the hill is the analog of the aphelion and the bottom (where the car moves fastest) is that of perihelion.

For Kepler's third law, have the students guess the simplest possible relation between the variables p and a. You can steer them into the equation $p = a$. The make a table on the chalkboard using the data for the Earth (i.e., $p = 1$ year, $a = 1$ AU). Ask the students to decide whether this is the correct equation since it is satisfied by the Earth numbers. After some prompting, one of them will suggest looking at the data for one of the other planets. Have a student look this up in Appendix 1 and come to the conclusion that $p = a$ is not correct. Ask them why it works for the Earth. Point out that the ultimate equation must work for the Earth as well as other planets. Ask for a function that will always work for the Earth. Eventually a student will suggest powers of p and a. Try different integer powers having a student who has a calculator do the calculation while another student looks up the data in the text. Point out that Kepler used trial and error to produce his final equation. Also point out that the constant of proportionality is unity because of the choice of units used.

Galileo's telescopic observations (section 4-5) provide an excellent example of the use of observations to test the validity of a scientific theory. The discovery of the Galilean satellites of Jupiter removed the uniqueness of the Earth as a second center of motion in the Copernican system. The changing phases and sizes of Venus were shown by Galileo to be consistent with a heliocentric, but not with a geocentric, cosmology. This should be demonstrated with a light source and sphere being careful to identify the configuration, apparent size and phase of Venus at various positions in both systems.

It is useful to describe Galileo's experiment of dropping objects from the leaning tower and discovering that they hit the ground simultaneously regardless of the difference in their mass. Point out that Galileo's method of doing science required him to perform the eqperiment despite the intuitive notion that the heavier one should strike the ground first. You can introduce the concept of inertia as a resistance to a change in motion which exactly compensates for the difference in weight. This then becomes a lead-in to Newton's first law. I give the students Galileo's statement of the hammer and feather and then show a video clip of one of the Apollo astronauts performing the experiment on the Moon literally. I then ask the class what the result would be if I dropped a hammer and feather in the classroom. They know the hammer would strike the floor first because of the effect of the air. I point out that it is not sufficient to answer the question intuitively. Finally a student will suggest actually dropping a hammer and feather and then I produce (with a fanfare) a hammer and feather from a drawer and we do the experiment several times. Of course, the hammer hits first but we then have a discussion of the results which is useful because the students understand what is happening. I point out to them at this point that despite their difficulties with mathematics, the fundamental physics is comprehensible.

Newton's laws of motion (section 4-6) require careful attention as most students cling to Aristotelian physics even after lengthy discussion. Be certain that the students appreciate the difference between velocity and acceleration. The concept of a force is relatively easy for students to grasp. However, the idea of forces acting without contact is somewhat foreign at first. They will accept this idea due to their previous exposure to gravity as the accepted explanation for falling bodies. It is helpful to review Aristotle's mechanics to present a formal structure for their general experiences. The idea that forces are required for motion to persist is very difficult to overcome.

Aristotle believed that the natural state of an object is at rest and that a moving object will always come to the natural state on its own. *Newton's first law* states that the natural state is that of uniform motion which requires an external influence to change. Aristotle's idea came from observations such as a block sliding along a table seeming to come to rest when one stops pushing it. The frictional force which causes the change is actually an external force directed against the motion. Since it is impossible to avoid the multitude of forces present near the Earth's surface, an object here never is able to exhibit the natural state. Far out into intergalactic space, external forces can be almost eliminated and uniform motion will result. It is the inertia of the object that keeps the motion uniform. This is the logical tie-in to Galileo's experiment.

Two subtle aspects of *Newton's second law* are the vector nature of forces and the concept of a net force. Accelerations always occur in the direction of the net force. All forces don't produce accelerations, only net forces do. This is an essential element in the recognition of the third law of motion.

The form of the second law can be used in a symbolic way that gives the students some insight. In $F = ma$, point out that m is another way of expressing inertia and that for a given force, some is used to overcome the inertia and the rest that is left over produces acceleration. Then, the larger the mass (inertia), the smaller the acceleration for the same force.

Newton's third law is very difficult for students to accept. They are frequently confused about the concept of net force and the equal and opposite forces required by the third law. If all forces are equal and opposite, how can one ever achieve a net force? Point out that the equal and opposite forces of the third law act on two different objects and not on a single object. It is helpful to discuss two general problems in relation to this law. The first has to do with hidden forces. Frequently, the presence of the action and/or reaction force can be masked by friction. The second has to do with mass differences. The third law says that the forces are equal in magnitude. Two objects having very different masses undergo very different accelerations. (Apply the second law.) Remember to point out that acceptance of this law requires that "no" celestial object is fixed and at the center of the universe. This is the first implication that the concept of the "center" of the universe may not be as straightforward as it once appeared.

A good example is that of a shotgun. The gun and the pellets receive the same force but since the mass of the pellets is less, their acceleration is greater. Also, by placing the stock of the gun against your shoulder, you essentially add the mass of your body to the mass of the gun. Which force is the action and which the reaction depends on whose point of view you are taking, the pellets or the gun. The important thing is that they are equal and oppositely directed.

Universal gravitation (section 4-7) can be presented as a logical extension of Kepler's laws of planetary motion and Galileo's experimental results with falling bodies. (The Moon and the apple!) The inverse-square nature of gravitational forces should be emphasized as it appears later in other contexts. The derivation of Kepler's laws is an excellent example of the ability of a successful scientific theory to explain the known and to predict the unknown. The derived forms of Kepler's laws are much more detailed than Kepler's empirical forms and are shown to be just special cases which result from the three basic laws of motion and universal gravitation.

The various factors in the law of gravitation originate from different aspects of our universe. The inverse square of the distance is simply geometry and comes from the fact that the surface area of a sphere increases as the square of the radius. The universal gravitational constant G owes its numerical value to the choice of units used. The real physics is buried in the masses and we know little if anything at all about how or why mass generates gravity. Perhaps when a quantum theory of gravity is developed we will understand this.

It is also interesting to note here that the *Newtonian Synthesis*, the use of the same laws to explain celestial and terrestrial motion, is a turning point in the development of astronomy. The ancient (Aristotelian) view held that the Earth and the heavens were governed by different rules and made of different materials. If this were the case, then our experiences in the Earthly realm could not be used to understand celestial objects. Newton explained the motion of apples and moons using the same rules. The assumption that celestial bodies are subject to the same laws of nature as the Earth is an essential ingredient in modern astronomy and removes the Earth forever from the unique position it held in ancient science.

It might be useful for you to repeat the calculation of perihelion and aphelion speeds (Box 4-3) using data for Halley's Comet. The difference in these speeds is much greater than for the Earth and illustrates the reason why the comet is visible for such a short fraction of its orbital period.

Generally, students do not know how to use their calculators when scientific notation is used. Students invariably use the multiply key rather than the exponent key when they are entering the numbers. I have them rewrite the equations explicitly replacing the "× 10" part of the numbers by the word "exp." This helps a little. Also, students are unsure when to multiply and when to divide when the equation is written as a "fraction" (i.e., numerator and denominator) because sometimes the denominator is a product of numbers. It works to say "if it's above the line, you multiply and if it's below the line, you divide."

Presentation of the *special and general theories of relativity* (section 4-8) challenge a basic assumption about science held by most students. The idea of reproducibility in science is quite fundamental. The idea that basic measured quantities such as length, time and mass are altered by relative motion of the objects with respect to the observer, and by the presence of matter seems to violate the concept of reproducibility. It must be emphasized that the laws of nature are more basic than the measurements. While the measurements can be relative, the laws of nature must be the same for all observers.

The relative merits of Einstein's and Newton's descriptions of the nature of space and time can be very misleading for students when they read about Newton's laws of motion and gravitation and Einstein's theories of relativity. Be sure to emphasize that relativity has been demonstrated to be an essential improvement in our understanding of nature. While Newton's "laws" are easier to remember and are adequate for most of our normal experiences in life, the theories of relativity are necessary to explain observations involving high rates of speed and strong gravitational fields.

Point out the irony in the fact that "theories" have superseded "laws" and that actually they are all theories. Sometimes students will say "It's only a theory." They need to realize that theories are all we have other than long list of numerical facts. Theories allow us to organize the facts in a way that makes some sense. All theories are open to revision and improvement since they are approximations to the truth. We will never know the "absolute truth" in science and if we did have such knowledge we would not be capable of realizing it.

The concepts of *conservation laws* (Box 4-4) serve to provide students with a fuller understanding of two central issues in the history of physical science. Newtonian celestial mechanics transformed Kepler's empirical laws into examples of much broader laws which govern our everyday experiences here on Earth as well as throughout the universe. One should also note the central role of these laws in our understanding of the transformation from Newtonian mechanics to relativity. The fundamental assumption upon which special relativity is based is that the laws of nature must be constant and the same for all observers, even if fundamental measured quantities cannot.

The role of *supercomputers* in modern astronomy (Box 4-5) is certainly worth noting. It is difficult to think of any area of modern astronomy which has not been affected by the applications of supercomputer models and simulations. Results of supercomputer simulations are displayed in Figures 7-18, 9-21, 9-22, 24-12, 24-13, 26-23, 26-26, 27- 31, 27-31, 27-32, and 29-15.

Answers to Chapter 4 Computational Questions

7. According to Kepler's law of equal areas, if 5.2 AU^2 is swept out in one particular year, then 5.2 AU^2 is swept out every year. So, 5.2 AU^2 is swept out in 1991, and 5×5.2 AU^2 = 26.0 AU^2 is swept out in five years.

8. Use Kepler's third law ($P^2 = a^3$) with $P = 1000$ yr.

$P^2 = 1000^2 = 10^6 = a^3$, and so $a = (10^6)^{1/3} = 10^{6/3} = 10^2 = 100$ AU

The average distance from the Sun is 100 AU. The maximum distance is 200 AU, which corresponds to the limiting case of a long, skinny elliptical orbit that grazes the Sun.

9. The average distance from the Sun is $\frac{1}{2}(0.5 + 3.5) = 2$ AU. Using Kepler's third law

$$P^2 = a^3 = 2^3 = 8 \text{ yr}^2, \text{ so } P = \sqrt{8} = 2.8 \text{ yr}$$

12. $F = G \dfrac{m_1 m_2}{r^2}$

$= 6.67 \times 10^{-11} \left(\dfrac{(5.98 \times 10^{24})(7.35 \times 10^{22})}{(3.844 \times 10^5 \times 10^3)^2} \right)$ N

$= 6.67 \times 10^{-11} \left(\dfrac{43.95 \times 10^{46}}{14.78 \times 10^{16}} \right)$ N

$= 6.67 \times 10^{-11} \times 2.974 \times 10^{30}$ N

$= 19.8 \times 10^{19}$ N

$\approx 2 \times 10^{20}$ N

As shown on page 69, the force between the Earth and the Sun is 3.53×10^{22} N. Thus the force between the Earth and the Moon is about $3.53 \times 10^{22} \div 2 \times 10^{20} \approx 180$ times weaker than the force between the Earth and the Sun.

13. The gravitational force is inversely proportional to the square of the distance, so the gravitational pull of the Sun on the Earth would be reduced by a factor of $10^2 = 100$.

18. Synodic period = 115.88 days $\left(\dfrac{1 \text{ yr}}{365.25 \text{ days}} \right)$ = 0.317 yr

From Box 4-1 on page 61 we have:

$$\frac{1}{P} = \frac{1}{E} + \frac{1}{S}$$

$$= 1 + \frac{1}{0.317} = 4.15 = \frac{1}{0.24}$$

The sidereal period of Mercury is therefore about a quarter of a year.

19. Refer to Box 4-1 and set $P = S$ and $E = 1$ yr. If you substitute these values into the equation for an inferior planet, you get the bizarre result that

$$\frac{1}{S} = 1 + \frac{1}{S}$$

There is no non-zero value for S that can satisfy this equation, because no finite number equals itself plus one. Thus it is impossible for an inferior planet to have equal sidereal and synodic periods.

Substituting $P = S$ and $E = 1$ yr into the equation for a superior planet we obtain:

$$\frac{1}{S} = 1 - \frac{1}{S}$$

Multiplying through by S and then solving for S, we get

$$1 = S - 1$$

$$S = 2 \text{ yr}$$

Referring to Appendix 1 at the back of the textbook we find that the sidereal period of Mars is 1.88 years and its synodic period is 779.87 days, which equals 2.14 years. So Mars is close to having its sidereal period equal its synodic period.

20. (a) The easiest way to do this problem is to realize that both the Moon and the satellite obey Kepler's third law as they orbit the Earth.

Referring to Appendix 3 at the back of the textbook, we find that the Moon's orbital period (P_m) is 27.322 days and its average distance from Earth (a_m) is 384,404 km. Let P_s be the satellite's orbital period and a_s be the semimajor axis of its orbit. We can then write the proportionality

$$\frac{P_m^2}{P_s^2} = \frac{a_m^3}{a_s^3}$$

which is Kepler's law for the Moon divided by Kepler's law for the satellite. Note that the constants in the two versions of Kepler's law are approximately equal and so have cancelled.

Substituting in the lunar data and the requirement that $P_s = 1$ day, we find

$$\frac{(27.322 \text{ days})^2}{(1 \text{ day})^2} = \frac{(384{,}404 \text{ km})^3}{a_s^3}$$

$$a_s^3 = \frac{(384{,}404)^3}{(27.322)^2}$$

By taking the cube root of both sides, we get

$$a_s = \frac{384{,}404}{(27.332)^{2/3}} = \frac{384{,}404}{(747.04)^{1/3}} = \frac{384{,}404}{9.0736} = 42{,}365 \text{ km}$$

which must be the satellite's average distance from the Earth's center.

(b) The center of the Earth must lie at the center of the satellite's orbit. Thus if the satellite were north of the equator at some point of its orbit, it would have to be south of the equator one-half orbit later. Such an orbit cannot be geosynchronous.

Furthermore, the orbit must be directly over the equator so that the orbit is least affected by asymmetries in the Earth's shape such as the equatorial bulge. If the orbit were inclined to the Earth's equator, the gravitational pull of the equatorial bulge would cause the orbit to precess. In that case, the sub-satellite point would drift very slowly back and forth across the equator in a north–south direction.

21. The semimajor axis of Pluto's orbit is:

$$a = 39.529 \text{ AU} = 5.914 \times 10^9 \text{ km}$$

The orbital period is:

$$P = 248.6 \text{ yr} \left(\frac{3.156 \times 10^7 \text{ s}}{1 \text{ yr}} \right) = 7.846 \times 10^9 \text{ s}$$

$$\frac{2\pi a}{P} = \frac{2(3.142)(5.914 \times 10^9)}{7.846 \times 10^9} = 4.74 \text{ km/s}$$

Using the equations in Box 4-3:

At perihelion

$$v = 4.74 \left(\frac{1+e}{1-e} \right)^{1/2} = 4.74 \left(\frac{1+0.248}{1-0.248} \right)^{1/2}$$

$$= 4.74 \left(\frac{1.248}{0.752} \right)^{1/2} = 4.74 \sqrt{1.650} = 6.11 \text{ km/s}$$

At aphelion

$$v = 4.74\left(\frac{1-e}{1+e}\right)^{1/2} = 4.74\left(\frac{1-0.248}{1+0.248}\right)^{1/2}$$

$$= 4.74\left(\frac{0.752}{1.248}\right)^{1/2} = 4.74\sqrt{0.6026} = 3.68 \text{ km/s}$$

22. If the orbital period (P) is measured in years, and the length of the semimajor axis (a) is measured in AUs, then the last equation in Box 4-3 may be written as:

$$P^2 = \frac{a^3}{M}$$

where M is measured in solar masses ($M_\odot$). So, if $P = 2$ yr and $a = 4$ AU, then

$$M = \frac{a^3}{P^2} = \frac{64}{4} = 16 \text{ M}_\odot$$

The star is 16 times more massive than the Sun.

23. From Box 4-4, we have the relation:

$$V_{escape} = \sqrt{\frac{2GM}{R}}$$

where M is the mass of the planet, R is the radius of the planet, and $G = 6.67 \times 10^{-11}$ newton-m^2/kg^2.

(a) For the Moon (Appendix 3 and Box 9-1):

$$M = 7.35 \times 10^{22} \text{ kg}$$

and

$$R = 1738 \text{ km} = 1.738 \times 10^6 \text{ m}$$

$$V_{escape} = \sqrt{\frac{2(6.67 \times 10^{-11})(7.35 \times 10^{22})}{(1.738 \times 10^6)}} = 2375 \text{ m/s}$$

or 2.375 km/s

(b) For Jupiter (Box 13-1):

$$M = 1.90 \times 10^{27} \text{ kg}$$

and

$$R = \tfrac{1}{2}D_{average} = \tfrac{1}{2}\left(\frac{142{,}800 + 133{,}500}{2}\right) = 6.91 \times 10^4 \text{ km}$$

$$E = 6.91 \times 10^7 \text{ m}$$

$$V_{escape} = \sqrt{\frac{2(6.67 \times 10^{-11})(1.90 \times 10^{27})}{(6.91 \times 10^7)}} = 6.06 \times 10^4 \text{ m/s}$$

or 60.6 km/s

(c) For the Sun (Box 18-1):

$$M = 1.99 \times 10^{30} \text{ kg}$$

and

$$R = 6.96 \times 10^5 \text{ km} = 6.96 \times 10^8 \text{ m}$$

$$V_{escape} = \sqrt{\frac{2(6.67 \times 10^{-11})(1.99 \times 10^{30})}{(6.96 \times 10^8)}} = 6.18 \times 10^5 \text{ m/s}$$

or 618 km/s

Physical reason: The gravitational force that an object needs to overcome in order to escape is determined more by the nearby mass than by the more remote mass because of the inverse square nature of the force. The Sun is much more massive than Jupiter but most of the Sun's mass is farther away from the escaping object than all of Jupiter's mass would be. Therefore, the escape velocity is less than the total mass number would suggest.

Mathematical reason: In the equation, M is larger for the Sun than for Jupiter, but so is R; and since you divide by R, part of the increase due to M is cancelled.

24. The mass is unchanged, but the distance is doubled. The gravitational force is inversely proportional to the square of the distance, so your weight would be reduced by a factor of $2^2 = 4$. You would weigh one-quarter of your Earth weight on that planet.

25. From Appendix 2 in the textbook we see that:

Mars mass = 0.107 × Earth mass

Mars radius = 0.53 × Earth radius

A decrease in planet mass would tend to decrease your weight, while a decrease in planet size would tend to increase your weight. These two effects work together according to the universal law of gravitation: your weight is proportional to the planet's mass and inversely proportional to the square of the planet's radius. Thus your weight would change by a factor of

$$\frac{0.107}{(0.53)^2} = 0.38$$

On Mars you would weigh 38 percent of your Earth weight, so astronomers say that the *surface gravity* on Mars is 0.38. Similar calculations for the other planets give the values listed in the "Surface gravity" column in Appendix 2 in the textbook.

CHAPTER 5

THE NATURE OF LIGHT AND MATTER

Chapter Summary

This chapter presents descriptions of the wave and particle models of light, the electromagnetic spectrum, blackbody radiation laws and temperature systems as well as discussions of the principles of spectral analysis and modern concepts of atomic structure. It concludes with an explanation of the Doppler effect and a brief note about stellar radial and tangential velocities.

5-1 Light travels through empty space at a speed of 300,000 km/s

The experimental methods and values of the speed of light provided by Galileo, Roemer, Fizeau and Foucault are discussed. The speed of light in a vacuum is contrasted in a qualitative sense with its speed in other materials.

5-2 Light is electromagnetic radiation and is characterized by its wavelength

The wave model of light is developed with reference to the experimental evidence provided by Newton and to the interference experiments by Young. The contribution of Maxwell to our modern understanding of electromagnetic radiation is noted. The regions of the electromagnetic spectrum are listed and discussed in terms of wavelength and examples of each region are supplied. The mathematical relationship involving wavelength, frequency and the speed of light is discussed.

Box 5-1 Temperatures and temperature scales

The Celsius, Kelvin, and Fahrenheit temperature systems are contrasted and the respective freezing and boiling points of water and a few other materials are noted. Transformation relations are derived which permit conversions between the Celsius and Fahrenheit systems.

5-3 An object emits electromagnetic radiation according to its temperature

The concepts of energy units, temperature and absolute zero are introduced and discussed in the context of the Stefan–Boltzmann law and Wien's law.

Box 5-2 The Sun's luminosity and surface temperature

The concept of luminosity is introduced and the solar constant is defined. The Stefan–Boltzmann law is used to compute the surface temperature for the Sun from luminosity and size measurements.

Box 5-3 Applications of the radiation laws

Examples are given for finding the temperature of the Sun from Wien's law, the dominant wavelength (l_{max}) for Sirius, and the flux ratio for Sirius compared to the Sun using the Stefan–Boltzmann law.

5-4 In a full explanation of blackbody radiation, light is assumed to have particlelike properties

The particle model of light is examined in the context of the photoelectric effect and Planck's law. The relationship involving energy per photon, frequency and wavelength is noted and Planck's constant is identified.

5-5 Each chemical element produces its own unique set of spectral lines

The introduction of the concepts of spectral lines and chemical elements leads to a discussion of Kirchoff's laws of spectral analysis. The concept of the atom is introduced.

Box 5-4 Photon energy and the electron volt

The concept of the amount of energy per photon is expanded and typical energies for each region of the electromagnetic spectrum are computed and displayed with corresponding wavelengths and temperatures computed from Wien's law.

5-6 An atom consists of a small, dense nucleus surrounded by electrons

The Rutherford model of the atom is developed and the three subatomic particles are identified.

5-7 Spectral lines are produced when an electron jumps from one energy level to another within an atom

The Bohr model of the hydrogen atom is introduced as having certain allowed orbits. Photons absorbed or emitted correspond to electron transitions. The photon wavelengths are in agreement with Balmer's formula.

Box 5-5 Atomic structure and energy-level diagrams

The discrete energy level structure of the hydrogen atom proposed in the Bohr model of the atom is examined in the context of the Balmer, Lyman and Paschen spectral line series. Spectral lines are described as resulting from electron transitions. Ionization is defined.

5-8 The wavelength of a spectral line is affected by the relative motion between the source and the observer

The Doppler effect is described and examples are provided of its application to the determination of the radial velocity of a star. The concept of transverse or tangential velocity is introduced.

Teaching Hints and Strategies

This chapter presents radiation laws and principles of spectral analysis and modern concepts of atomic structure. It provides the necessary background for an understanding of modern astrophysics and can be used to expand on the use of models in modern science.

The transition from Chapter 4 can be affected by discussing the experimental determination of the constancy of the *speed of light* (section 5-1) by Michelson and Morley and its implications in terms of relativity. The element common to all forms of radiation is the speed of light in a vacuum.

The need for more than one model of light provides some insight into the limitations inherent in the use of physical models to help us to understand and explain unusual phenomena in science. The quantitative aspects of the *wave model of light* (section 5-2) should be noted at this point. A demonstration of Young's double slit experiment or diffraction of a laser beam as it passes a human hair is very useful in demonstrating the wave nature of light. It is helpful to describe constructive and destructive interference of waves in some detail at this point.

The role of *temperature* (section 5-3) in the blackbody radiation laws should be given special emphasis here. These laws are the astronomer's link to the temperatures of celestial objects. It can be informative to illustrate the nature of the Stefan–Boltzmann law by comparing the flux emitted by three stars having the same size but three different typical temperatures. Use 3000, 6000 and 30,000 kelvins to represent low, medium (the Sun) and high temperature stars. Use relative units to eliminate the constants and keep the algebra simple. Be sure to point out that these laws work reasonably well for planets also. You might compare the Earth (300 kelvins) with the three stars above in terms of total emitted flux.

Since examples are also worked out in the textbook in Box 5-3, it is a good idea to use the same solution format in the class examples. Students at this level are more comfortable if they have a set pattern to follow. Have them work the problems at the end of the chapter using this format.

The concept of the *kinetic definition of temperature* (Box 5-1) is important later in discussing gas pressure and atmospheric retention. The need for an absolute temperature system is imperative when attempting to compare two temperatures or when making calculations involving temperature. Be sure to stress that, although all temperature systems are equally accurate, only an absolute temperature system can be used for calculations without modification for zero point differences.

The *temperature and luminosity of the Sun* (Box 5-2) can lead to a discussion of the effects of different temperatures for stars on their planetary systems. What would happen to the energy reaching the Earth if the Sun were to become 10% hotter or cooler?

Most students find it confusing that neither the wave nor the *particle model of light* (section 5-4) is "right." Be sure to point out that our conceptual models are just analogies to permit easier understanding, but which often are limited. Light possesses both wavelike and particlelike properties so both models are correct but neither is complete. Light will manifest itself as particles or waves depending on which of these is being revealed by the particular apparatus being used. In other words, light knows how to behave but we have trouble describing it. The general concept of the photoelectric effect can be illustrated with a photocell, a galvanometer and a flashlight.

Kirchoff's laws of spectral analysis (section 5-5) are empirical in origin. The observation of a continuous spectrum indicates the presence of a luminous or incandescent solid, liquid or dense gas. Note that the radiation laws describe quantitatively the radiation from an ideal object of this type. A few moments should be spent in clarifying what a dense gas is and contrast it to a low-density or rarefied gas. Some mention should be made about the relationship between density and pressure as well. Be certain that students understand that spectral lines can only be produced in low density or rarefied gases. Many students think that spectral lines are images of something on the light source. They should be reminded that the lines are in fact images of the slit in the spectrograph at wavelengths which are more, or less, intense than the continuum. The lines in spectra are unique for each ionization state of each type of atom and, therefore, can be used to identify the types of atoms in low-density gases. Always remind students that the absence of lines of a certain element does not necessarily indicate that the element is not there, as different physical conditions can also affect the line intensities.

A classroom demonstration is very useful at this point. Inexpensive diffraction gratings and spectrum tubes along with a power supply can be obtained from any of several vendors of laboratory equipment. Hand out the gratings and then hold into view an incandescent light bulb followed by several spectrum tubes. Especially important here is the hydrogen tube. Ask the class to describe their observations in as much detail as possible and then make conclusions such as:

1. the light bulb produces a continuous spectrum

2. the filament of the bulb, being solid, is approximating a dense gas

3. all gas tubes produce bright line spectra

4. the pattern of lines is different for each gas

5. the pattern for each element observed is unique

6. the pattern for elements not observed is probably unique.

It is not possible to demonstrate a dark line spectrum in the classroom but the results of a hypothetical demonstration can be given. Be sure that the students understand that the same lines would be present for a given element as they saw in the bright-line spectrum. An analogy to a photographic print and its negative is useful here. Point out that both contain the same information but the appearance is reversed.

The use of *energy units for photons* (Box 5-3) and conversions between wavelength, frequency and energy per photon are important in the education of the student interested in X-ray and gamma ray astronomy. Remind the students that while optical astronomers typically refer to wavelengths, radio astronomers prefer frequencies, and X-ray and gamma ray astronomers usually indicate the energy per photon.

Clearly our understanding of the nature of light was only possible in conjunction with a revolution in our knowledge about the nature of matter. The early Rutherford model of the *structure of the atom* (section 5-6) is easily envisioned as a tiny solar system. It is useful to exploit this physical model which is readily adopted by most students to point out the limitations of physical models. An introduction to the electrical force here is a natural to indicate that gravitation is not the only force which acts at a distance and that it cannot explain the stability of atoms. This is also an excellent time to apply Newton's laws of motion to the model of the stable atom to indicate their application in the microscopic realm as well as the macroscopic realm. Note the relative strengths of the electric and gravitational forces.

The energy changes of an atom accompanied by *electron transitions* (section 5-7) should be clearly indicated as the cause of emission or absorption of photons (light) by matter. The quantization of the orbits in the theory is imposed by the observation of line spectra. Without quantized orbits, energy could be absorbed or emitted gradually and continuously and some kind of continuous spectrum would result. The line spectra are telling us that the orbits (or energy levels) are analogous to a staircase. It is impossible to exist except on a step. Sometimes students feel that in a theory, we are trying to impose the rules on nature. They need to understand that in a theory, we are only attempting to describe the rules that nature follows even in the absence of human interpreters. The application of quantum theory in atomic structure to explain spectral lines may be a little bit easier for students to grasp than the statistical predictions of Planck's law. The limitations of the solar system model of atomic structure are obvious at this point.

A small buzzer attached to the end of a string and whirled above the head can provide an excellent demonstration of the *Doppler effect* (section 5-8) Have students note the tone or frequency of the buzzer when at rest to that when it is only moving laterally and not along the line of sight. The concepts of tangential and radial velocities are treated in greater detail in section 19-1 of the textbook.

An interesting conclusion to this subject might be a discussion of the wave nature of the elementary particles of which material objects are made. Most students are willing to accept the progressive reduction in size of the particles of which matter is made, but have difficulty accepting the fact that these "particles" have wave properties. The realization that electrons, protons and neutrons have both wave and particle properties makes the "conversion of matter to energy" concept a little more acceptable.

Ask the students why they think the wave nature of matter is not apparent except on the atomic scale. Then work out the wavelength for a baseball traveling at 100 mph (a Nolan Ryan fastball). Point out that the resulting extreme smallness of the wavelengths of macroscopic matter makes the wave nature unobservable.

Answers to Chapter 5 Computational Questions

1. The Earth's circumference is π times its diameter (listed in Appendix 2 in the textbook), so the total distance around the Earth is

$$3.1416 \times 12{,}756 \text{ km} \approx 40{,}000 \text{ km}$$

Light travels 300,000 km in one second, which corresponds to 300,000/40,000 = $7\frac{1}{2}$ trips around the world.

5. $\nu\lambda = c$

$$c = 3 \times 10^8 \text{ m/s}$$

$$\lambda = 825 \text{ nm} = 8.25 \times 10^{-7} \text{ m}$$

$$\nu = \frac{c}{\lambda} = \frac{3 \times 10^8 \text{ m/s}}{8.25 \times 10^{-7} \text{ m}} = 0.364 \times 10^{15} \text{ Hz}$$

$$= 3.64 \times 10^{14} \text{ Hz}$$

6. $\nu = 103.3 \text{ MHz} = 1.033 \times 10^8 \text{ Hz}$

$$\lambda = \frac{c}{\nu} = \frac{3 \times 10^8 \text{ m/s}}{1.033 \times 10^8 \text{ Hz}} = 2.90 \text{ m}$$

8. Wien's law states that:

$$\lambda_{max} T = 2.9 \times 10^{-3} \text{ mK}$$

So,

$$\lambda_{max} = \frac{2.9 \times 10^{-3} \text{ mK}}{12,200 \text{ K}} = \frac{2.9}{1.22} \times 10^{-3-4} \text{ m}$$

$$= 2.38 \times 10^{-7} \text{ m}$$

$$= 238 \text{ nm}$$

9. Wien's law states that:

$$\lambda_{max} T = 2.9 \times 10^{-3} \text{ mK}$$

So,

$$T = \frac{2.9 \times 10^{-3} \text{ mK}}{\lambda_{max}}$$

$$\lambda_{max} = 445 \text{ nm} = 4.45 \times 10^{-7} \text{ m}$$

Thus, the surface temperature of Procyon is :

$$T = \frac{2.9 \times 10^{-3} \text{ mK}}{4.45 \times 10^{-7} \text{ m}} = 0.652 \times 10^{7-3} \text{ K}$$

$$= 6520 \text{ K}$$

10. Assume the book is at room temperature, about 295 K. Using Wien's law we obtain

The Nature of Light and Matter 41

$$\lambda_{max} = \frac{2.9 \times 10^{-3} \text{ mK}}{T}$$

$$= \frac{2.9 \times 10^{-3} \text{ mK}}{295 \text{ K}} \approx 10^{-5} \text{ m}$$

$$= 10 \text{ μm}$$

18. The Sun's surface temperature is

$$T_\odot = 5800 \text{ K}$$

First you must convert to degrees Celsius by subtracting 273:

$$T_\odot = (5800 - 273)°C$$

$$= 5527°C$$

To convert to Fahrenheit, use the equation given in Box 5-1 on page 87:

$$T_F = \frac{9}{5} T_C + 32$$

$$= \frac{9}{5}(5527) + 32$$

$$= 9949 + 32 = 9981°F$$

$$\approx 10{,}000°F$$

19. You must first convert 98.6 °F into kelvin. Your body temperature in degrees Celsius is:

$$T_C = \frac{5}{9}(T_F - 32)$$

$$= \frac{5}{9}(98.6 - 32) = \frac{5}{9}(66.6)$$

$$= 37°C$$

So your body temperature in kelvin is 37 + 273 = 310 K

Using Wien's law, you can now calculate the wavelength of maximum intensity (λ_{max}):

$$\lambda_{max} = \frac{2.9 \times 10^{-3} \text{ mK}}{T}$$

$$= \frac{2.9 \times 10^{-3} \text{ mK}}{310 \text{ K}} = 9.35 \times 10^{-6} \text{ m}$$

$$= 935 \text{ μm}$$

This is infrared radiation.

20. The luminosity of a star (L_*) is equal to the energy flux from its surface times its surface area. Since its energy flux is proportional to its surface temperature to the fourth power and the area of a sphere is proportional to the square of its radius, we can write

$$L \propto R^2 T^4$$

Thus we can write the proportion:

$$\frac{L_*}{L_\odot} = \left(\frac{R_*}{R_\odot}\right)^2 \left(\frac{T_*}{T_\odot}\right)^4$$

For Sirius $L_* = 25 L_\odot$ and $R_* = 1.67 R_\odot$, and so

$$25 = (1.67)^2 \left(\frac{T_*}{T_\odot}\right)^4$$

$$\left(\frac{T_*}{T_\odot}\right)^4 = \frac{25}{(1.67)^2} = 8.964$$

$$\frac{T_*}{T_\odot} = \sqrt[4]{8.964} = 1.73$$

Since $T_\odot = 5800$ K

$$T_* = 1.73 \times 5800 = 10{,}000 \text{ K}$$

21. 511 keV = 511×10^3 eV

$$= 511 \times 10^3 \text{ eV} \left(\frac{1.60 \times 10^{-19} \text{ J}}{1 \text{ eV}}\right)$$

$$= (511)(1.60) \times 10^{2+3-19} \text{ J}$$

$$= 8.18 \times 10^{-14} \text{ J}$$

According to Planck's law: $E = h\nu$, thus

$$\nu = \frac{E}{h} = \frac{8.18 \times 10^{-14} \text{ J}}{6.63 \times 10^{-34} \text{ Js}}$$

$$= \left(\frac{8.18}{6.63}\right) \times 10^{-14+34} \text{ Hz}$$

$$= 1.23 \times 10^{20} \text{ Hz}$$

Wavelength is related to frequency by

$$\lambda = \frac{c}{\nu} = \frac{3 \times 10^8 \text{ m/s}}{1.23 \times 10^{20} \text{ Hz}}$$

$$= \left(\frac{3}{1.23}\right) \times 10^{8-20} \text{ m} = 2.4 \times 10^{-12} \text{ m}$$

Such a photon belongs to the gamma ray region of the electromagnetic spectrum.

22. We use Balmer's formula from section 5-7:

$$\frac{1}{\lambda} = R\left(\frac{1}{4} - \frac{1}{n^2}\right)$$

with $\quad R = 1.09737 \times 10^7 \text{ m}^{-1}$

and $\quad n = 6$ for Hδ

Then:

$$\frac{1}{\lambda} = (1.09737 \times 10^7)\left(\frac{1}{4} - \frac{1}{36}\right) = 2.44 \times 10^6 \text{ m}^{-1}$$

So:

$$\lambda = 4.10 \times 10^{-7} \text{ m} \quad \text{or} \quad 410 \text{ nm}$$

23. The energy flux is proportional to the fourth power of the temperature. Procyon's temperature (6500 K) is greater than the Sun's (5800 K) by a factor of 6500/5800 = 1.12. The flux from Sirius' surface is therefore $(1.12)^4 = 1.57$ times that from the Sun's surface.

24. The wavelength of maximum intensity is inversely proportional to temperature. For the Sun, λ_{max} is 500 nm. For a star twice as hot as the Sun, λ_{max} is half as large, or 250 nm.

The energy flux is proportional to the fourth power of the temperature. Both stars have the same size, so the difference in brightness depends only on the difference in their surface temperatures. The star that is twice as hot as the Sun emits $2^4 = 16$ times as much energy as the Sun, and therefore is sixteen times brighter than the Sun.

25. The star's surface temperature is half that of the Sun, so the star's energy flux is $(1/2)^4$, or one-sixteenth that of the Sun. In other words, each square meter of the cooler star's surface emits only 1/16 as much energy as a square meter of the Sun's surface. But the cooler star's diameter (and therefore

its radius) is 10 times larger than the Sun's. Since the area of a sphere is proportional to the square of its radius, the cooler star's surface area is 100 times that of the Sun. Thus the total energy output of the cooler star is

$$1/16 \times 100 = 6.25$$

times that of the Sun. At the same distance, the cooler star would be 6.25 times brighter than the Sun.

26. As explained in section 5-8, the Doppler effect relates a wavelength shift $\Delta \lambda$ to radial velocity v by:

$$\frac{\Delta \lambda}{\lambda_o} = \frac{v}{c}$$

Thus,

$$v = \left(\frac{\lambda - \lambda_o}{\lambda_o}\right) c$$

$$v = \left(\frac{486.112 - 486.133}{486.133}\right) c$$

$$= \left(\frac{-0.021}{486.133}\right)(3 \times 10^5 \text{ km/s})$$

$$= -\frac{(2.1)(3)}{(4.86133)} \times 10^{-2+5-2} \text{ km/s}$$

$$= -1.296 \times 10^1 \text{ km/s} \approx -13 \text{ km/s}$$

Since v is negative, the star is moving toward the Earth.

27. In the laboratory, the wavelength of

$$H\alpha \text{ is } 656.285 \text{ nm} = \lambda_o.$$

Therefore,

$$\Delta \lambda = \lambda - \lambda_o = 656.331 - 656.285 = 0.046 \text{ nm}$$

Since this quantity is positive, the star is moving away from us. Its speed, according to the Doppler effect, is

$$v = \frac{\Delta \lambda}{\lambda_o} c = \left(\frac{0.046}{656.285}\right)(3 \times 10^5 \text{ km/s})$$

$$= 21 \text{ km/s}$$

CHAPTER 6

OPTICS AND TELESCOPES

Chapter Summary

This chapter describes laws of optics and astronomical telescopes. The major components of refracting and reflecting telescopes are presented with a description of basic optical imperfections. The roles of CCDs and spectrographs in astronomy are discussed. Radio astronomy instruments are described and a survey of the more important satellite observatories is given with some discussion of their advantages and impact on our understanding of the nature of celestial objects.

6-1 A refracting telescope uses a lens to concentrate incoming starlight at a focus

The phenomenon of refraction is presented and the concept of the focal length of a lens is introduced. The differences between objective and eyepiece lenses are identified. The magnifying power of a telescope is explained and chromatic aberration is examined.

Box 6-1 Major refracting telescopes

A list is provided which contains the observatories, year completed, locations, apertures and focal lengths of fourteen of the largest refracting telescopes in the world.

6-2 A reflecting telescope uses a mirror to concentrate incoming starlight at a focus

The law of reflection is presented and principal components of reflecting telescopes are identified. The Newtonian, Cassegrain, prime and coudé-focus locations are contrasted. The concepts of light-gathering power and resolving power are introduced. The optical limitations of reflecting telescopes are discussed and the development and role of Schmidt telescopes for survey work are reviewed.

Box 6-2 Major reflecting telescopes

A list is provided which contains the observatories, year completed, locations, apertures and focal lengths of eleven of the largest reflecting telescopes in the world.

Box 6-3 New Technology Telescopes

Telescopes using thin mirrors and adaptive optics are described. Various NTT's under design or construction are listed.

Box 6-4 The Palomar Sky Survey

The two Palomar Sky Surveys and the extension into the southern hemisphere are discussed.

6-3 An electronic device is often used to record the image at a telescope's focus

The advantages of the CCD over the photographic plate in image detection are noted.

6-4 Spectrographs record the spectra of astronomical objects

The functions of prisms and gratings to produce spectra in spectrographs are reviewed.

6-5 A radio telescope uses a large concave dish to reflect radio waves to a focus

A brief history of the origins of radio astronomy is followed by a discussion of the limited resolving power of radio telescopes. Interferometry and VLBI techniques are described.

6-6 Telescopes in orbit around the Earth detect radiation that does not penetrate the atmosphere

Absorption of different types of electromagnetic radiation by the Earth's atmosphere is outlined. Advantages of the space-based observatories are noted and the characteristics and value of the IRAS, IUE, and three HEAO satellites are enumerated. The potential of the Hubble Space Telescope is examined along with some of its early problems. A brief introduction is provided for the AXAF and GRO observatories.

Teaching Hints and Strategies

When discussing *refracting telescopes* (section 6-1) like the one shown in Figure 6-3, point out that incoming parallel rays indicate light coming from a single point in space due to the great distances of celestial objects. Note also that these parallel rays are bent so as to form the image. Light from a second point in space is detected as a second set of parallel rays which approach the telescope from a different direction. Consideration of the two sets of parallel rays can show that all points of the objective lens, or primary mirror in the case of a reflecting telescope, contribute light to all points in the image. Obstructions in and near the telescope do not block a specific part of the field of view, but only decrease the amount of light which ultimately forms the image. The role of the eyepiece in a telescope is demonstrated by allowing the converging rays from a point source to diverge and refract at the eyepiece to emerge as a parallel group of rays again. Our eyes would normally see light from this point source as a parallel group of rays without the telescope. The role played by the eyepiece in determining the magnification should be explained here as well. The variation in the amount of refraction with variation in wavelength, or dispersion, leads naturally to a discussion of chromatic aberration and to a general discussion of the technical and economic limitations of refracting telescopes.

The precise nature of *refraction and reflection* (sections 6-1 and 6-2) can be demonstrated using a low power laser and a fish tank half filled with water. A cover for the tank permits the trapping of smoke in the top to facilitate viewing of the refracted laser beam as it approaches and leaves the water. Place a mirror on the bottom to produce clear, simple reflection.

It should be noted here that modern *reflecting telescopes* (section 6-2) are designed with several secondary mirrors so that different focus locations can be used. Changing the focal length of a telescope changes its basic characteristics and permits greater flexibility in its use for astronomical research. The limitations on resolving power of large telescopes on the Earth are generally not determined by the optical quality or diameter of the telescope, but are controlled by the disturbances in the atmosphere due to variable refraction.

The information about *major reflecting and refracting telescopes* (Boxes 6-1 and 6-2) provides for some interesting observations. Compare the completion dates of the refractors with those of the reflectors. How many major new refractors are being designed or constructed? How about major new reflectors? Why are there so few refractors in the southern hemisphere compared to reflectors? Explain why having telescopes in the southern hemisphere is important?

The adaptive optics of NTT's (Box 6-3) can eliminate the effects of slowly varying seeing because the frequency of the corrections is high enough. Rapid variations cannot be followed. Fortunately, most distortion is from the slower

48 CHAPTER 6

variations due to the movement of large cells of air having different optical characteristics.

The *Palomar Sky Survey* (Box 6-4) is one of the most valuable references that astronomers have. Every observatory has a set of sky survey prints along with transparent overlays which have coordinate grid lines marked and the location and identification of thousands of objects. The new sky survey plates are being scanned and digitized for computer use. Because of improvements in photographic emulsions and processing, the new sky survey extends to much fainter levels.

CCD cameras and spectrographs (sections 6-3 and 6-4) are examples of the auxiliary instruments used by astronomers. Most students visualize the modern astronomer looking through the telescope. Emphasize that the telescope merely gathers light and forms an image. Most of the advances in the study of faint objects which have taken place since completion of the 200" telescope on Mount Palomar have been due to improvements in the auxiliary instruments and detectors. Small telescopes today can do many projects which required much larger telescopes in the past. It is interesting to discuss the problems associated with storage of large volumes of digital data compared to the simplicity of photographic plates.

The wavelength dependence of resolving power should be noted and discussed in the context of the sizes of *radio telescopes* (section 6-5). Compute and compare typical resolving powers for the largest radio and optical telescopes. Make it clear that radio dishes working together have a resolving power related to the separation distance between the antennas, while the light gathering power is determined by adding the areas of the antennas involved.

The need for *space astronomy* is discussed in the last section (6-6) of this chapter. Note that gamma ray, X-ray, ultraviolet and infrared observatories in space are required to overcome the absorption of these types of radiation by the Earth's atmosphere, while orbiting optical telescopes are needed to improve resolving power which is limited by atmospheric seeing.

Students should learn that despite the often publicized defect in the optics of the HST, this telescope is providing an unprecedented view of the universe. Point out that the defect means that the telescope is not performing at its intended potential, but it is still better than any other telescope in the world in some respects. The news media seem to prefer failures rather than successes.

Answers to Chapter 6 Computational Questions

15. The area of a circle is proportional to the square of its diameter. The 5-m Palomar mirror therefore has an area $5^2 = 25$ times larger than the area blocked by the observing cage. Only 1/25 of the incoming light is obstructed.

16. Light-gathering power is proportional to the area of the mirror or lens that collects the light. The diameter of the Palomar mirror is 5000 mm, which

is 1000 times larger in diameter than the dark-adapted human pupil. The light-gathering power of the Palomar telescope, which is proportional to the square of its diameter, is 1000^2 or one million times greater than that of the human eye.

18. You must use the small-angle formula discussed in Box 1-1 to convert the angular size of the smallest discernable surface features into a linear size. The distance (d) to Jupiter is 628×10^6 km, and the angle (α) subtended by the smallest features is 0.1 arc sec. Thus, the actual size (D) of the smallest visible features is

$$D = \frac{\alpha d}{206{,}265} = \frac{(0.1)(628 \times 10^6 \text{ km})}{206{,}265}$$
$$\approx 300 \text{ km}$$

The distance from the surface of the Earth to the surface of the Moon is about 376,300 km, and the resolution of the human eye is 60 arc sec. Again using the small-angle formula, we find that the smallest discernable lunar features have a size of

$$D = \frac{\alpha d}{206{,}265} = \frac{(60)(376{,}300 \text{ km})}{206{,}265} = 110 \text{ km}$$

Thus, the HST can resolve features larger than 300 km across on Jupiter's moons, whereas only those features larger than about 110 km across on our nearby Moon can be distinguished with the unaided human eye.

19. Magnifying power equals the focal length of the mirror divided by the focal length of the eyepiece. The telescope's focal length is 2000 mm. Consequently,

(a) a 9 mm eyepiece gives a magnification of $2000/9 = 222 \times$

(b) a 20 mm eyepiece gives a magnification of $2000/20 = 100 \times$

(c) a 55 mm eyepiece gives a magnification of $2000/55 = 36 \times$

The smallest angular separation that can be resolved by a telescope is inversely proportional to the telescope's diameter. In the text, we saw that the resolving power of the 4-m telescopes at Kitt Peak and Cerro Tololo is about 0.03 arc sec. The primary mirrors of these telescopes are 20 times larger in diameter than that of a 20-cm Newtonian reflector. Consequently, the Newtonian's resolving power is about 20 times 0.03 = 0.6 arc sec. In actuality, under excellent observing conditions, you would probably be able to resolve two stars that are about 1 arc sec apart.

20. At opposition, the distance to Pluto is about 38.5 AU = 5.8×10^9 km. As we saw in Question 14, the resolution of the HST is 0.1 arc sec. using the

small-angle formula (Box 1-1), we find that the size of the smallest discernable features at Pluto's distance from Earth is

$$D = \frac{\alpha d}{206{,}265}$$

$$= \frac{(0.1)(5.8 \times 10^9 \text{ km})}{206{,}265}$$

$$\approx 2800 \text{ km}$$

Pluto's diameter is 2,290 km, and so the HST would not be able to distinguish any features on that planet.

CHAPTER 7

OUR SOLAR SYSTEM

Chapter Summary

This chapter summarizes the principal observational evidence regarding the current structure of the solar system which is the basis for our understanding of its formation and evolution. The physical characteristics of the most massive solar system objects other than the Sun are presented with an explanation of the classification criteria for these objects. The atomic nature of the different chemical elements is described and the information acquired from spectroscopic chemical abundance determinations is reviewed. The processes responsible for the formation of the elements are noted. The theory of retention of planetary atmospheres is summarized. The accretion theory of the formation and evolution of the solar system is discussed.

7-1 The planets are classified as either terrestrial or Jovian by their physical attributes

Section 7-1 describes the classification of planets based on their physical properties. The differences between orbit sizes and temperatures for Jovian and terrestrial planets are noted. The common orbital planes for all of the planets is revealed. The members of the two classes of planets are identified with Pluto remaining the exception. The sizes, masses and average densities of the two classes are contrasted.

7-2 Seven giant satellites are as big as terrestrial planets

A brief summary of planetary satellites is provided with size and average density data listed for the seven major planetary satellites.

7-3 Spectroscopy reveals the chemical composition of the Sun, planets, and stars

The role of spectroscopy in determining the chemical compositions of celestial bodies is explained. The nature of chemical elements is described and the periodic table of the elements is presented. Atoms and molecules are defined and examples of common molecules are given.

Box 7-1 Atoms and isotopes

The constituent parts of atoms are discussed in the context of atomic structure. The role of the electrical force in atomic structure is revealed and the concept of atomic number is introduced. Isotopes are described.

7-4 Hydrogen and helium are abundant on the Jovian planets, whereas the terrestrial planets are composed mostly of heavy elements

The chemical composition of the Jovian and terrestrial planets is given and the role of temperature in determining the composition is explained.

Box 7-2 Thermal motion and the retention of an atmosphere

The concepts of kinetic energy and thermal energy are defined and combined with the concept of the escape velocity developed in Newtonian mechanics to form the basis for the theory of the retention of planetary atmospheres.

7-5 Small chunks of rock and ice also orbit the Sun

The existence of asteroids and comets as rocky and icy bodies respectively is established. The division of these objects into two families according to their distances from the Sun is made in an analogous fashion to the two planetary families.

7-6 The relative abundances of the elements are the result of cosmic processes

The cosmic abundances of the elements are briefly reviewed with references to those elements which are formed in the Big Bang, in normal stellar evolution and in supernova explosions.

7-7 The solar system formed in the solar nebula during the birth of the Sun

The solar nebula is described. The differences between the chemical compositions of the two classes of planets are examined in the context of condensation temperatures of materials commonly thought of as gases, ices and rocks. The flattening of the nebula is attributed to its rotation. Angular momentum is mentioned but is not formally defined. (See Box 23-1 for formal definition.) Helmholtz contraction is briefly discussed.

7-8 The inner planets formed by the accretion of planetesimals, whereas the outer planets formed by an accumulation of gases in the solar nebula

Some of the details of the formation of the solar system, such as the nature of planetesimals, protoplanets, accretion, and stellar winds, are investigated in the context of computer simulations. Chemical differentiation is defined in connection with the homogeneous accretion theory. The heterogeneous accretion theory is also presented.

Teaching Hints and Strategies

The *terrestrial and Jovian classifications* (section 7-1) for planets are immediately suggested by inspection of Figure 7-2. The planets divide naturally into two groups based on size. Examination of Appendix 2 in the textbook shows that most of their physical properties (mass, size, average density, chemical composition, rotation periods) divide in the same way. Ask how such diverse objects can come from a common source? Review the distances of the planets from the Sun. It is common for many students to overlook the uneven distribution of planets in the solar system. The diagrams used to describe the geocentric and heliocentric cosmologies often show the planetary orbits being rather uniformly spaced. Be sure to call attention to the fact that the terrestrial planets are all very close to the Sun while the Jovian planets are much more spread out, as is clearly shown in Figure 7-1 in the textbook. It is instructive to use the inverse square law of light to compute the solar flux at several planets relative to that at the Earth. In contrast to the differences between these two classes, note that 1) the planets have orbital planes which are very closely aligned in space as opposed to a random distribution, 2) all planets revolve in the same direction (counter clockwise as seen from the north) in their orbits around the Sun, and 3) most planets rotate in this same direction, with Venus and Uranus being the exceptions. These facts seem to imply a common origin for the solar system objects. The trick is to explain the similarities and differences in one single theory.

The presentation of the *seven giant satellites* (section 7-2) as a group can be used to dispel a common misconception held by students that all planets have solid surfaces like the Earth. Point out that this is obviously not the case for the Jovian planets which may have no solid surfaces or have such surfaces only in their deep interiors. Emphasize that the large moons in the outer solar system are more like the Earth than are the planets themselves. Note that the difference between the average densities of the Jovian and terrestrial planets is a clear indicator of composition differences. Call attention to the difference in density between terrestrial planets and the major satellites.

When discussing the application of *spectroscopy* (section 7-3) to the determinations of the chemical compositions of atmospheres, remind students that the spectrum of a planet consists of sunlight reflected from some layer in the atmosphere or from the surface of the planet. It will contain absorption lines produced in the solar atmosphere, in the planetary atmosphere and the Earth's atmosphere. The composition of the atmosphere gives little information about the vast majority of the mass in the planet. The average density provides some insight into the interior composition.

A careful presentation of the defining characteristics and abbreviations for *atoms and isotopes* (Box 7-1) at this time can prevent total confusion later in the alphabet soup of physical science. Be sure to cover the composition of molecules and the concept of isotopes as the variation of neutron numbers for a given number of protons. Show students how a periodic table, which shows the atomic weights of the elements, can be used to derive the numbers of neutrons in the most common isotopes of an element. Round the atomic weight to the nearest integer and subtract the atomic number to obtain the number of neutrons. Provide examples of commonly cited isotopes such as carbon 12 and 14 and uranium 235 and 238. Review the definition of ions as variations of the number of electrons and the definition of elements as the variation of the number in protons to put the definition of an isotope in context.

To introduce the data on *cosmic abundances of the elements* (section 7-6) you might list the common gases found in the atmospheres of the planets and the principal constituents of the sun and stars. Discuss why the elements Li, Be and B are not very abundant even though they are rather simple in structure. These elements tend to be fused into more massive nuclei in stellar interiors.

Some insight into the compositions of planetary atmospheres can be developed by applying the *theory of retention of atmospheres* (Box 7-2) to common gases found in planetary atmospheres. Make a table of the number of nucleons (neutrons plus protons) in the common gases found in the atmospheres of the planets (H_2, He, N_2, O_2, H_2O, CO_2, NH_3 and CH_4). Consider these molecules as potential artificial satellites. Discuss the role of temperature in controlling the motions of the particles and the fact that the low mass particles move faster and the more massive particles move more slowly. This leads to a better understanding of the differences in the compositions between terrestrial and Jovian atmospheres as indicated by their spectra.

The *accretion theory* (section 7-7) is accepted by most astronomers and explains the current conditions in the solar system, such as a common orbital plane, common orbital and rotational directions and general compostion differences, with almost no arbitrary assumptions. It is important to point out that the best scientific theories involve very few special conditions. The general rules which describe the high degree of order in the motions of the planets and satellites are probably caused by the natural process of gravitational collapse and the conservation of angular momentum. Astronomers have observed spherical dark clouds from which stars form for decades and have just recently started to observe flattened disks surrounding stars in which planets may have formed or may form in the future. The evolution of a large, slowly rotating spherical cloud can be described using a rotating stool and two hand weights. The conservation of angular momentum requires that the disk rotate more rapidly as the weights move in toward the axis of rotation. This increase in angular velocity enhances the "centrifugal" forces and will ultimately stabilize the collapse in the plane of rotation. Demonstrate that collapse parallel to the axis of rotation produces no change in angular velocity and will not be stabilized until the material has formed a flattened disk. Thus the common orbital planes and directions of the planets result from the laws of nature. The fact that Venus and Uranus rotate in the retrograde direction is not explained by this theory, nor by any other simple theory.

The choice between the nebular (accretion) and catastrophe theories for the formation of the solar system have a bearing on the potential number of planetary systems which might be expected in the Galaxy. If the nebular theory is correct, then planetary systems should be quite common. No special conditions are required to form a planetary system. If the solar system formed as the result of a collision of the Sun with another star, then similar systems would be much less common because stellar collisions are quite likely rare events in the Milky Way. Make clear that the Big Bang which is thought to have created the universe is not the event which started the formation of the solar system. The Big Bang took place 15–20 billion years ago while the solar system was formed about 5 billion years ago.

Computer simulations (section 7-8) are very helpful in solving the mystery of the differences between the terrestrial and Jovian planets. Most of the differences derive from the different temperatures which existed in the inner and outer solar nebula. Volatile materials did not accumulate into the objects which formed in the high temperature regions. Thus the terrestrial planets were deprived of a vast majority of the material in the inner solar nebula and have small masses and sizes. The differences in average density are also explained in this manner. You might note here that perhaps as much as 1 or 2 solar masses were ejected from the solar system when the Sun completed its formation. The different rotation rates of the planets are due to the conservation of angular momentum and the temperature differences. A much higher fraction of the material in the outer solar nebula accumulated into the Jovian planets and hence they are rapid rotators. In the case of terrestrial planets, much of the material failed to fall inward which resulted in much more modest rotation rates. Students often overlook the fact that Jovian planets have much shorter rotation periods than the terrestrial planets.

General teaching hints and strategies for the solar system

Students can come to view the next several chapters as an exercise in memorization if they are not given adequate direction. Develop a framework for organization which can assist in mastery of the subject matter while illustrating the scientific method. Discuss such scientific activities as data organization, classification, analysis and generation of hypotheses.

Data acquisition

Have students organize the data from Chapters 8–16 in tabular form. Point out where limitations in the observations present problems. Recommend that students develop a table in which they can record the key elements for each of the terrestrial planets and a similar table for the Jovian planets. Make a column for each of the planets and make one or more rows for each of the structural elements to be discussed similar to the table below:

	Mercury	Venus	Earth	Mars
Diameter (Earth = 1)				
Atmosphere				
composition				
pressure				
temperatures				
Surface				
composition				
oceans				
topography				
plate tectonics				
erosion				
craters				
volcanoes				
Interior				
mean density				
core				
Magnetic field				
Satellites				
Rings				

As students read the chapters, have them fill in the appropriate information in the table. Use the Earth as a standard and use relative values rather than absolute units when appropriate.

Classification

Concentrate on comparing and contrasting the information in the tables. Clearly identify the criteria for classification and discuss the possibilities of other criteria. Ask the students to decide whether Mercury and Pluto are more similar to the other planets or to planetary satellites. Have them justify their conclusions. Make a transparency which shows the relative sizes of all terrestrial planets, the major satellites, the largest minor planets and Pluto.

Analysis and Generation of Hypotheses

When all chapters relevant to a certain class of objects have been read, provide continuity by emphasizing the explanations and interpretations of the entries rather than the entries themselves. What do magnetic fields tell us about planetary interiors? How are volcanoes used to derive information about planetary plate tectonics?

Make an effort to synthesize the material in Chapter 9 on the Moon and Chapter 14 on the Galilean satellites of Jupiter with similar material in section 15-8 dealing with Titan and in section 16-7 dealing with Triton.

For more advanced students you can provide data about the periods of rotation and revolution for a planet, its orbital eccentricity, and the inclination of its axis of rotation to its orbital plane (e.g., Appendixes 1 and 2). Have them determine what seasonal cycles should be like. What are the primary causes of seasons for the planet. How many "days" are there in a "year" for the planet? Emphasize how much of the knowledge we have is based on space exploration compared to Earth-based observations.

Answers to Chapter 7 Computational Questions

17. First you need to calculate the mass of the asteroid. Its mass (m) is related to its density (ρ) and volume (V) by

$$m = \rho V$$

Since the asteroid is spherical,

$$V = \frac{4}{3} \pi r^3$$

where r is the asteroid's radius.

$$\rho = 5 \text{ g/cm}^3 = 5000 \text{ kg/m}^3$$

$$r = \frac{1}{2} \text{ km} = 500 \text{ m}$$

Thus the mass of the asteroid is

$$m = \rho V = \frac{4}{3}\pi\rho r^3$$

$$= \frac{4}{3}(3.1416)(5000)(500)^3 \text{ kg}$$

$$= \frac{4}{3}(3.1416)(5)(5)^3 \times 10^9 \text{ kg}$$

$$= 2.62 \times 10^{12} \text{ kg}$$

The asteroid's speed is

$$v = 25 \text{ km/s} = 2.5 \times 10^4 \text{ m/s}$$

The asteroid's kinetic energy (KE) is

$$KE = \frac{1}{2}mv^2 = \frac{1}{2}(2.62 \times 10^{12})(2.5 \times 10^4)^2 \text{ J}$$

$$= \frac{1}{2}(2.62)(2.5)^2 \times 10^{20} \text{ J}$$

$$= 8.2 \times 10^{20} \text{ J}$$

Since 1 atom bomb = 20 kilotons

$$= 20(4.2 \times 10^{12}) \text{ J}$$

$$= 8.4 \times 10^{13} \text{ J}$$

$$KE = 8.2 \times 10^{20} \text{ J}\left(\frac{1 \text{ atom bomb}}{8.4 \times 10^{13} \text{ J}}\right)$$

$$\approx 10^7 \text{ atom bombs}$$

Thus the impact of the asteroid would release ten million times more energy than the bomb that destroyed Hiroshima.

18. From Box 7-2 on page 134, we have the relation

$$v = \left(\frac{3kT}{m}\right)^{1/2} \text{ where } k = 1.38 \times 10^{-23} \text{ J/K}$$

Our Solar System 59

$$v = \left(\frac{3(1.38 \times 10^{-23})(5800)}{1.673 \times 10^{-27}} \right)^{1/2}$$

$$= \left(\frac{3(1.38)(5.8)}{1.673} \times 10^7 \right)^{1/2}$$

$$= (1.435 \times 10^8)^{1/2} = \sqrt{1.435} \times 10^4 \text{ m/s}$$

$$= 1.2 \times 10^4 \text{ m/s} = 12 \text{ km/s}$$

19. From Box 7-2 on page 135 we have the relation

$$v_{escape} = \left(\frac{2GM}{R} \right)^{1/2}$$

where $\quad M = M_\odot = 1.989 \times 10^{30}$ kg
and $\quad R = R_\odot = 6.96 \times 10^8$ m
and $\quad G = 6.67 \times 10^{-11}$ Nm^2/kg^2

$$v_{escape} = \left(\frac{2(6.67 \times 10^{-11})(1.989 \times 10^{30})}{6.96 \times 10^8} \right)^{1/2} \text{ m/s}$$

$$= \left(\frac{2(6.67)(1.989)}{6.96} \times 10^{11} \right)^{1/2} \text{ m/s}$$

$$= (38.1 \times 10^{10})^{1/2} \text{ m/s} = \sqrt{38.1} \times 10^5 \text{ m/s}$$

$$= 6.17 \times 10^5 \text{ m/s} = 617 \text{ km/s}$$

From the solution to problem 18, we know that the velocity of hydrogen atoms at the Sun's surface is about 12 km/s. From Box 7-2, we know that the escape velocity must be at least six times greater than the average speed of the atoms, if they are to be retained. The Sun's escape velocity is about 50 times the speed of the hydrogen atoms. Therefore the Sun retains the hydrogen.

20. Using the formula for escape velocity given in Box 7-2, we have

$$v_{escape} = \left(\frac{2GM}{R} \right)^{1/2}$$

$$= \left(\frac{2(6.67 \times 10^{-11})(1.48 \times 10^{23})}{2.63 \times 10^6} \right)^{1/2} \text{ m/s}$$

60 CHAPTER 7

$$= \left(\frac{2(6.67)(1.48)}{2.63} \times 10^6\right)^{1/2} \text{ m/s}$$

$$= (7.51 \times 10^6)^{1/2} \text{ m/s} = \sqrt{7.51} \times 10^3 \text{ m/s}$$

$$= 2.74 \times 10^3 \text{ m/s} = 2.74 \text{ km/s}$$

The spacecraft must attain a velocity of 2.74 km/s to leave Ganymede.

But alas, THIS IS A TRICK QUESTION, because Jupiter's powerful gravity would still prevent the spacecraft from returning to Earth.

To see what speed the spacecraft needs to return to Earth, we must calculate the escape velocity from Jupiter at the distance of Ganymede's orbit.

According to Appendix 3, the distance from Jupiter to Ganymede is

$$d = 1.07 \times 10^9 \text{ m}$$

Jupiter's mass is $M = 1.90 \times 10^{27}$ kg and thus the escape speed at Ganymede's orbit is

$$v_{escape} = \sqrt{\frac{2(6.67 \times 10^{-11})(1.90 \times 10^{27})}{1.07 \times 10^9}}$$

$$= \sqrt{2.37 \times 10^8} = 1.54 \times 10^4 \text{ m/s}$$

$$= 15.4 \text{ km/s}$$

22. First of all, you must calculate Phobos' orbital radius:

$$a = 5980 + \frac{1}{2}(6794) = 9377 \text{ km}$$

$$= 9.375 \times 10^6 \text{ m}$$

From Box 4-3 on page 68 we see that the appropriate form of Kepler's third law is

$$P^2 = \left(\frac{4\pi^2}{GM}\right)a^3$$

were M is the mass of Mars.

We are told that the orbital period of Phobos is

$$P = 0.31891 \text{ days} = 2.755 \times 10^4 \text{ s}$$

Thus

$$M = \frac{4\pi^2 a^3}{GP^2} = \frac{4\,(3.1416)^2(9.375 \times 10^6)^3}{(6.67 \times 10^{-11})(2.755 \times 10^4)^2}\ \text{kg}$$

$$= \frac{4\,(3.1416)^2(9.375)^3}{(6.67)(2.755)^2} \times 10^{18+11-8}\ \text{kg}$$

$$= 6.43 \times 10^{23}\ \text{kg}$$

The average density of Mars ($\bar{\rho}$) equals its mass divided by its volume.

$$\bar{\rho} = \frac{M}{V} = \frac{M}{\tfrac{4}{3}\pi r^3}$$

where r is the radius of Mars (3397 km)

$$\bar{\rho} = \frac{3M}{4\pi r^3} = \frac{3\,(6.43 \times 10^{23}\ \text{kg})}{4\pi(3.397 \times 10^6\ \text{m})^3}$$

$$= \frac{3\,(6.43)}{4\,(3.1416)(3.397)^3} \times 10^{23-18}\ \text{kg/m}^3$$

$$= 3900\ \text{kg/m}^3$$

23. In Box 5-1 on page 85 we saw that the Celsius and Fahrenheit scales are related by

$$T_F = \frac{9}{5} T_C + 32$$

Set $T_F = T_C = T$ and solve for T

$$T = \frac{9}{5}T + 32$$

$$\frac{4}{5}T = -32$$

$$T = -\frac{5}{4}\,(32) = -40°$$

Thus $-40°C = -40°F$

Temperature on the Kelvin scale (T_K) is related to temperture on the Celsius scale by the simple equation

$$T_K = T_C + 273$$

and the Celsius scale is related to the Fahrenheit scale by

$$T_C = \frac{5}{9}(T_F - 32)$$

Combining these two equations and setting

$$T_K = T_F = T$$

we obtain

$$T = \frac{5}{9}(T - 32) + 273$$

Solving for T, we get

$$T = \frac{9}{4}(273) - \frac{5}{4}(32) = 574\frac{1}{4}$$

So, $574\frac{1}{4}°F = 574\frac{1}{4}$ K.

24. Approximately 98 percent of the matter in the universe is hydrogen and helium; heavier elements account for only 2 percent. The Earth is composed almost entirely of heavy elements. To see what the Earth's mass would have been if it had retained hydrogen and helium in proportion to its heavy elements, we must ask "Two percent of what mass (X) equals the Earth's mass (M_e)?"

In other words,

$$M_e = 0.02\ X\ \text{or},\ X = 50 M_e.$$

Thus the Earth would have been 50 times more massive than it is today, if it had been able to keep the hydrogen and helium that came along with its heavier elements.

For comparison, from Appendix 2 we note that Saturn, Uranus, and Neptune have masses between 15 and 94 times that of Earth. These Jovian planets may therefore have rocky cores roughly comparable to the mass of the Earth.

25. It is reasonable to assume that a planet 50 AU from the Sun would be composed primarily of ice, which has a density of 1000 kg/m^3. There would probably be an admixture of heavier elements, however, and gravity would compress the planet's ices somewhat. So let's assume that the density of the hypothetical planet is ρ_p = 1500 kg/m^3. The density of the Earth is ρ_e = 5500 kg/m^3. Density is mass divided by volume (V_e). Since the mass of the Earth and hypothetical planet are the same, we can write

$$\rho_e V_e = \rho_p V_p$$

where the subscripts e and p refer to the Earth and hypothetical planet, respectively. The volume of a sphere is proportional to the cube of its radius, and so the volumes of the planets are proportional to the cubes of their diameters (d). Consequently,

$$d_p = \left(\frac{\rho_e}{\rho_p}\right)^{1/3} d_e = \left(\frac{5500}{1500}\right)^{1/3} d_e = 1.54\, d_e$$

Thus the hypothetical planet would have a diameter about 1.54 times larger than Earth's.

CHAPTER 8

OUR LIVING EARTH

Chapter Summary

This chapter begins with a discussion of the composition of the surface of the Earth and plate tectonics followed by a description of its interior and applications of seismology to determine the nature of the mantle and core. The composition and structure of Earth's atmosphere are reviewed. The Earth's magnetic field and magnetosphere are related to aurorae. Life on Earth is discussed in the context of global warming and the greenhouse effect.

8-1 Earth rocks contain clues about the history of our planet's surface

Basic elements of geology are introduced with the definitions of the terms rock and mineral. Descriptions are provided for the three classes of rocks, igneous, sedimentary and metamorphic. Some examples of different rock types are discussed.

Box 8-1 Earth data

This table contains the orbital and physical data for the Earth. Oblateness is defined.

8-2 Studies of earthquakes reveal the Earth's layered interior structure

The use of seismic waves to study the Earth's interior is explained. Earthquakes are described and their epicenters and foci are defined. S waves and P waves are described and their role in mapping the run of density with depth in the Earth and in detecting the molten core of the Earth are discussed. The physical properties and chemical compositions of the mantle, outer core and inner core are reviewed.

8-3 Plate tectonics produce earthquakes, mountain ranges and volcanoes that shape the Earth's surface

The theory of plate tectonics is discussed with reference to seafloor spreading, the asthenosphere, oceanic rifts, subduction zones and the lithosphere. Some geographic regions on the Earth are explained in terms of the theory of plate

Our Living Earth   65

tectonics. Convection is defined. The role of earthquake mapping in defining plate boundaries is noted. Hot-spot volcanism on Earth is discussed in the context of volcanoes on Mars and Venus.

8-4 Absorption of sunlight and the Earth's rotation govern the behavior of our nitrogen–oxygen atmosphere

The large abundances of O_2 are attributed to the existence of life on Earth and the role of the oceans in removing CO_2 from the atmosphere is noted. The chemical composition of the Earth's atmosphere is described and the temperature variations with altitude are used to define the four layers of the atmosphere. The role of ozone in absorption is identified and problems related to ozone depletion are noted.

Box 8-2   Atmospheric pressure

The concept of atmospheric pressure is developed and units for expressing pressures are reviewed.

8-5   The Earth's magnetic field produces a magnetosphere that traps particles from the solar wind

The dynamo effect is presented. The nature of the magnetosphere is described and the interaction of magnetic fields with the solar wind to form bow shocks is discussed. The structure of the Earth's magnetic field is examined in terms of the magnetopause and shock waves. The trapping of charged particles from the solar wind is explained and aurorae are discussed.

8-6   A burgeoning human population is profoundly altering Earth's biosphere

The biosphere is defined. A brief note is made of the population increases on Earth. The impact of civilization on the greenhouse effect and on global warming due to increasing abundance of CO_2 is reviewed.

Teaching Hints and Strategies

[Refer to the Teaching Hints and Strategies section of Chapter 7 of this Instructor's Guide for general strategies applicable to solar system objects.]

The discussion of various types of rocks (section 8-1) provide an opportunity to mention once again that ages for celestial bodies such as the Earth, the Moon, and meteorites are based on the times of solidification of the rocks. Any time rocks become molten, their radiological clocks are reset at the time of resolidification. The rocks on the surfaces of the Earth and Moon show a wide variety of ages depending upon local geological history.

The limitations on direct observations of the *Earth's interior* (section 8-2) which lead to use of seismology by geophysicists is rather similar to the investigative restrictions encountered by astronomers. It is important to stress the time variable nature of the Earth's interior and the fluid motions which occur. The chemical differentiation, as indicated by the increase in average density with increasing depth, is an important clue to the past thermal history of the Earth. The role of the fluid iron–nickel core in the generation of the Earth's magnetic field should be noted as well.

The fact that the Earth has a plastic mantle and liquid core is foreign to many students. The theory of *plate tectonics* (section 8-3) has revolutionized modern planetary science and added new life to the subject of planetary geology. The great success of this theory, which has been developed over the past two decades, requires that we consider that planets may be geologically active like the Earth. The geologic structures on Earth, and their explanation in terms of our knowledge of its internal structure, allow us to speculate about the physical processes in the interiors of the other terrestrial planets based on geologic structures there. The role of convection in Earth's mantle should be clarified. Figure 8-10 indicates that the mantle is a solid while Figure 8-14 implies convective flow. Compare this material with wavey glass in an old window. We think of the glass as a solid over short periods of time. But over long times it acts as a very thick or viscous fluid.

The temperature profile of *Earth's atmosphere* (section 8-4) shown in Figure 8-17 makes for an interesting analysis. Describe the heating mechanisms which result in the rather complex temperature variations with height in the atmosphere. The absorption of visible light by the Earth's surface and of infrared radiation by carbon dioxide and water vapor molecules in the lowest layers of the troposphere cause the high temperatures near the surface of the Earth. The absorption by ozone in the stratosphere contributes to the the temperature peak there. Absorption by individual atoms and ions in the upper thermosphere helps to produce rising temperatures there. Remind students that similar graphs are presented for Venus and Mars. Suggest that they compare and contrast them. The impact of man-made gases on the ozone layer discussed in this section should not go without mention. One might prefer to discuss this topic along with global warming and the ozone hole in the last section of this chapter.

Elaborate on the *origins and evolutions of the atmospheres of three of the terrestrial planets* (Box 7–2). While there are about 100 chemical elements and literally millions of compounds, only three chemical elements account for at least 98% of the gases in the atmospheres of these planets. The major difference is the dominance of carbon dioxide in the atmospheres of Venus and Mars while it is only a trace gas in Earth's atmosphere. The reason for this is the extraction of this gas from the atmosphere due to chemical reactions with dissolved salts in the liquid water oceans on the Earth's surface. The carbonate rocks found on the Earth contain what might otherwise have been in the atmosphere. The greenhouse effect on Venus provides an extreme example of what might have happened on Earth if liquid water had not been present.

It is helpful to use a compass to review the properties of the *Earth's magnetic field* (section 8-5). A horizontal compass shows the direction of the field in a plane parallel to the surface of the Earth. A compass suspended in a vertical plane containing the north magnetic pole can illustrate the inclination of magnetic field lines relative to the normal to the Earth's surface. A pair of bar magnets can be used to explain the interaction of magnetic forces and their similarity to electrostatic forces. One bar magnet can be used to show the magnetic influence on the compass, which is just a magnet. Iron filings can be placed on top of a plastic sheet with a bar magnet below to show the magnetic field lines. This demonstration can be done using an overhead projector to permit viewing by a large class. A few minutes should be devoted to a discussion of the concept of a "force field" as a general vector diagram. The field lines merely demonstrate the direction and magnitude of the force.

The rising gobal population and its impact on *global warming* (section 8-6) provide an excellent opportunity to discuss applications of astronomy, geophysics and space science. The drought of 1987 focused attention on the potential impact of the greenhouse effect. Discuss the attention received by theories of global warming by the media before and since this unusual year. Be sure to discuss the role of the space program in monitoring many of the variables in this complex planet on which we exist. An interesting application of the results of planetary exploration program would be to test the reliability of our global climatic models on the atmospheres of other planets to evaluate their relative merits. It is worth noting the effects of CFC's on the ozone layer as noted in section 8-4 as another important interface between modern society and the tenuous atmosphere.

Answers to Chapter 8 Computational Questions

17. A cursory examination of any map of the world reveals that the distance between Africa and South America is about 6600 km.

The time required for Africa and South America to separate by 6600 km equals that distance divided by the speed of separation (3 cm/yr):

$$6600 \text{ km} \div 3 \text{ cm/yr} = \frac{6.6 \times 10^3}{3 \times 10^{-5}} = 2.2 \times 10^8 \text{ yr}$$

68 CHAPTER 8

So, Africa and South America began moving apart about 220 million years ago.

19. The volume of a sphere is proportional to the cube of its radius. In the text we saw that the radius of the core (r_c) is 3500 km, the average radius of the Earth (r_e) is 6350 km, and the thickness of the crust is 30 km.

The fraction of the Earth occupied by the core is:

$$\frac{r_c^3}{r_e^3} = \frac{3500^3}{6350^3} = 0.17$$

So the core occupies 17 percent of the Earth's volume. The mantle's inner radius is 3500 km and its outer radius is 6320 km. Thus the fraction of the volume occupied by the mantle is:

$$\left(\frac{6320^3 - 3500^3}{6350^3}\right) = 0.82$$

So the mantle occupies 82 percent of the Earth's interior. The core and mantle together occupy 17 + 82 = 99 percent of the Earth's volume. That leaves 1 percent for the Earth's crust.

20. According to Box 8-1, the Earth's diameter is 12,756 km, and so its volume (V) is:

$$V = \frac{4}{3}\pi(6378)^3 = 1.087 \times 10^{12} \text{ km}^3 = 1.087 \times 10^{21} \text{ m}^3$$

The mass of the Earth (M_e) is listed as 5.976×10^{24} kg, so the average density ($\bar{\rho}$) is:

$$\bar{\rho} = \frac{M}{V} = \frac{5.976 \times 10^{24}}{1.087 \times 10^{21}} = 5500 \text{ kg/m}^3$$

The mass of the Earth (M_e) equals the mass of the core plus the mass of the mantle and crust. The mass of the core is its density (ρ_c) times its volume (V_c). Similarly, the mass of the mantle and crust is their density ($\rho_m = 3500$ kg/m^3) times their volume (V_m). So we can write:

$$M_e = \rho_c V_c + \rho_m V_m$$

From Problem 16, we know that the core occupies 17 percent of the Earth's volume, and so the mantle and crust occupy 100 − 17 = 83 percent of the volume. We can therefore write

$$V_c = 0.17 V_e \quad \text{and} \quad V_m = 0.83 V_e$$

where V_e is the volume of the Earth. Substituting these equalities in the formula for the Earth's mass, and dividing through by the Earth's volume, we get

$$\frac{M_e}{V_e} = \bar{\rho} = 0.17\, \rho_c + 0.83\, \rho_m$$

Substituting in the values for the Earth's average density ($\bar{\rho}$) and the density of the mantle and crust (assumed to be 3500 kg/m^3), we get

$$5500 = 0.17\, \rho_c + 0.83 \times 3500$$

and solving for ρ_c

$$\rho_c = \left(\frac{5500 - 2905}{0.17}\right) = 15{,}300 \text{ kg/m}^3$$

which is about four times denser than typical crustal rock.

21. Make a two-column table listing altitude and atmospheric pressure, halving the pressure for every 5 1/2 km of height above the Earth's surface as follows:

Altitude (km)	Pressure (bars)
0	1
5.5	1/2
11	1/4
16.5	1/8
22	1/16
27.5	1/32
33	1/64
38.5	1/128
44	1/256
49.5	1/512
55	1/1024 = 1 millibar

CHAPTER 9

OUR BARREN MOON

Chapter Summary

Lunar surface features, the absence of a lunar magnetic field and the nature of the lunar interior are reviewed. Tidal forces in the Earth–Moon system are described. The method of radioactive dating of rocks is explained. The composition and ages of lunar rocks are discussed in the context of the origin and evolution of the Moon. The strengths and weaknesses of the various theories of the origin of the Moon are examined.

9-1 The Moon's surface is covered with plains and craters

Earth-based observations of the Moon are described with reference to synchronous rotation and libration. The nature and probable origins of lunar maria, craters, highlands and mountain ranges on the lunar surface are delineated.

Box 9-1 Moon data

This table contains the orbital and physical data for the Moon.

9-2 Manned exploration of the lunar surface was one of the greatest adventures of human history

The efforts to reach and explore the lunar surface which took place in the 1960s and early 1970s are reviewed. The contributions of the Ranger, Surveyor, Orbiter, Zond, Luna and Apollo satellite series are discussed.

9-3 The Moon has no magnetic field but may have a small, solid core beneath a thick mantle

The Apollo experiments to measure seismic waves and magnetism are described. The results are reported and interpreted in terms of the Moon's structure.

9-4 Gravitational forces between the Earth and the Moon produce tides

Tidal forces on the extended bodies of the Earth and the Moon are described. Spring and neap tides are defined in the context of the phases of the Moon. The changing rotation rate of the Moon and increasing Earth–Moon distance are explained in terms of tidal interaction.

9-5 Lunar rocks were formed 3 to 4.5 billion years ago

The minerals found in lunar rock samples returned to Earth by the Apollo astronauts are identified. The composition of the lunar surface is examined and contrasted with that of the Earth.

Box 9-2 Radioactive age dating

The process of radioactive age dating of rocks is examined in some detail. Radioactive decay, isotopes and half-lives are defined and discussed. The calibration of abundance ratios of original to final stable isotope abundances using nonradiogenic isotopes is described.

9-6 The Moon probably formed from debris cast into space when a huge planetesimal struck the proto-Earth

Four major theories dealing with the origin of the Earth–Moon system are described in some detail. Evidence is presented in support of and against the various theories. Volatile and refractory elements are defined. A brief scenario for the formation and evolution of the Moon which is consistent with most of the data is outlined.

Teaching Hints and Strategies

[Refer to the Teaching Hints and Strategies section of Chapter 7 of this Instructor's Guide for general strategies applicable to solar system objects.]

The differences between the general appearance of *the surface of the near and far sides of the Moon* (section 9-1) in terms of highlands and maria should be noted. Be sure to explain the use of crater surface density to estimate ages of surface features on planets and satellites. The references to volcanism during various stages in the evolution of the Moon can be discussed in the context of the past theories of the possible volcanic origin of craters. Such theories have been abandoned in recent years, but were given serious consideration just a few years ago. The detailed picture we have of the lunar surface today stands in stark contrast to the diverse and uncertain theories which were not uncommon in the 1960's. Some discussion of those theories might illustrate the explosion of information which has resulted from the space program and permit greater appreciation of our current state of knowledge of many other solar system objects.

The *composition and ages of lunar rocks* (section 9-2) brought back by Apollo astronauts have proven to be invaluable in generating information which has led to a rather detailed scenario regarding the origin and evolution of the Moon. The fact that the Moon has no atmosphere is one reason that astronomers are interested in the lunar surface. Discuss the role of wind erosion and water erosion on the Earth and their impact on the ages of rocks and surface features. Stress the dramatic effect of the freezing and thawing cycle for water and the fact that it expands as it freezes. This process is very destructive as potholes in most highways in northern states clearly illustrate. The lack of these processes on the Moon permit much older rocks to exist there. The lunar surface is generally much older than the surface of the Earth. The dating of lunar rocks is much more informative about early periods in the evolution of the solar system than the same process for Earth rocks due to the lack of erosion on the Moon.

Emphasize that the *experiments left on the Moon* (section 9-3) as part of the Apollo exploration have supplied data long after the astronauts left the lunar surface. The seismometers have generated data which provides insight into the interior structure of the Moon just as seismic data does for the Earth. The study of the Moon has provided valuable information regarding the early history of the Earth about which no record remains on the Earth. Understanding the Earth is of direct benefit to the tax-paying public. Also, the Apollo landings had high "entertainment" value. The cost to advertisers to get the viewing ratings achieved then would have paid for much of the space program.

Students have great difficulty understanding why there are two *tides* (section 9-4) each day rather than one. You might want to discuss the Roche limit at this point and note the role of tidal forces in maintaining ring systems. It would be wise to refer back to Box 3-1 for relevant information about the orbit of the Moon to clarify the effects of the varying distance of the Moon from the Earth and the precessional cycle to note their impact on the tides. Also see the corresponding comments in this Instructor's Guide.

The analysis of the chemical abundances of lunar rocks as compared to Earth rocks is essential to our *understanding of the origin of the Moon* (section 9-5). Some reference to theories other than the collisional ejection theory should be made, as the theory of the origin of the Moon is much less certain than the general theory for the formation of the solar system.

When discussing *theories of the origin of the Earth and Moon* (section 9-6) it is wise to remind students yet again that the solar system formed long after the origin of the universe. Many students, if not a majority, attribute the origin of the solar system to events directly related to the Big Bang. Remind students that the solar system is only about 5 billion years old while the origin of the universe is dated at 15–20 billion years ago. Call attention to the definitions of volatile and refractory substances here.

The method of *radioactive age dating* (Box 9-2) described here is the primary source of data about the ages of such diverse celestial objects as the Earth, the Moon and meteorites. This might be an appropriate time to discuss the connection between radioactive dating techniques, the theory of evolution, and other methods of determining ages. If you have discussed the definition of isotopes, you can explain how certain isotopes of lead can be traced to specific isotopes of uranium. The ages determined by this process are usually the most recent solidification ages of the rocks and minerals being analyzed. The uncertainty associated with this process is now quite small. Meteorites especially are very consistent in age. The upper limit for the ages of meteorites is very clearly defined at 4.5–4.6 billion years ago for many of the oldest meteorites (carbonaceous chondrites). This presents a clear dilemma for persons believing in a relatively recent creation. While the evidence for evolution might be somewhat less than totally convincing, the ages of meteorites and rocks from the Earth and the Moon have been clearly established. This is an excellent time to review the essay by Sandra Faber again to emphasize the value of astronomy.

You might want to discuss the impact of so-called Creation "Science" on modern science education at this point. The effort to require the teaching of a theory of creation along with the theory of evolution by state legislatures seems to be a recurring problem for modern biology and for science in general. The exclusion of the theory of evolution from many high school biology texts due to this requirement by the state of Texas clearly demonstrates how a concentrated effort by a small group of persons can directly affect science education throughout the country due to economic considerations. Recent complaints about "bland" textbooks in reports about the crisis in education are of interest here as well. This might be an opportune time to discuss briefly how the ages of stars are determined from theories of stellar evolution and how the age of the universe is estimated from the expansion of stellar systems. The consistency of the results of these three independent methods of age determination are reassuring.

Answers to Chapter 9 Computational Questions

16. The average density of the Moon is its mass (M) divided by its volume (V). From Box 9-1 we see that the Moon's mass is 7.35×10^{22} kg, and its radius (r) is 1/2 (3476 km) = 1.738×10^6 m.

$$\bar{\rho} = \frac{M}{V} = \frac{M}{\frac{4}{3}\pi r^3} = \frac{7.35 \times 10^{22} \text{ kg}}{\frac{4}{3}\pi (1.738 \times 10^6)^3} = 3340 \frac{\text{kg}}{\text{m}^3}$$

Crustal rocks from the Earth and Moon have densities in the range of 3000 to 3500 kg/m³. The Moon cannot have a significant iron core because, if it did, its average density would be higher than the density of typical crustal rock.

17. First calculate the weight of the 80-kg person on Earth, which is mass times the acceleration due to Earth's gravity (9.8 m/s²):

Earth weight = 80 × 9.8 = 784 newtons = 176 pounds

This weight is the force with which gravity pulls the person down on the surface of the Earth. According to Newton's law of gravitation, that force (F_E) is given by

$$F_E = G \frac{mM_E}{R_E^2}$$

where G is the universal constant of gravitation, m is the person's mass, and M_E and R_E are the mass and radius of the Earth, respectively.

Similarly, the person's weight on the Moon (F_M) is given by

$$F_M = G \frac{mM_M}{R_M^2}$$

where M_M and R_M are the mass and radius of the Moon, respectively.

Now find the ratio of the person's Moon-weight to his Earth-weight by dividing the equation for F_M by the equation for F_E:

$$\frac{F_M}{F_E} = \left(\frac{M_M}{M_E}\right)\left(\frac{R_E}{R_M}\right)^2$$

We know that M_M/M_E = 0.012, and R_M/R_E is (1738 km/6378 km) = 0.27. Consequently

$$\frac{F_M}{F_E} = (0.012)(0.27)^{-2} = 0.16$$

Therefore a person standing on the Moon weighs 16 percent of his/her Earth weight. An 80-kg person presses down on the lunar surface with a force of

0.16 × 784 newtons = 125 newtons = 28 pounds

18. As we saw in section 9-3 of the text, seismometers left on the Moon by Apollo astronauts detected up to 150 impacts per year by meteoroids with masses up to 1000 kg. Let's use these upper limits in the hope that they will adequately include matter added to the Moon by meteoroids too small to be detected by the seismometers. Thus, the rate at which the Moon is gaining mass is:

$$150 \times 1000 = 1.5 \times 10^5 \frac{\text{kg}}{\text{yr}}$$

The Moon's mass is 7.35×10^{22} kg. To see how long it would take to add this amount of matter to the Moon, we divide the Moon's mass by the accretion rate:

$$\frac{7.35 \times 10^{22}}{1.5 \times 10^5} = 4.9 \times 10^{17} \text{ years}$$

which is more than ten million times the age of the universe. Clearly, meteoroid impacts are increasing the Moon's mass at an insignificant rate.

18. The Moon obeys Kepler's third law: the square of its period is proportional to the cube of its semimajor axis. Today, the Moon's period and semimajor axis are 27.32 days and 384,400 km, respectively. To find the Earth–Moon distance (d) when its orbital period is 47 days, we construct the following proportion:

$$\frac{(27.32)^2}{(384,400)^3} = \frac{(47)^2}{d^3}$$

Rearranging terms, we obtain

$$d = \left(\frac{47}{27.32}\right)^{2/3} 384,400 \text{ km} = 552,000 \text{ km}$$

CHAPTER

10

SUN-SCORCHED MERCURY

Chapter Summary

This chapter reviews the current state of knowledge of the planet Mercury. The limitations of Earth-based optical observations are discussed and the advantages of radio and radar observations are noted. The discovery and explanation of the 3-to-2 spin-orbit coupling are addressed and the peculiar cycles of days and seasons are examined. The results of the Mariner 10 mission form the basis of comparisons of the surface features with those of the Moon. The interior structure and magnetic field are compared with those of the Earth.

10-1 Earth-based optical observations of Mercury are difficult to make and often prove disappointing

The difficulty in observing Mercury from the Earth's surface is explained. The time difference between the rising or setting of Mercury and the Sun at greatest elongation is computed. Favorable and unfavorable elongations are contrasted and a table of maximum elongations through 1999 is presented. Albedo is defined and solar transits by inferior planets are discussed.

Box 10-1 Mercury data

This table contains the orbital and physical data for the planet Mercury.

10-2 Mercury rotates slowly and has an unusual spin–orbit coupling

Synchronous rotation is defined and the evidence leading to the discovery that Mercury is not in synchronous rotation is reviewed. Integer spin–orbit coupling is described and the relationship between the orbital period and rotation period is discussed in the context of daily and seasonal cycles.

10-3 Photographs from Mariner 10 revealed Mercury's heavily cratered, lunarlike surface

The Mariner 10 photographic revelation of extensive cratering, intercrater plains and scarps provide for excellent comparisons with related features on the lunar surface. Differences are discussed in terms of the different evolutionary histories of the two bodies due to different masses and chemical compositions.

10-4 Mercury has an iron core and a magnetic field, like Earth

The average density of Mercury is used to develop a simple model of the planet's interior which is qualitatively similar to that of the Earth. Mercury's magnetic field is contrasted with that of the Earth. The nature of the magnetosphere is described and the interaction of magnetic fields with the solar wind to form bow shocks and magnetosheaths is discussed.

Teaching Hints and Strategies

[Refer to the Teaching Hints and Strategies section of Chapter 7 of this Instructor's Guide for general strategies applicable to solar system objects.]

When discussing *Earth-based observations of Mercury* (section 10-1), it is important that students understand that Mercury is almost never seen in total darkness. It is seen during morning and evening twilight when the atmospheric temperatures are undergoing rapid change, which causes turbulence. This makes studies of Mercury from the Earth very difficult at best. If you look up the rising and setting times of Mercury, you will find only a few nights each year when Mercury rises or sets before or after astronomical twilight begins or ends. It is interesting to note that Mercury has the second most eccentric and second most inclined orbit after Pluto, as is clearly shown in Appendix 1 in the textbook.

The peculiar *rotation rate of Mercury* (section 10-2) leads to the very strange situation that one solar day is two years long. It is important that students learn to translate the periods of astronomical motions such as rotation and revolution into meaningful cycles which we experience here on Earth. The variety of these motions in the solar system is quite enlightening. Try to get your students to consider the effects of rotation period, revolution period, orbital eccentricity and tilt of the axis of rotation into physically observable phenomena. What is the period from one sunrise to the next? How long is one set of seasons? How does the altitude of the Sun vary throughout the year? What is the dominant cause of seasons? What is the range of temperature throughout one daily cycle and one annual cycle? How would these phenomena influence colonies on these bodies?

The *Mariner 10 photographs of Mercury* (section 10-3) provide an excellent opportunity to involve students in comparative planetology. Have them compare images of Mercury and the Moon as shown in Figures 10-7 to 10-14 for Mercury to Figures 9-3 to 9-7 for the Moon. These images clearly display the similar surfaces. Point out the differences between the front and back sides of the Moon shown in Figure 9-6. As a reinforcing activity you might ask students to look ahead to Figures 12-4, 12-23, 14-14, 14-15, 14-17, 15-20, 15-24, 16-6 and 16-16 to show that cratering is a common surface feature on the surfaces of many solar system objects. You might ask why images of some solar system objects are dominated by impact craters and others are not. Note the roles of geologic activity and of atmospheres in limiting or eliminating craters. The lack of any substantive atmosphere on Mercury (section 10-2) is related to it high surface temperature combined with its low escape velocity. Refer back to Box 7-2 to show that the escape velocity on Mercury is much lower than on Venus. Thus Venus can retain gases quite readily which are lost on Mercury even though Venus has a higher surface temperature.

While the size and surface features of Mercury are like those of the Moon, the *interior of Mercury* (section 10-4) is much more similar to the interiors of the Earth and Venus. Mercury has a high mean density which indicates that it is much more iron rich than any other planet or moon in the solar system.

When discussing the *magnetic field of Mercury* (section 10-4), note that the slow rotation was expected to preclude a magnetic field due to the dynamo effect, yet it has a weak field similar to that of Earth. Compare the scales for Figures 10-17 and 8-19 to be certain that students realize that the magnetic field of Mercury is much weaker than that of Earth. Also compare Figure 10-16 to Figure 9-20. Detailed differences in the collisional parameters resulted in the retention of stripped matter from the Earth as a satellite, and the loss of the stripped matter by Mercury. It would be good to review Box 4-5 here.

Answers to Chapter 10 Computational Questions

14. We must first calculate the distance to Mercury when it is at greatest elongation, at an angle of about 25° from the Sun. It is helpful to draw a diagram, as shown below:

Since 1 AU = 1.5×10^{11} m, the distance (r) between Earth and Mercury is

$$r = (1.5 \times 10^{11} \text{ m})(\cos 25°)$$

$$= (1.5 \times 10^{11})(0.906) = 1.36 \times 10^{11} \text{ m}$$

We can now turn to the small-angle formula (see Box 1-1):

$$D = \frac{\alpha r}{206,265}$$

with α = 1 arc sec. Thus,

$$D = \frac{1 \times 1.36 \times 10^{11} \text{ m}}{2.06 \times 10^{5}} = 0.66 \times 10^{6} \text{ m} = 660 \text{ km}$$

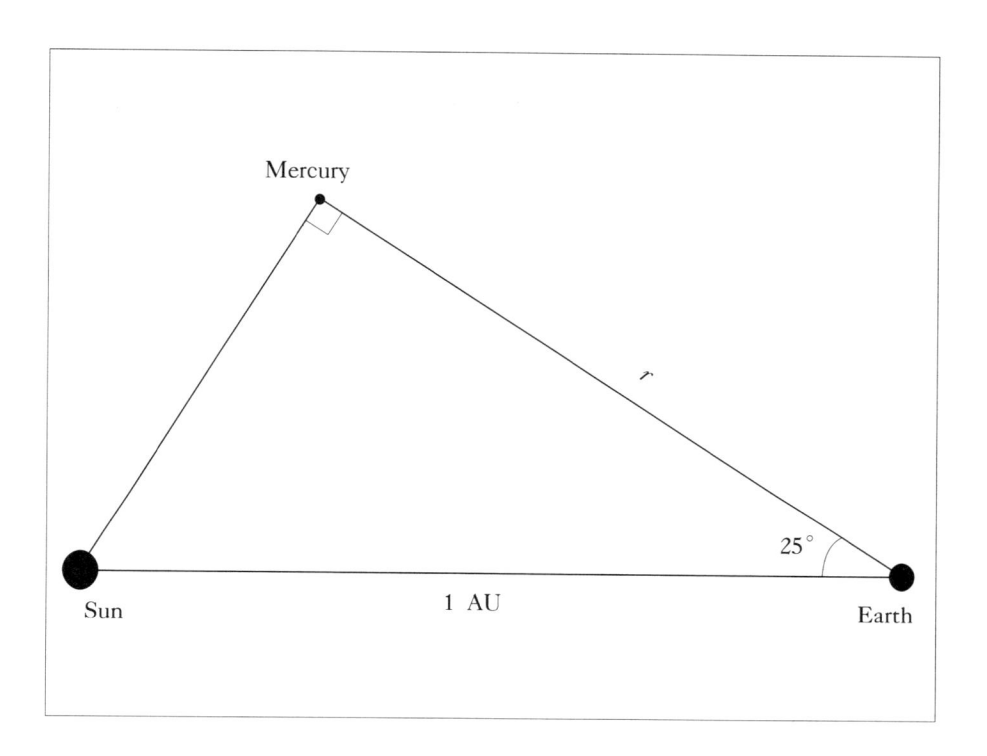

As mentioned in the text (section 10-3), the diameter of the Caloris Basin is about 1300 km, so this feature is resolved.

16. As mentioned in section 10-2 of the text, the temperature of Mercury's sunlit side reaches a maximum of about 700 K. Using Wien's law we have

$$\lambda_{max} = \frac{2.9 \times 10^{-3} \text{ mK}}{T}$$

$$= \frac{2.9 \times 10^{-3}}{700} \text{ m} = 4.14 \times 10^{-6} \text{ m}$$

$$\approx 4.1 \text{ μm}$$

Note that if you had used the average daytime temperature of 623 K given in Box 10-1, you would have calculated that

$$\lambda_{max} = 4.6 \text{ μm}$$

18. The velocity (v) of a point on Mercury's equator due to the planet's rotation is Mercury's circumference ($2\pi R$) divided by its period of rotation (P):

$$v = \frac{2\pi R}{P}$$

$$= \frac{2\pi(2439 \text{ km})}{58.6 \text{ days}}$$

$$= \frac{2\pi(2439 \text{ km})}{(58.6 \text{ d})(24 \frac{\text{hr}}{\text{d}})(3600 \frac{\text{s}}{\text{hr}})}$$

$$= 3.03 \times 10^{-3} \text{ km/s}$$

We can now make use of the formula for the Doppler shift

$$\frac{\Delta\lambda}{\lambda_o} = \frac{v}{c}$$

with $\lambda_o = 500$ nm. Thus,

$$\Delta\lambda = \left(\frac{v}{c}\right)\lambda_o$$

$$= \left(\frac{3.03 \times 10^{-3} \text{ km/s}}{3 \times 10^5 \text{ km/s}}\right)(500 \text{ nm})$$

$$= 5.05 \times 10^{-6} \text{ nm}$$

Thus light reflected from the approaching side of Mercury is blueshifted by about 5×10^{-6} nm while light reflected from the receding side of the planet is redshifted by that same amount. The total wavelength shift across the planet is therefore $2\Delta\lambda$ or about 10^{-5} nm.

19. As we saw in Box 7-2, the requirement for the retention of a gas is:

$$6\left(\frac{3kT}{\mu m_H}\right)^{1/2} < \left(\frac{2GM}{R}\right)^{1/2}$$

Squaring both sides of this inequality we get

$$36\left(\frac{3kT}{\mu m_H}\right) < 2G\frac{M}{R}$$

and so

$$\mu > \frac{54kTR}{GM m_H}$$

Inserting the mass, radius, and temperature of Mercury along with the values of the physical constants we get

$$\mu > \frac{54\,(1.38 \times 10^{-23})(620)\,(2.439 \times 10^6)}{(6.67 \times 10^{-11})\,(3.3 \times 10^{23})\,(1.67 \times 10^{-27})} \approx 31$$

Thus the molecular weight must be greater than 31. Oxygen ($\mu = 32$), carbon dioxide ($\mu = 44$), and argon ($\mu = 40$) are possible atmospheric constituents. But the Sun's ultraviolet radiation breaks O_2 and CO_2 molecules into atoms that easily escape from the planet.

20. From Newton's second law (see section 4-6) we know that the weight (W) of an object equals its mass (m) times the accelertion due to gravity (g):

$$W = mg$$

On the Earth, g = 9.8 m/s² and so an 80-kg person weighs

$$80 \times 9.8 = 784 \text{ N} = 176 \text{ pounds}$$

In Box 10-1 we read that the surface gravity on Mercury is 0.39 times that of Earth. Thus an 80-kg person standing on Mercury weighs

$$0.39 \times 784 \text{ N} = 306 \text{ N} = 69 \text{ pounds}$$

Similarly, in Box 9-1 we read that the Moon's surface gravity is 0.166 of that on Earth. So an 80-kg person on the Moon would weigh

$$0.166 \times 784 \text{ N} = 130 \text{ N} = 29 \text{ pounds}$$

21. From Kepler's third law we know that

$$P^2 = a^3$$

For Mercury P = 87.969 days = 0.2408 yr

For *Mariner 10* P = 2 × 0.2408 yr = 0.4816 yr

$$a^3 = (0.4816)^2 = 0.2319$$

$$a = (0.2319)^{1/3} = 0.614 \text{ AU}$$

The semi-major axis of *Mariner 10*'s orbit is therefore about 0.61 AU.

CHAPTER 11

CLOUD-COVERED VENUS

Chapter Summary

This chapter describes the thick atmosphere of Venus which limits our observations of the surface, the retrograde rotation of the planet, the greenhouse effect, the lack of a magnetic field and what the data collected by the Pioneer Venus satellites tell us about the nature of the atmosphere. The impact of volcanoes on atmospheres are described and the radar mapping results are examined.

11-1 The surface of Venus is hidden beneath a thick, highly reflective cloud cover

The Earth-based observations of Venus as a morning and evening "star" are described. Its albedo, apparent magnitude, changing phases and apparent size are reviewed. Table 11-1 contains the dates of the conjunctions and elongations of Venus for the decade of the 1990s.

Box 11-1 Venus data

This table contains the orbital and physical data for the planet Venus.

11-2 Venus's rotation is slow and retrograde

The spectroscopic determination of the retrograde rotation of Venus is discussed. Early radar observations of the accuracy of the rotation rate are detailed and the integer relationship between daily cycles of Venus and the interval from one inferior conjunction to the next is noted.

11-3 The surface of Venus is very warm because of the greenhouse effect

The determination of the surface temperature of Venus using radio observations is described. The concept of the greenhouse effect, discussed in section 8-6 for Earth, is reviewed. The role of the Venusian atmosphere in driving up the temperature is explained.

11-4 Spacecraft descending into Venus's dense, corrosive atmosphere found several sulfur-rich cloud layers

The *Venera* spacecraft of the Soviet Union are described as are the *Pioneer Venus Orbiter* and *Multiprobe* of the United States. The upper, middle and lower cloud layers of the Venusian atmosphere are identified and their vertical extents defined. The chemical composition of the Venusian atmosphere is examined in detail. The roles of the Pioneer satellites in the investigation of the atmosphere are examined. Convection is defined and circulation patterns are described. The rapid rotation of the atmosphere is contrasted to the slow rotation of the planetary surface.

Box 11-2 Venus's ionosphere

The failure of the *Mariner 2* satellite to detect a magnetic field is noted. The slow rotation of Venus is identified as the probable reason for the lack of a magnetic field. The influence of the Venusian ionosphere on the solar wind is reviewed.

11-5 Active volcanoes are probably responsible for Venus's clouds

The circumstantial evidence that active volcanoes may be contributing to the Venusian atmosphere is reviewed. A brief history of the evolution of the Venusian atmosphere is suggested.

11-6 The absence of water on Venus suggests a hellish past

Possible reasons for the depletion of carbon dioxide in the Earth's atmosphere and of water in the Venusian atmosphere are reviewed. The theory that there may have been a large amount of water in Venus's atmosphere at one time is described. The mechanism for the destruction of water molecules is given. Images by the Venera probes are displayed

11-7 Venus is covered with gently rolling hills and two "continents"

The detailed mapping of the Venusian surface by the Magellan spacecraft is discussed. The two Venusian continents are identified and the general topography is described and compared with that of the Earth. Volcanoes on Venus are compared with those on Earth and on Mars. The role of plate tectonics on Venus is contrasted with its counterpart on Earth.

84 CHAPTER 11

Teaching Hints and Strategies

[Refer to the Teaching Hints and Strategies section of Chapter 7 of this Instructor's Guide for general strategies applicable to solar system objects.]

It should be emphasized that the dense atmosphere of Venus has made visual *observations of its surface* from Earth (section 11-1) impossible. This dense atmosphere results in a very high albedo which is why Venus is the third brightest celestial source behind the Sun and Moon.

The *retrograde rotation of Venus* (section 11-2) has some interesting facets. Note the duration of a Venusian day of about 116 Earth days. You might discuss how this might effect the daily temperature variation there. How much does the Earth's temperature change over a typical daily cycle? What factors other than rotation period influence the range of temperature? Point out that water has a very high specific heat and acts as a thermal storage medium on Earth. Also be sure to discuss the slightly different meaning of the term retrograde here. Retrograde motion in the context of cosmology is the periodic westward motion of the planets relative to the background stars. This is in contrast to their more typical eastward motion called direct motion. The retrograde rotation of Venus refers to the fact that nearly all planets rotate in a counter-clockwise direction as seen from the north, while Venus rotates in the clockwise direction.

The *greenhouse effect for Venus* (section 11-3) can be compared to that on Earth. Remember to point out that the atmosphere of Venus is dominated by carbon dioxide which is just a trace gas in Earth's atmosphere. Without liquid water on its surface, Earth might have had much more carbon dioxide in its atmosphere and a much higher surface temperature and atmospheric pressure. Many students erroneously believe that Mercury must have the highest surface temperature because it is the planet closest to the Sun.

The *absence of a magnetic field on Venus* (Box 11-2) can be used to review the nature of the dynamo effect on Earth. Although Venus has a similar size and density, it rotates much too slowly to generate a magnetic field. Be sure to note the very different explanations for the Box figure and Figure 8-19 and the very different scales involved.

It is interesting to note that the *clouds in the atmosphere of Venus* (sections 11-4 and 11-5) are dominated by sulfur compounds while sulfur is not a major constituent of the atmosphere. This is similar to Earth's atmosphere in which water vapor clouds are common while water vapor is not abundant. Have students look for similar examples in the atmospheres of other planets.

The *origins of the atmospheres of the terrestrial planets* (section 11-6) should be emphasized. The role of volcanic activity in the injection of gases into the atmospheres of terrestrial planets should be clearly identified. *Volcanoes* (section 11-5) have been found on the surfaces of nearly all terrestrial planets and on many of the major satellites as well. Many students tend to believe that a planetary atmosphere is permanent and unchanging. Try to reinforce the ideas of evolution of planetary atmospheres introduced in section 8-6 for the Earth. You might have students compare the sizes of the largest volcanoes found in the inner solar system. Ask them to comment on why such structures are found on other bodies and why they have such different sizes. Russian spacecraft have been landed on Venus's surface (one of only two planets studied in this manner to date) and provided a few photographs.

Point out that *radar mapping* (section 11-7) by radar from Earth and by space probes has provided virtually all of our information about the surface topography of Venus due to its dense atmosphere. Note also that Venus is only one of three celestial objects to date upon which space probes have made soft landings and returned images and other scientific data. You might also note that the former Soviet Union has been more successful in experimenting on the surface of Venus while the United States has been more successful on Mars. Emphasize the radar maps of Venus generated by the Magellan spacecraft. New Magellan pictures are being released periodically.

Answers to Chapter 11 Computational Questions

14. In Box 11-1 on page 195 we read that the surface temperature on Venus is 750 K. Making use of Wien's law we find that

$$\lambda_{max} = \frac{2.9 \times 10^{-3} \text{ mK}}{T}$$

$$= \frac{2.9 \times 10^{-3}}{750} \text{ m}$$

$$= 3.87 \times 10^{-6} \text{ m}$$

$$\approx 3.9 \text{ μm}$$

This infrared wavelength could be easily detected telescopically, were it not for Venus's thick cloud cover that efficiently absorbs such radiation.

15. The velocity (v) of a point on Venus's equator due to the planet's rotation equals Venus's circumference ($2\pi R$) divided by its period of rotation (P):

$$v = \frac{2\pi R}{P}$$

$$= \frac{2\pi (6.057 \times 10^3 \text{ km})}{243.01 \text{ days}}$$

$$= \frac{2\pi(6.057 \times 10^3 \text{ km})}{(243.01 \text{ d})(24 \text{ hr/d})(3600 \text{ s/hr})}$$

$$= 1.813 \times 10^{-3} \text{ km/s}$$

We can now make use of the formula for the Doppler shift

$$\frac{\Delta \lambda}{\lambda_0} = \frac{v}{c}$$

with $\lambda_0 = 550$ nm. Thus

$$\Delta \lambda = \left(\frac{v}{c}\right)\lambda_0$$

$$= \left(\frac{1.813 \times 10^{-3} \text{ km/s}}{3 \times 10^5 \text{ km/s}}\right)(550 \text{ nm})$$

$$= 3.32 \times 10^{-6} \text{ nm}$$

Thus light reflected from the approaching side of Venus is blueshifted by about 3.32×10^{-6} nm while light from the receding side of the planet is redshifted by that same amount. The total wavelength shift across the planet is therefore $2\Delta\lambda$ or about 6.64×10^{-6} nm.

19. You need to be careful in interpreting this problem. Pressure is properly measured in newtons per square meter (not kg/m^2). You must multiply the density (ρ) by the acceleration of gravity (g) to find the pressure exerted by a 1-m-tall column of water. For a column h meters tall, the resulting pressure (P) is:

$$P = \rho g h$$

thus

$$h = \frac{P}{\rho g}$$

$$P = 90 \text{ atm} = 90 \times 1.013 \times 10^5 \text{ N/m}^2$$

$$g = 9.8 \text{ m/s}^2$$

$$\rho = 1000 \text{ kg/m}^3$$

$$h = \frac{90 \times 1.013 \times 10^5}{1000 \times 9.8} = 930 \text{ m}$$

or nearly 1 km beneath the sea!

CHAPTER 12

THE MARTIAN INVASIONS

Chapter Summary

This chapter begins with a brief history of Earth-based observations of Mars. Martian surface features, physical properties and composition of the Martian atmosphere, and geological information about surface materials are reviewed as revealed by an active space exploration program. Properties of the two Martian satellites Phobos and Diemos and details about the results of Viking experiments, which searched for evidence of life forms on the surface of Mars, are also presented.

12-1 The best Earth-based observations of Mars are made during favorable oppositions

The value of favorable oppositions is pointed out. Opposition dates through 1999 are presented. The apparent brightness and size variations are related.

Box 12-1 Mars data

This table contains the orbital and physical data for the planet Mars.

12-2 Earth-based observations originally suggested that Mars might have some form of extraterrestrial life

A brief history of Earth-based investigations of Martian surface features by Cassini, Herschel, Schiaparelli and Lowell is given.

12-3 Space probes to Mars found craters, volcanoes, and canyons—but no canals

Photographic information from early space probes that revealed very old impact craters, but no evidence of canals, is discussed. The volcanoes revealed by *Mariner 9* are compared with volcanoes on Earth and on Venus. The Valles Marineris is described and comments are provided about how these features are interpreted in the context of plate tectonics.

12-4 Surface features indicate that water once flowed on Mars

Evidence for the existence of water, such as residual polar caps and old stream channels similar to river beds, is discussed in the context of the thin Martian atmosphere. Ancient floods on Mars are contrasted with typical rivers on Earth and with ancient Earth flooding.

12-5 Earth and Mars began with similar atmospheres that evolved very differently

The evolution of the Martian atmosphere is described. The role of water in converting carbon dioxide from the atmosphere into carbonate rocks on both Earth and Mars is discussed.

12-6 The Viking landers sent back close-up views of the Martian surface

The missions of the two Viking spacecraft are detailed. Explanations of the seasonal changes in the appearance of the surface are presented.

12-7 Instruments on the Viking landers revealed a highly seasonal Martian climate

The composition and physical properties of the Martian atmosphere are compared and contrasted with those of the Earth's atmosphere. Daily and seasonal variations in climate are examined.

12-8 The Viking landers found abundant iron in the Martian regolith but failed to detect living organisms

The chemical composition of the Martian surface as determined by Viking, the average density for Mars, and absence of a magnetic field are reviewed to suggest that chemical differentiation is quite different on Mars than on Earth. The three experiments performed during the Viking mission to search for evidence of life forms are described and their results are discussed. An explanation is provided for the early reports that suggested a positive result.

12-9 The two Martian moons resemble asteroids

The physical and orbital properties of Phobos and Diemos are considered. The accretion and capture theories for the origin of these satellites are presented.

12-10 Future missions to Mars will give us important new insights about the planet

The loss of the Mars Observer spacecraft is mentioned and its mission objectives are outlined. The inevitability of a Mars Observer-like mission and some of its possible observations are discussed. Russian plans to explore Mars in the near future are described.

Essay by Marcia Neugebauer
Exploring the Planets

Essay Summary

This essay presents many of the questions about the Earth and planets that drive the scientists involved in the unmanned space exploration programs. The various stages in planetary exploration ranging from flybys to manned landings are listed and discussed.

    This essay provides a good introduction to a class discussion of the motivations of the scientists actually involved in the space program compared to those of the general public. The benefits to the scientists are knowledge, satisfaction, employment, career advancement, etc. The benefits to the public are an increasing awareness of the universe we live in and the considerable fruits of the new technologies developed to support the space program. These often provide economic benefits that pay for the exploration many times over.

Teaching Hints and Strategies

[Refer to the Teaching Hints and Strategies section of Chapter 7 of this Instructor's Guide for general strategies applicable to solar system objects.]
    Man's long investigation of the *Martian surface* (sections 12-1 and 12-2) with optical telescopes has led to much speculation about the existence of life on Mars. A great deal of science fiction literature concerning Mars has resulted. In fact, the scientific evidence thus far still makes Mars a possible source of life forms in the past if not at present. The results of the Viking spacecraft in this context should be emphasized.
    The spectacular *spacecraft images of the Martian surface* (section 12-3) provide clues about the nature of the Martian interior. Have students look first for similarities with surface features on other bodies and then try to find differences. Stress the explanations and interpretations which are derived from the pretty pictures.
    The *stream channels on Mars* (section 12-4) are clear evidence of an atmosphere that has changed with time. The dendritic pattern of tributaries to the

dried river beds indicates a widespread distribution of surface water that could only have been produced by rainfall. On the other hand, the large flood channels that drain into low-lying plains were caused by melting permafrost producing a different form of water erosion. Neither form of water can exist in the contemporary Martian surface environment.

Some of the dynamics of *atmospheric evolution* (section 12-5) on Earth and Mars help to develop an understanding of the importance of the planetary exploration program. The ability to examine planets having different physical conditions from Earth permits astronomers to develop a better appreciation for the details of how atmospheres may change with time. This information will be helpful for scientists trying to predict how Earth's atmosphere will evolve in the future.

The value of the *Viking landers* (sections 12-6 through 12-8) is reflected by the wealth of information they have provided about the surface and atmosphere of Mars as summarized in several sections of this chapter. Emphasize again that Mars is only one of three celestial objects beyond Earth on which we have been able to achieve soft landings followed by data collection and transmission. Remind students that space exploration in the outer solar system has just begun and should not be abandoned.

The *small moons of Mars* (section 12-9) are the first very small objects in the solar system for which direct images were available. Point out that they are not spherical like planets and satellites. Also draw attention to the cratering on their surfaces and note the fact that they may not be natural satellites. The similarity in appearance of Phobos and Deimos to the asteroid Gaspra as photographed by the Galileo spacecraft (Figure 17-6), supports the idea that the former may be captured asteroids.

When discussing the *Mars Observer* (section 12-10) be sure to emphasize that the loss of the spacecraft in 1993 is only a temporary setback. The correct next step is still to place a long-lived orbiter at Mars. A spoiled public and congress have come to expect flawless missions in the unmanned space exploration missions. Like any piece of hardware subject to a hostile environment, we are bound to lose one occasionally. Follow-ons include a remotely manipulated Mars rover and eventually direct exploration by humans. Perhaps even a permanent station will be a reality. The Biosphere II experiment in Arizona is a prelude to enclosed, self-sustaining environments. When students ask when all this is possible, you can tell them that it would be reasonable that it could be within their lifetimes. What seems to be lacking is not the technology but the will to use it in this way.

Answers to Chapter 12 Computational Questions

15. The linear diameter (D) of the smallest feature visible through a telescope that can resolve an angle α is given by

$$D = \frac{\alpha d}{206{,}265}$$

where d is the distance from the object to the observer (see Box 1-1).

During the 1995 opposition $d = 10^8$ km

For $\alpha = 1''$ we have

$$D = \frac{1 \times 10^8}{206{,}265} \text{ km}$$

$$= 480 \text{ km}$$

For the Hubble Space Telescope with $\alpha = 0.1''$ we get

$$D = \frac{0.1 \times 10^8}{206{,}265}$$

$$= 48 \text{ km}$$

16. You must first calculate the distance from Mars to Earth at greatest elongation. At greatest elongation, the Earth, Mars, and the Sun form a right triangle.

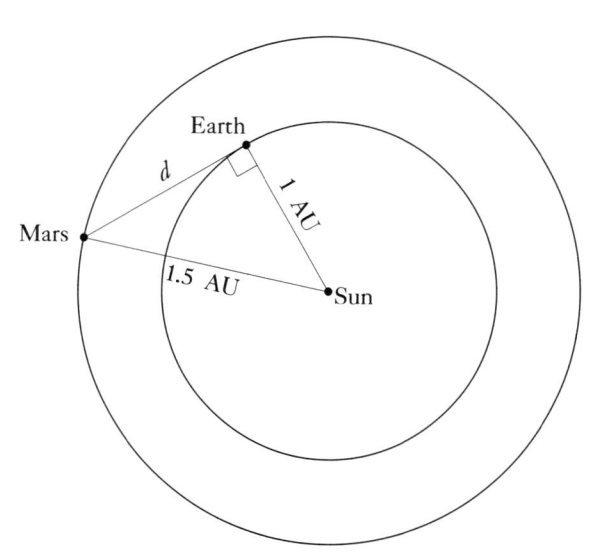

We can therefore use the Pythagorean Theorem to calculate d, as follows:

$$d^2 + 1^2 = (1.5)^2 = 2.25$$
$$d^2 = 2.25 - 1 = 1.25$$
$$d = \sqrt{1.25} = 1.12 \text{ AU}$$
$$d = 1.12 \times 1.496 \times 10^8 \text{ km}$$
$$= 1.67 \times 10^8 \text{ km}$$

The average distance between the Earth and the Moon is 384,400 km = D.

Using the small-angle formula in Box 1-1, we can calculate the largest angular separation between the Earth and Moon, as seen from Mars:

$$\alpha = \frac{206{,}265 D}{d}$$
$$= \frac{206{,}265 \times 384{,}400}{1.67 \times 10^8}$$
$$= \frac{2.06 \times 3.84 \times 10^8}{1.67 \times 10^8} = 474 \text{ arc sec}$$
$$= 7.9 \text{ arc min}$$

Since this is much larger than the resolving power of the human eye, you would indeed see the Earth and Moon as a "double star."

17. From Appendix 3 we learn that the orbital radius (*a*) of Phobos is 9380 km and its orbital period (*P*) is 0.319 day = 2.76×10^4 seconds.

From Box 4-3 on page 68 we see that the appropriate form of Kepler's third law is:

$$P^2 = \left(\frac{4\pi^2}{GM}\right) a^3$$

where *M* is the mass of Mars.

Thus,

$$M = \frac{4\pi^2 a^3}{GP^2} = \frac{4\,(3.1416)^2\,(9.38 \times 10^6)^3}{(6.67 \times 10^{-11})(2.76 \times 10^4)^2} \text{ kg}$$
$$= 6.41 \times 10^{23} \text{ kg}$$

This is quite close to the value of 6.42×10^{23} kg listed in Box 12-1 on page 210.

CHAPTER 13

JUPITER: LORD OF THE PLANETS

Chapter Summary

The differences between terrestrial and Jovian planets are reviewed in the context of temperature differences in the early solar nebula. The features visible in the Jovian atmosphere are described. Differential rotation, radio observations and the oblate shape of the planet are related to qualitative conclusions about the interior structure of the planet. The liquid molecular hydrogen and liquid metallic hydrogen layers of the interior are described and related to the large, complex Jovian magnetic field. The detailed observations of the atmospheric features provided by the *Pioneer* and *Voyager* spacecraft are reviewed and explained. The temperature variations with altitude are displayed and circulation patterns are discussed in some detail.

13-1 Huge, massive Jupiter is composed largely of lightweight gases

The mass and average density of Jupiter are recounted and the abundances of hydrogen and helium in the planet are reviewed. Typical Earth-based observations of Jupiter are described with belts, zones and the Great Red Spot denoted. Differential rotation is discussed in the context of optical and radio observations. Table 13-1 lists the apparent angular size and apparent magnitudes of Jupiter at the oppositions between 1993 and 1999.

Box 13-1 Jupiter data

This table contains orbital and physical data for the planet Jupiter.

13-2 Pictures taken during flybys show many details in Jupiter's clouds

Photographic data provided by the *Pioneer* and *Voyager* satellites are discussed and the nature of circulation patterns around the Great Red Spot and general wind conditions are elucidated.

13-3 Because of its rapid rotation, Jupiter's weather patterns are stretched around the planet

The vertical structure of the Jovian atmosphere is illustrated and the temperature and pressure profiles with depth are presented. The variations in the chemical composition of different cloud layers are discussed. High- and low-pressure systems are described in the context of terrestrial weather patterns. The Coriolis effect is discussed and cyclones and anticyclones are defined and related to white and brown ovals in the Jovian atmosphere. The zonal wind patterns are displayed graphically and computer modeling results are introduced. Potential chemical compounds which might be responsible for the colors of the clouds are discussed. Sulfur, phosphorus and organic compounds are considered. The future of space investigations of Jupiter by the Galileo satellite are described.

13-4 Jupiter's oblateness divulges the planet's rocky core

Oblateness is defined and the structure of the interior and approximate physical conditions in Jupiter's core are described.

13-5 A metallic hydrogen interior endows Jupiter with a powerful magnetic field

Thermal and nonthermal radiations are noted. The role of the magnetosphere of Jupiter in contributing to the decametric radio radiation is related. The liquid molecular hydrogen and metallic hydrogen shells of the interior of the planet are described. The magnetic field is explained as due to the rotation of the metallic liquid hydrogen shell. The mapping of the Jovian magnetosphere by the instruments on the *Pioneer 10* and *11* and the *Voyager 1* and *2* spacecraft is reported. The extremely high temperatures detected in the plasma in the magnetosphere are noted.

Teaching Hints and Strategies

[Refer to the Teaching Hints and Strategies section of Chapter 7 of this Instructor's Guide for general strategies applicable to solar system objects.]

The *features visible in the atmosphere of Jupiter* (section 13-1) can be discussed in the context of the major differences between physical conditions which exist on the Jovian planets as opposed to the terrestrial planets. The solid surfaces of the Jovian planets, if they exist at all, are buried under thick, dense atmospheres and deep layers of liquids. An observing session in which Jupiter can be viewed with a telescope can be very helpful in providing some appreciation for this material. Some astrophotography can record the belts and zones if suitable instrumentation is available.

For those students wanting to observe *oppositions of Jupiter*, (Table 13-1) you might point out that the Jovian planets are much more systematic in their oppositions than are the terrestrial planets. The inferior planets and Mars are close to the Earth and have quite erratic apparent motions. Note that the oppositions of Mars occur in alternate years. The Jovian planets are much farther away and appear to move more uniformly. Jupiter has oppositions about every 13 months. The synodic periods of the other Jovian planets approach 1 year as their distances from the Sun increase. (See Appendix 1 in the textbook.)

The *Great Red Spot* (section 13-2) has been visible for more than three hundred years and provides insight into the longevity of structures in Jupiter's atmosphere. The discussion of these structures in the context of meteorological terms familiar to the students will help students to understand.

When discussing the *vertical structure of Jupiter's atmosphere* (section 13-3), caution students to note the zero point in Figures 13-9 and 15-13. For the terrestrial planets, the altitude in the atmosphere is expressed with zero representing the surface of the planet. Without an observable surface for Jovian planets, some arbitrary reference level must be adopted as the "top" of the atmosphere. Note the the zero level seems to correspond to the point of reversal of the temperature profile. The numbers below this zero level represent "depth" in the atmosphere rather than "altitude or height."

The *computer simulations of Jovian clouds* might be discussed, after a brief introduction to the vocabulary, in the context of meteorological events on the Earth. This approach might provide a sense of familiarity with the terminology.

The mention of the *Galileo* spacecraft leads naturally to a discussion of the need for a systematic planetary exploration program. The terrestrial planets have been investigated by series of increasingly more sophisticated spacecraft. The first of these were passing observatories. Second generation versions were designed to study more detailed problems. The most sophisticated spacecraft were more experimental than observational in nature. The *Venera* landers on *Venus* and the *Viking* landers on Mars are excellent examples of this last class. The spacecraft which have explored the outer solar system are of the first- and early second-generation type. The *Galileo* spacecraft, designed to visit Jupiter, is a much more sophisticated piece of hardware than either *Pioneer* or *Voyager*. It is important to convey the idea that the exploration of the Jovian planets has just begun and it is essential to aggressively pursue this program in the future.

The *oblateness* (section 13-4) of Jovian planets provides direct data about their interior structures. It is interesting to point out also that the determination of a rotation period, simple for solid planets with visible surface features, becomes very complicated by the fluid nature of these objects. The description of Jupiter as a "failed star" should be noted as well. The concept of energy

96   CHAPTER 13

should be reviewed and applied to the excess energy emitted by Jupiter and other Jovian planets.

Be sure to note that the *magnetic field of Jupiter* (section 13-5) is produced by liquid metallic hydrogen rather than by iron, which is thought to be responsible for the magnetic fields of the terrestrial planets.

The *magnetosphere of Jupiter* shown in Figure 13-14 can be compared and contrasted with those of Mercury, shown in Figure 10-17 and of the Earth, shown in Figure 8-19.

Answers to Chapter 13 Computational Questions

14. Let $M_\odot$ and M_J be the masses of the Sun and Jupiter respectively. Let $r_\odot$ be the distance from the Sun to the center of mass of the Sun–Jupiter system. Let r_J be the distance from Jupiter to the same center of mass. Then, from the definition of center of mass, we have

$$M_\odot r_\odot = M_J r_J$$

Note that r_J is just the semimajor axis of Jupiter's orbit. Rearranging terms and inserting values from the appendixes we get

$$r_\odot = \left(\frac{M_J}{M_\odot}\right) r_J$$

$$= \left(\frac{1.90 \times 10^{27} \text{ kg}}{1.99 \times 10^{30} \text{ kg}}\right) \times 7.783 \times 10^8 \text{ km}$$

$$= 7.43 \times 10^5 \text{ km}$$

Thus the center of mass is 743,000 km from the Sun's center. Since the Sun's radius is 696,000 km, the center of mass is 47,000 km above the solar surface.

15. Let's use Io whose orbital parameters (see Appendix 3) are:

$$a = 421{,}600 \text{ km} = 4.216 \times 10^8 \text{ m}$$

$$P = 1.769 \text{ days} = 1.528 \times 10^5 \text{ s}$$

Assuming that the mass of Jupiter is much greater than the mass of Io, Kepler's third law becomes:

$$P^2 = \left(\frac{4\pi^2}{GM}\right) a^3$$

where M is the mass of Jupiter. Thus

$$M = \frac{4\pi^2 a^3}{GP^2}$$

$$= \frac{4\,(3.1416)^2\,(4.216 \times 10^8)^3}{(6.67 \times 10^{-11})(1.528 \times 10^5)^2} \text{ kg}$$

$$= 1.90 \times 10^{27} \text{ kg}$$

This is the same as the value given in Box 13-1 on page 230.

16. In Box 13-1 on page 230 we find that Jupiter's surface gravity is 2.54 times that of Earth. Therefore a 150-pound person would weigh 2.54 times as much on Jupiter, or

$$\text{weight} = 2.54 \times 150 \text{ pounds}$$

$$= 381 \text{ pounds}$$

17. The dimensions of the Great Red Spot are roughly 40,000 km by 14,000 km. Assume that the outer edge of the Great Red Spot can be approximated by a circle whose radius is:

$$\frac{1}{2}\left(\frac{40{,}000 + 14{,}000}{2}\right) = 13{,}500 \text{ km}$$

The circumference of a circle is 2π times its radius and the time it takes to cover this distance is 6 days = 6 × 24 hours = 144 hours. Thus the wind velocity is:

$$v_{\text{wind}} = \frac{2\pi \times 13500 \text{ km}}{144 \text{ hr}}$$

$$= \frac{2\,(3.1416)(13500)}{144} \text{ km/hr}$$

$$= 590 \text{ km/hr}$$

$$\approx 370 \text{ mi/hr}$$

CHAPTER 14

THE GALILEAN SATELLITES OF JUPITER

Chapter Summary

This chapter describes the discovery of the four major satellites of Jupiter by Galileo. The appearance of these objects from Earth is related. Photographs and other data provided by spacecraft observations are reviewed and are interpreted in the context of the formation and evolution of the Jovian system. A detailed description of each of the satellites is presented.

14-1 The Galilean satellites are easily seen with Earth-based telescopes

The changing positions of the Galilean satellites with time and their orbital periods are described. Eclipses of the satellites are noted as are occultations of stars and other satellites. The synchronous rotation of all of these moons is pointed out.

14-2 Data from spacecraft yielded accurate sizes, masses, and densities of the Galilean satellites

The determinations of diameters and masses for the satellites by the *Voyager* spacecraft are related. The use of these data to compute average densities is described.

Box 14-1 The Galilean satellites

This table contains mean distance, sidereal period, diameter, mass and mean density for each of the Galilean satellites and comparison to the Moon and Mercury is made.

14-3 The formation of the Galilean satellites probably mimicked the formation of our solar system

The similarities of general trends in size and average density which exist for both the two classes of planets in the solar system and the Galilean satellite system are pointed out. The theoretical explanations for these trends are recounted for the Galilean satellites. The spectroscopic identification of water-ice on the three outermost satellites is reviewed.

14-4 The *Voyager* spacecraft discovered several small moons and a ring around Jupiter

The precise size and irregular shape of Amalthea revealed by *Voyager 1* are discussed. The number of satellites in the Jovian system is related with descriptions of major orbital groups of satellites. The physical properties of the ring discovered by the *Voyager* spacecraft are discussed.

Box 14-2 Jupiter's family of moons

This table lists the size, period and year of discovery for each of 16 satellites of Jupiter.

14-5 Io is covered with colorful deposits of sulfur compounds ejected from active volcanoes

The theoretical prediction of strong tidal forces on Io due to Jupiter and the other satellites is discussed. The manifestation of these forces in the form of volcanic activity is related. The discovery of volcanic activity on Io is recounted and the surface conditions and composition are described in detail. The roles of the volcanic eruptions in continuously transforming the surface and in contributing material to the Io torus are examined. The flow of electric current between Io an Jupiter is noted.

14-6 Europa is covered with a smooth layer of ice that is crisscrossed with numerous cracks

The smooth, streaked surface of Europa is discussed in the context of tidal forces on the 100 km thick, water-ice crust. The paucity of craters is interpreted to indicate a young surface.

14-7 Ganymede and Callisto have heavily cratered, icy surfaces

The extensive cratering on the two outermost Galilean satellites is described and compared with the lunar surface in terms of age. The thick water or water-ice mantles of these two moons are illustrated. The cratering record on Ganymede is described in some detail. The possible role of plate tectonics in producing the grooved terrain on Ganymede is presented. The absence of grooved terrain on Callisto is explained as evidence of a very old surface which has not been altered since the heavy bombardment early in the history of the solar system. The appearance and age of the Valhalla Basin on Callisto are discussed.

Teaching Hints and Strategies

[Refer to the Teaching Hints and Strategies section of Chapter 7 of this Instructor's Guide for general strategies applicable to solar system objects.]

When *introducing the Galilean satellites of Jupiter* (section 14-1), you might review their role in reducing public resistance to the heliocentric theory of Copernicus. The Copernican system required two centers of motion in the universe. The Sun was the center of the orbits of the planets while the Earth was the center of the orbit of the Moon. Many people found the concept of having two centers of motion to be unsatisfactory. Galileo noted, upon his discovery of these moons, that Jupiter must be a center of motion in either the geocentric or heliocentric system.

The *formation of the Galilean satellites of Jupiter* (section 14-3) can be described as a miniature solar system. The general trends to be noted here are the density, mass and size variations with distance from the planet. The energy released by Jupiter during collapse should be noted as an analog to the solar energy released into the solar nebula. The excess energy radiated by Jupiter is of special interest in this context.

The *many moons of Jupiter* (Box 14-2) are well suited for illustrating the difference between natural satellites and captured satellites. The data in Appendix 3 in the textbook are more helpful than those in Box 14-2 for this activity. Point out the two groups of satellites beyond the orbit of Callisto. The first group has direct orbits and eccentricities ranging from 0.13 to 0.21. The second group has retrograde orbits and eccentricities ranging from 0.17 to 0.37. Discuss the capture of asteroids as an alternative to simultaneous accretion and review the satellites of Mars in section 12-9 and the theories about the origin of the Moon in section 9-6.

The *geologic activity of Io* (Section 14-5) is due to gravitational interactions between Io and its neighbors as opposed to the internal sources which are, or were, responsible for such activity in the terrestrial planets and the Moon. The overwhelming alteration of the surface of Io in a very brief period of time is also quite different from volcanic activity elsewhere in the solar system. The volcanic activity on Io was an accidental discovery by the Voyager spacecraft and is an example of their value in investigating the solar system.

In contrast to Io, the *surface of Europa* (section 14-6) is very smooth and inactive. Tidal forces are manifested on Europa in the form of long cracks in the surface but lack the intensity to produce volcanism. The average physical properties of Io and Europa are rather similar otherwise. The lack of numerous impact craters implies that the surface must be of modest age.

Show why the tidal forces produced by Jupiter on *Ganymede* and *Callisto* (section 14-7) are much smaller than those on Io or Europa. The intense cratering of the surfaces of these two bodies indicates older surfaces. The lower densities indicate much greater abundances of volatile materials than are present in the inner two Galilean satellites. Tectonic activity also tends to decrease with increasing distance from Jupiter. Europa shows strong evidence of such activity, Ganymede shows evidence of past activity and Callisto shows no evidence of such activity. As this activity is probably due in large part to tidal action by Jupiter, this is not surprising.

Answers to Chapter 14 Computational Questions

11. Any calculation which shows that the ratio a^3/P^2 is a constant is acceptable. For instance, using data from Box 14-1 on page 244 we can construct the following table:

SATELLITE	a	a^3	P	P^2	a^3/P^2
Io	4.216×10^5	74.94×10^{15}	1.77	3.13	2.39×10^{16}
Europa	6.709×10^5	301.8×10^{15}	3.55	12.6	2.40×10^{16}
Ganymede	1.07×10^6	1.23×10^{18}	7.16	51.3	2.40×10^{16}
Callisto	1.88×10^6	6.64×10^{18}	16.69	278.6	2.38×10^{16}

The fact that the numbers in the right-hand column are virtually the same constitutes a demonstration of Kepler's third law.

13. Under conditions of excellent seeing $\alpha \approx 1$ arc second. Near opposition the distance to Jupiter is about

$$d = 7.78 \times 10^8 \text{ km} - 1.50 \times 10^8 \text{ km}$$
$$= 6.28 \times 10^8 \text{ km}$$

Using the small-angle formula (recall Box 1-1) we obtain

$$D = \frac{\alpha d}{206,265}$$

$$= \frac{1 \times 6.28 \times 10^8 \text{ km}}{206,265} = 3.04 \times 10^3 \text{ km}$$

$$\approx 3000 \text{ km}$$

The smallest visible features seen through your telescope are about 3000 km in diameter.

CHAPTER 14

14. We can use the small-angle formula with

$$D = 150 \text{ km}$$

$$d = 4.45 \text{ AU} = 4.45 \times 1.496 \times 10^8 \text{ km}$$

$$= 6.66 \times 10^8 \text{ km}$$

$$\alpha = \frac{206{,}265 D}{d} = \frac{2.06 \times 10^5 \times 1.50 \times 10^2}{6.66 \times 10^8}$$

$$= 0.46 \times 10^{-1} \approx 0.05 \text{ arc sec}$$

15. The curvature of Io in Figure 14-7 on page 247 can be fitted with a circle whose radius is about 30 cm. The height of the plume in the picture is about 1 cm. Io's radius is 1/2 (3630 km) = 1815 km. We can therefore write the proportion:

$$\frac{1 \text{ cm}}{30 \text{ cm}} = \frac{x}{1815 \text{ km}}$$

$$x = \frac{1815}{30} \text{ km} \approx 60 \text{ km}$$

16. We must first convert 1 ton/s into kilograms per second as follows:

$$1 \frac{\text{ton}}{\text{s}} \times 2000 \frac{\text{lbs}}{\text{ton}} \times \frac{1 \text{ kg}}{2.2 \text{ lb}} = 909 \text{ kg/s}$$

The mass of Io is 8.92×10^{22} kg (see Box 14-1 on page 244). One-tenth of this mass is 8.92×10^{21} kg.

The time needed to eject one-tenth of Io's mass at a rate of 909 kg/s is

$$\frac{8.92 \times 10^{21} \text{ kg}}{909 \text{ kg/s}} = 9.81 \times 10^{18} \text{ s} \left(\frac{1 \text{ yr}}{3.16 \times 10^7 \text{ s}} \right)$$

$$= \left(\frac{9.81}{3.16} \right) \times 10^{18-7} \text{ yr} = 3.1 \times 10^{11} \text{ yr} = 310 \text{ billion years}$$

This answer is about 60 times greater than the age of the solar system, which is about 5 billion years old.

17. Assume that Ganymede's orbit is a circle whose circumference is $2\pi r$. From the data in Box 14-1 we can take $r = 1.07 \times 10^6$ km. Furthermore, Ganymede's orbital period is

$$P = 7.16 \text{ d} = (7.16 \text{ d})(24 \text{ hr/d})(3600 \text{ s/h})$$

$$= 6.186 \times 10^5 \text{ seconds}$$

The average speed ($\bar{v}$) of Ganymede along its orbit is

$$\bar{v} = \frac{2\pi r}{P} = \frac{2(3.1416)(1.07 \times 10^6 \text{ km})}{6.186 \times 10^5 \text{ s}}$$

$$= 10.87 \text{ km/s}$$

The diameter of Ganymede is 5262 km, and so the time needed to enter or leave Jupiter's shadow is

$$\frac{5262 \text{ km}}{10.87 \text{ km/s}} = 484 \text{ seconds}$$

$$\approx 8.1 \text{ minutes}$$

CHAPTER 15

THE SPECTACULAR SATURNIAN SYSTEM

Chapter Summary

This chapter covers Earth- and spacecraft-based observations of Saturn, the rings of Saturn and its satellites. The structure of Saturn's rings are discussed. The interior of Saturn and its excess energy emission are examined and explained. The atmosphere is compared and contrasted with that of Jupiter. Details of the ring system and its connection with satellite orbits are reviewed. The nature of the atmosphere of Titan is described and specific features of the moderate-sized moons are presented.

15-1 Earth-based observations reveal three broad rings encircling Saturn

The appearance of the ring system from Earth is described. The changing view of the ring plane is illustrated and the thickness of the rings is noted. The major rings and gaps are identified. Table 15-1 contains the dates of oppositions of Saturn and its apparent magnitudes through the remainder of the 1990s.

Box 15-1 Saturn data

This table contains orbital and physical data for the planet Saturn as well as the range of apparent angular size, the apparent magnitude at maximum brightness and the apparent size of the Sun as seen from the planet.

15-2 Saturn's rings are composed of numerous fragments of ice and ice-coated rock

The observational evidence supporting the individual small particle model for the Saturnian ring system is reviewed. The spectroscopic evidence of water-ice from the ring particles is noted. The concepts of tidal forces and of the Roche limit are described and examined in the context of planetary rings.

15-3 Saturn's rings consist of thousands of narrow, closely spaced ringlets

The vast numbers of ringlets revealed by satellite observations are related. The braided structure detected for the F ring is noted. An analysis of the makeup of the various regions of the rings based on scattering observations made by satellites is presented. In Figure 15-8 a brief review is provided of the extent and distance from the center of the planet for each of the major rings which make up the ring system. The rings are presented in order from the planet out and the orbits of the inner satellites are shown to the same scale.

15-4 Saturn's innermost satellites affect the appearance and structure of its rings

The existence of the gaps is explained in terms of orbital resonance. The role of small shepherd satellites in producing the broad bands and gaps in the ring system is explained. The grouping of the satellites of Saturn by size is described.

15-5 Saturn's atmosphere extends to a greater depth and has higher wind speeds than Jupiter's atmosphere

The chemical composition and temperature profiles with altitude in the Saturnian atmosphere are compared with those of Jupiter as are the patterns of wind speed with latitude. The discovery of the white spot by the HST in 1990 is mentioned.

15-6 Saturn's internal structure is quite similar to Jupiter's and includes a layer of liquid metallic hydrogen

A model of the internal structure deduced from measurements of Saturn's oblateness is discussed. The interior liquid molecular hydrogen and liquid metallic hydrogen shells and rocky core are described and compared with those of Jupiter. The magnetic field of Saturn is discussed.

15-7 Like Jupiter, Saturn emits more radiation than it receives from the Sun

The excess energy emission by Saturn is compared with that of Jupiter. The hypothesis that helium may be responsible for this excess is considered.

15-8 Titan has a thick, opaque atmosphere rich in methane, nitrogen, and hydrocarbons

The retention and evolution of an atmosphere by Titan are examined. The composition and structure of the atmosphere of Titan is illustrated and described. The possibility of liquid ethane condensing and evaporating similar to the behavior of water on the Earth is noted.

15-9 The features of the icy surfaces of Saturn's six moderate-sized moons provide clues to their histories

The common densities and orbital characteristics of the six moderate-sized moons of Saturn are noted. The surface features of each of these moons are described and potential explanations of their significance provided.

Box 15-2 Saturn's satellites

This table contains the distance of the satellite from the planet, orbital period, size, and density for each of 18 moons of Saturn.

Teaching Hints and Strategies

[Refer to the Teaching Hints and Strategies section of Chapter 7 of this Instructor's Guide for general strategies applicable to solar system objects.]

The *rings of Saturn* (sections 15-1 and 15-2) were the first planetary rings to be discovered. If you have access to a telescope, make every effort to let your students observe Saturn. It is invariably the most appreciated celestial object during public observing sessions. Ring systems provide an excellent opportunity to discuss tidal forces and the Roche limit. Were these rings once material in a satellite which has been destroyed or do they represent material too close to the planet to form into a satellite? The discovery of the ring systems of Jupiter by the *Voyager* spacecraft and of Uranus and Neptune by stellar occultation indicates that, while the rings of Saturn are the most spectacular in the solar system, they represent a common structure in the environments of Jovian planets. The complicated distribution of the particles which make up the rings demonstrate the action of gravitational forces of the satellites of the planets. It should be noted also that the explanation of the existence and structure of these ring systems is provided by Newtonian mechanics and clearly demonstrates the value of Newton's laws. It is important to stress that the Roche limit is related to the density of a moon. Higher density objects have stronger internal gravitational forces and can remain stable at distances closer to the parent planet than can low density material.

The *gaps in the rings of Saturn* (section 15-3 and 15-4) are similar to the Kirkwood gaps in the asteroid belt which is discussed in Chapter 17. It might be helpful to have students read and discuss that section at this time. The situation with ring systems is always much more complex due to the effects of several satellites at the same time.

Emphasize the comparison of the *atmosphere of Saturn* (section 15-5) with that of Jupiter. The principal differences arise due to differences in temperature resulting from their different distances from the Sun and to surface gravity differences. A quick calculation of the relative fluxes of sunlight reaching these two planets will illustrate the expanse of the outer solar system and reinforce the use of the inverse square law of light.

Be sure to stress the similarities between the *internal structure of Saturn* (section 15-6) and that of Jupiter and relate that structure to its magnetic field. Point out the rather different relative sizes of the primary zones in Figure 15-14.

The *excess emission* by Saturn (section 15-7) is quite interesting. Be sure that students understand that the net radiant flux for Saturn is even more out of balance than that of Jupiter although Saturn should not be contracting like Jupiter is. Stress the very different explanations for these two cases.

Although *Titan* (section 15-8) is as large as the largest Galilean satellites, its dense atmosphere makes it less spectacular visually than the other major moons in the solar system. Be sure to compare and contrast it with the other large moons and refer to Box 14-1 and the photograph on the cover page for Chapter 14. Note that is is almost identical in size and density to Ganymede. The Voyager data regarding the atmosphere on Titan is obviously of greatest importance. It should be noted that the atmosphere was detected years ago by spectroscopic analysis using Earth-based telescopes. It is nitrogen rich like the Earth's atmosphere but the different temperatures on the two bodies result in quite different conditions. The possible surface conditions are rather interesting.

When discussing *Saturn's satellites* (section 15-9 and Box 15-2) you might refer to Appendix 3 in the textbook and note that while the Saturnian system has only one major moon, it has several intermediate-sized moons. Jupiter, on the other hand, has four major moons but no intermediate-sized moons. You might want to prepare a transparency showing the relative sizes of the intermediate-sized moons of Saturn and of Uranus. Note that Uranus has no large satellite. In Box 15-2 you should note the irregular shapes of the smaller moons.

Answers to Chapter 15 Computational Questions

12. A particle orbiting in the Enke division has a semimajor axis of

$$a = 1.335 \times 10^5 \text{ km}$$
$$= 1.335 \times 10^8 \text{ m}$$

We can calculate the particle's orbital period (P) with Kepler's third law:

$$P^2 = \left(\frac{4\pi^2}{GM}\right)a^3$$

where M = Saturn's mass = 5.69×10^{26} kg

$$P^2 = \frac{4\,(3.1416)^2\,(1.335 \times 10^8)^3}{(6.67 \times 10^{-11})\,(5.69 \times 10^{26})}$$

$$= \frac{4\,(3.1416)^2(1.335)^3}{(6.67)(5.69)} \times 10^{24+11-26}$$

$$= 2.475 \times 10^9 \text{ s}^2 = 24.75 \times 10^8 \text{ s}^2$$

and so

$$P = 4.975 \times 10^4 \text{ s } (1 \text{ hr}/3600 \text{ s})$$

$$= 13.82 \text{ hr}$$

Checking the data in Box 15-2 on page 266, we see that this value of P is approximately a 2:3 resonance with Mimas and 3:1 resonance with Tethys.

13. We can use the small-angle formula (Box 1-1 on page 8) to calculate the size of the smallest feature detectable at the distance of Saturn from Earth:

$$D = \frac{\alpha d}{206,265}$$

$\alpha = 0.05$ arc sec

d = distance to Saturn at opposition

$$= 1427 \times 10^6 \text{ km} - 150 \times 10^6 \text{ km}$$

$$= 1277 \times 10^6 \text{ km}$$

$$\approx 1.28 \times 10^9 \text{ km}$$

$$D = \frac{(0.05)(1.28 \times 10^9 \text{ km})}{2.06 \times 10^5} = \left(\frac{6.4}{2.06}\right) \times 10^{9-2-5} \text{ km}$$

$$\approx 300 \text{ km}$$

The Hubble Space Telescope is capable of resolving features as small as 300 km in size on Saturn.

14. As we saw in the previous question, the Hubble Space Telescope can resolve features as small as 300 km across at the distance of Saturn. Box 15-1 on page 266 lists five satellites that are substantially larger than this limit: Tethys, Dione, Rhea, Titan, and Iapetus.

Mimas and Enceladus are quite close to the limit of HST's ability to resolve details.

15. In Box 7-2 on pages 134-135 we saw that escape velocity (v_{escape}) is given by:

$$v_{escape} = \left(\frac{2GM}{R}\right)^{1/2}$$

We can calculate the mass of Titan (M) from its density (ρ) and radius (R) given in Box 15-2 on page 266 as follows:

$$\rho = \frac{M}{V} = \frac{M}{\frac{4}{3}\pi R^3}$$

so

$$M = \frac{4}{3}\pi \rho R^3$$

Since

$$\rho = 1880 \text{ kg/m}^3$$

and

$$R = \frac{1}{2}(5150 \text{ km}) = 2.575 \times 10^6 \text{ m}$$

we obtain

$$M = \frac{4}{3}(3.1416)(1880 \text{ kg/m}^3)(2.575 \times 10^6 \text{ m})^3$$

$$= \frac{4}{3}(3.1416)(1.88)(2.575)^3 \times 10^{3+18} \text{ kg}$$

$$= 1.34 \times 10^{23} \text{ kg}$$

Substituting the values for the mass and radius of Titan into the equation for the escape velocity, we get:

$$v_{escape} = \left(\frac{2(6.67 \times 10^{-11})(1.34 \times 10^{23})}{2.575 \times 10^6}\right)^{1/2}$$

$$= \left(\frac{2(6.67)(1.34)}{2.575} \times 10^{-11+23-6}\right)^{1/2}$$

$$= \left(6.94 \times 10^6\right)^{1/2}$$

$$= 2.63 \times 10^3 \text{ m/s}$$

From Box 7-2 we know that the requirement for the retention of a gas of molecular weight µ is:

$$6\left(\frac{3kT}{\mu m_H}\right)^{1/2} < \left(\frac{2GM}{R}\right)^{1/2} = v_{escape}$$

where m_H = mass of hydrogen atom = 1.67×10^{-27} kg and T = Titan's surface temperature.

Assuming that Titan's surface temperature is roughly the same as Saturn's ($-180°C$ = 93 K) we find:

$$6\left(\frac{3(1.38 \times 10^{-23})(93)}{\mu(1.67 \times 10^{-27})}\right)^{1/2} < 2.63 \times 10^3$$

$$6\left(\frac{3(1.38)(93)}{\mu(1.67)} \times 10^{-23+27}\right)^{1/2} < 2.63 \times 10^3$$

$$6\left(\frac{2.31 \times 10^6}{\mu}\right)^{1/2} < 2.63 \times 10^3$$

$$6\left(\frac{2.31}{\mu}\right)^{1/2} < 2.63$$

$$\frac{(36)(2.31)}{\mu} < (2.63)^2 = 6.92$$

$$\mu > \frac{(36)(2.31)}{6.92} = 12$$

Any gas whose molecular weight is greater than 12 will be retained.

17. The easiest way to do this problem is to use the fact that all of Saturn's satellites obey Kepler's third law as they orbit the planet. For instance, Titan's period (382.69 hr) and semimajor axis (1.222×10^9 m) are in accord with Kepler's third law. Thus, for any particle orbiting Saturn with a period P and a semimajor axis a, we can write the proportion:

$$\frac{P^2}{(382.69)^2} = \frac{a^3}{(1.222 \times 10^9)^3}$$

At the outer edge of the A ring, a = 136,600 km = 1.366×10^5 km = 1.366×10^8 m, and so

$$\frac{P^2}{(382.69)^2} = \left(\frac{1.366 \times 10^8}{1.222 \times 10^9}\right)^3 = (1.118 \times 10^{-1})^3$$

$$P = (382.69)[(1.118 \times 10^{-1})^3]^{1/2}$$
$$= 14.3 \text{ hr } (3600 \text{ s}/1 \text{ hr})$$
$$= 5.15 \times 10^4 \text{ seconds}$$

The circumference of the particle's orbit is

$$2\pi a = 2(3.1416)(1.366 \times 10^8 \text{ m})$$
$$= 8.58 \times 10^8 \text{ m}$$

and so, the particle's velocity is

$$v = \frac{2\pi a}{P} = \frac{8.58 \times 10^8 \text{ m}}{5.15 \times 10^4 \text{ s}} = 1.67 \times 10^4 \text{ m/s}$$
$$= 16.7 \text{ km/s}$$

At the inner edge of the B ring, $a = 92{,}000$ km $= 9.20 \times 10^7$ m, and so

$$\frac{P^2}{(382.69)^2} = \left(\frac{9.20 \times 10^7}{1.222 \times 10^9}\right)^3 = 4.267 \times 10^{-4}$$

$$P = 382.69(4.267 \times 10^{-4})^{1/2} = 7.91 \text{ hr}$$
$$= 2.85 \times 10^4 \text{ seconds}$$

At the inner edge of the B ring, the particle's average speed is:

$$v = \frac{2\pi a}{P} = \frac{2(3.1416)(9.20 \times 10^7 \text{ m})}{2.85 \times 10^4 \text{ s}}$$
$$= 20.3 \text{ km/s}$$

The maximum observable wavelength shift occurs when the rings are viewed nearly edge-on. The difference in velocity between the approaching and receding inner edge of the B ring is 2×20.3 km/s $= 40.6$ km/s. Using the formula for the Doppler shift, we get

$$\frac{\Delta \lambda}{\lambda_0} = \frac{v}{c}$$

$$\Delta \lambda = \lambda_0 \left(\frac{v}{c}\right) = (500 \text{ nm})\left(\frac{40.6 \text{ km/s}}{3 \times 10^5 \text{ km/s}}\right)$$
$$= 0.068 \text{ nm}$$

CHAPTER

16

THE OUTER WORLDS

Chapter Summary

This chapter covers the physical and orbital properties of Uranus, Neptune and Pluto. The discoveries of Uranus and Neptune are contrasted. The similarities and differences between Uranus and Neptune are discussed and contrasted with properties of the two larger Jovian planets, Jupiter and Saturn. The *Voyager* 2 observations of Uranus in 1986 and Neptune in 1989 are discussed in some detail and the physical properties of the Uranian and Neptunian satellites are presented. The unusual orbits of the satellites of Neptune are examined. The discoveries of Pluto and its moon Charon are described. The peculiar orbital properties of Pluto and its small size are examined in the context of hypotheses which suggest possible links between Neptune and Pluto in the past. The physical and orbital properties of Charon are discussed.

16-1 Uranus was discovered by chance, but Neptune's existence was predicted with Newtonian mechanics

The discovery of Uranus by William Herschel during routine telescopic observations is related. The subsequent predictions of the existence of another planet which led to the discovery of Neptune are outlined.

Box 16-1 Uranus data

This table contains orbital and physical data for Uranus.

Box 16-2 Neptune data

This table contains orbital and physical data for Neptune.

16-2 The large tilt of Uranus's axis of rotation produces extreme seasonal changes on this nearly featureless planet

The unusual alignment of the axis of rotation of Uranus and its effect on the atmospheric conditions are discussed. The differential rotation of the atmosphere is described and the rotation period based on the magnetic field is given.

16-3 Uranus is orbited by satellites that bear the scars of many shattering collisions

The Voyager observations of the five intermediate-sized satellites are described and interpreted in the context of massive collisions which may be related to the unusual orientation of the rotation axis of the planet.

Box 16-3 Uranus's satellites

This table contains the distance of the satellites from the planet, orbital period, size and mean density for each of the 15 moons of Uranus.

16-4 Uranus is circled by a system of thin, dark rings

The discovery of the more prominent rings of Uranus from Earth in 1977 is recounted. The rings are described and contrasted with the rings of Saturn.

16-5 Neptune is a cold, bluish world with Jupiterlike atmospheric features and a Uranuslike interior

The atmospheric features photographed by the *Voyager 2* satellite in 1989 are discussed. The Great Dark spot is described and compared with the Great Red Spot on Jupiter. Anticyclonic flows and cirrus clouds are also noted. The unusual magnetic axis is examined. The difference in composition between Uranus and Neptune and the other giant planets is explained.

16-6 The magnetic fields of both Uranus and Neptune are oriented at unusual angles

The large differences between the rotation axes and magnetic axes of Uranus and Neptune are given. Possible causes for the misalignments are examined. The sources of these fields are considered.

16-7 Triton is a frigid, icy world with a young surface and a tenuous atmosphere

The unusual orbital properties, retrograde and high inclination orbit, are reviewed followed by a capture theory to explain these unusual characteristics. A description of Triton's unique surface features observed by *Voyager 2* and their possible origin is discussed.

Box 16-4 Neptune's satellites

The orbit sizes, periods and the diameters are listed for the eight known moons of Neptune as of 1989.

16-8 Pluto was discovered after a laborious search of the heavens

The long search for and discovery of Pluto are described. Pluto's strange orbit is reviewed and its rotation period is revealed.

Box 16-5 Pluto data

This table contains orbital and physical data for the planet Pluto.

16-9 Pluto and its moon Charon may be typical of a thousand "ice dwarfs" that orbit the Sun at the outskirts of the solar system

The common rotation period for the planet and orbital period for Charon are discussed. The sizes, masses and mean densities of the two objects are reported. The possible origin of the Pluto–Charon system by collision is mentioned as well as the possibility of capture. The idea that the outer solar system is populated with many icy bodies is discussed.

Teaching Hints and Strategies

[Refer to the Teaching Hints and Strategies section of Chapter 7 of this Instructor's Guide for general strategies applicable to solar system objects.]

    The circumstances surrounding the *discovery of Uranus and Neptune* (section 16-1) are quite different and can be used to illustrate two different aspects of scientific discovery. While the discovery of Uranus by William Herschel could be regarded as accidental, because he was not specifically looking for a planet, it was detected by William Herschel in one of his systematic surveys of the heavens. Much of modern astronomy is related to astrophysical investigations. Many of the objects being studied were selected as a result of general surveys for specific types of objects. Many of the nonstellar objects listed in the *New General Catalogue* were first observed and listed by Herschel as were many double star systems. Survey work in astronomy is usually not terribly exciting, but it provides vast amounts of information about individual objects which can be 1) studied in detail at some later time, 2) used to identify specific types of objects or 3) studied by statistical analysis. You might point out that the catalogues listed in Box 2-1 are all the result of many years of meticulous work. The discovery of Neptune clearly reveals the benefits of a precise understanding of the laws of nature. The theoretical prediction that Neptune should exist should be noted as an example of the verification of a scientific theory by prediction.

    Be sure to discuss the strange consequences of the tilt of the axis of rotation (section 16-2) on the daily and seasonal changes in the altitude and azimuth of the Sun for observers on Uranus. Contrast these motions with the seasonal changes of the azimuth of sunrise and sunset due to the combination of the rotation and revolution of the Earth. Contrast these events for an observer at the pole and another at the equator. Stress the fact that the daily and seasonal cycles are quite different from one planet to another. You should have students review the material in Box 2-3 on Tropics and Circles.

The *satellites of Uranus* (section 16-3 and Box 16-3), as revealed by *Voyager 2* images, provide a very interesting system. A very useful summary of the results of this encounter is provided by the PBS Nova program titled "The Planet that Got Knocked on Its Side." The information has resulted in a number of interesting theories about the evolution of the Uranian system which includes an explanation of the tilt of the rotation axis of Uranus.

Contrast the *rings of Uranus* (section 16-5) with the rings of Jupiter and of Saturn. Explain the differences in scattering properties and the differences in the widths in each case. Discuss why these ring systems are so different from each other. It is interesting to note that the rings of Uranus were detected before the ring system of Jupiter and that they were detected from Earth-based observations by occultation techniques.

Many interesting features were revealed in the *atmosphere of Neptune* (section 16-5) by the *Voyager 2* images. The concept of differential rotation and zonal wind patterns can be clearly presented by some of the computer animation movies of these images.

The very unusual tilts and offsets of the magnetic fields of both Uranus and of Neptune (section 16-6) deserve special attention. The explanation of the strange magnetic properties of Uranus could easily be attributed to collisional events associated with its axis of rotation and its satellites until similar features were found on Neptune. The peculiar orbits of Triton and Nereid may be evidence of collisions in the Neptune system as well.

When discussing the *satellites of Neptune* (section 16-7 and Box 16-4), you should note that natural satellites normally orbit parent planets in the equatorial plane and have nearly circular orbits. The peculiar orbits of Triton and Nereid could have resulted from a possible gravitational interaction with Pluto in the past. The Voyager flyby in 1989 provided facinating images of Triton. The accepted size of Triton was reduced by almost a factor of two which signifcantly changed its ranking in the list of major satellites.

The history of the *discovery of Pluto* (section 16-8) is a rather interesting example of a serendipitous discovery which resulted from a well planned search initiated by irrelevant data. The calculations which led to the discovery of Neptune in 1846 were followed by more calculations and similar predictions of the existence of yet another planet. The determination of the mass of Pluto, subsequent to the discovery of Charon, reveals that Pluto has too little mass to produce any significant perturbations. The search for Pluto was initiated as the result of predictions of gravitational perturbations. It was discovered in spite of the fact that it does not produce those perturbations. Some astronomers believe that Pluto is not the real Planet X and that a tenth planet remains to be found. They have estimates for the position in the sky but lack enough time on sufficiently large telescopes to conduct the required search. In this respect, Tombaugh was lucky. He had an adequate telescope dedicated to the single purpose of finding Planet X.

Pluto and its moon Charon (section 16-9) now have replaced the Earth–Moon system as the solar system's double planet. Until the discovery of Charon in 1978, it was commonly noted that the Earth and Moon were closer in size than any other planet satellite combination in the solar system. Note that Charon is a little over half the size of Pluto while the Earth is about four times bigger than the Moon. Also emphasize that Pluto is smaller and less massive than our Moon. As Pluto was not included in the classification scheme for planets, it might be appropriate at this point to describe briefly why this is so. Review the common physical properties for the two classes of planets and compare with those of Pluto. Be sure to note that Pluto is currently closer to the Sun than Neptune and will remain so through 1999. The search for a planet beyond the orbit of Pluto is always a subject of great interest. Planet X (X might represent the traditional unknown in math and science or the Roman numeral for 10) is discussed in popular astronomy articles on a regular basis. It is also sometimes associated with possible perturbations of the comet cloud which might have been related to mass extinctions like that which resulted in the elimination of the dinosaurs. While these theories are quite speculative, they certainly capture the imaginations of the students. You might also discuss the series of eclipses which have been observed in the late 1980's for Pluto and Charon. These have greatly enhanced our knowledge about these two objects. It is rather interesting that such eclipses should occur about once every 125 years, but the first such series of eclipses occurred only about ten years after the discovery of Charon. If that discovery had been made in the early 1990's, we would have had to wait for a century to observe these events.

Answers to Chapter 16 Computational Questions

12. Once again we can turn to the small-angle formula

$$D = \frac{\alpha d}{206{,}265}$$

where $\alpha = 0.25$ arc sec

and d = distance from Earth to Pluto at opposition

$$= 5.91 \times 10^9 \text{ km} - 1.5 \times 10^8 \text{ km}$$

$$= 5.76 \times 10^9 \text{ km}$$

The largest diameter (D) that can be resolved is

$$D = \frac{(0.25)(5.76 \times 10^9)}{2.06 \times 10^5} \text{ km}$$

$$\approx 7000 \text{ km}$$

13. The Hubble Space Telescope can resolve features whose angular size is 0.05 arc sec. Neptune's distance from Earth at opposition is

$$d = 4.497 \times 10^9 \text{ km} - 0.150 \times 10^9 \text{ km}$$

$$= 4.347 \times 10^9 \text{ km}$$

We can use the small-angle formula to calculate the diameter (D) of the smallest resolvable feature on Neptune as follows:

$$D = \frac{\alpha d}{206{,}265}$$

$$= \frac{(0.05)(4.347 \times 10^9 \text{ km})}{206{,}265}$$

$$\approx 1000 \text{ km}$$

Since this value is much smaller than the Great Dark Spot's dimensions (12,000 km × 8,000 km) this feature is resolvable.

At the distance of Pluto ($d = 5.76 \times 10^9$ km) we find

$$D = \frac{\alpha d}{206{,}265}$$

$$= \frac{(0.05)(5.76 \times 10^9 \text{ km})}{2.06 \times 10^5}$$

$$\approx 1400 \text{ km}$$

Since Pluto's diameter is 2290 km, the Hubble Space Telescope is able to resolve the planet's disk.

14. You must first determine the angular speed (ω) of Uranus as it moves along its orbit. Uranus's orbital period is 84.01 yr and so:

$$\omega = \frac{360°}{P} = \frac{360°}{84.01 \text{ yr}}$$

$$= \frac{360°}{84.01 \text{ yr}} \left(\frac{1 \text{ yr}}{365.22 \text{ d}} \right) \left(\frac{1 \text{ d}}{24 \text{ hr}} \right) \left(\frac{1 \text{ hr}}{60 \text{ min}} \right) \left(\frac{3600 \text{ arc sec}}{1°} \right)$$

$$= 0.0293 \text{ arc sec/min}$$

The maximum duration of a stellar occultation equals the time it takes for Uranus to traverse an angle equal to its angular diameter. In Box 16-1 we read that the maximum angular diameter of Uranus is 3.7 arc sec. Thus the duration of the occultation is:

$$\frac{3.7 \text{ arc sec}}{0.0293 \text{ arc sec/min}} = 126 \text{ minutes}$$

$$= 2 \text{ hours } 6 \text{ minutes}$$

15. The maximum gravitational force of Neptune on Uranus occurs as the following configuration:

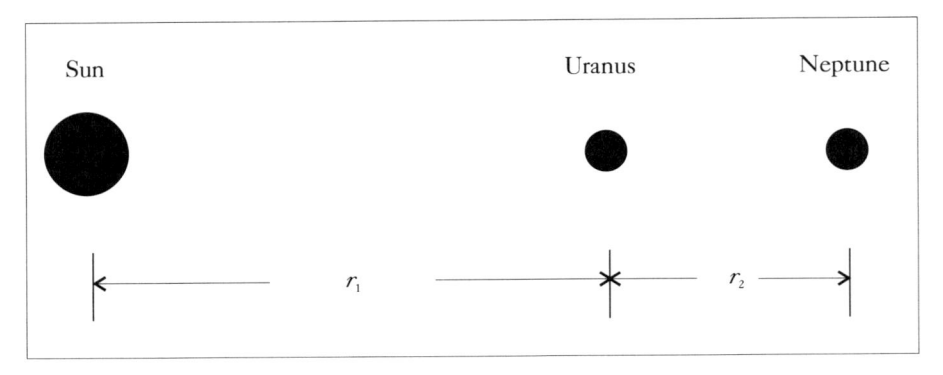

The gravitational force of the sun on Uranus is: $F_{\odot U} = \dfrac{GM_{\odot}M_U}{r_1^2}$

The gravitational force of Neptune on Uranus is:

$$F_{NU} = \frac{GM_N M_U}{r_2^2}$$

Taking the ratio of these two forces we see that

$$\frac{F_{NU}}{F_{\odot U}} = \frac{\dfrac{M_N M_U}{r_2^2}}{\dfrac{M_{\odot} M_U}{r_1^2}} = \frac{M_N}{M_{\odot}}\left(\frac{r_1}{r_2}\right)^2$$

From Boxes 16-1 and 16-2 and the appendixes at the back of the book we learn that

$$M_N = 1.03 \times 10^{26} \text{ kg}$$

$$M_{\odot} = 1.99 \times 10^{30} \text{ kg}$$

$$r_1 = 19.19 \text{ AU}$$

$$r_2 = 30.09 \text{ AU} - 19.19 \text{ AU} = 10.9 \text{ AU}$$

and so

$$\frac{F_{\odot U}}{F_{NU}} = \left(\frac{1.03 \times 10^{26}}{1.99 \times 10^{30}}\right)\left(\frac{19.19}{10.9}\right)^2$$

$$= (0.518 \times 10^{-4})(1.76)^2$$

$$= 1.61 \times 10^{-4}$$

Neptune exerts about 1.6×10^{-4} times as much gravitational force on Uranus as does the Sun.

18. The dimensions of the Great Red Spot are 40,000 km in longitude by 14,000 km in latitude. The average diameter of this feature therefore is:

$$\bar{D}_{GRS} = \frac{(40{,}000 \text{ km} + 14{,}000 \text{ km})}{2} = 27{,}000 \text{ km}$$

The dimensions of the Great Dark Spot are 12,000 km in longitude by 8,000 km in latitude. The mean diameter of this feature is:

$$\bar{D}_{GDS} = \frac{12{,}000 \text{ km} + 8{,}000 \text{ km}}{2} = 10{,}000 \text{ km}$$

The ratio of these diameters is:

$$\frac{\bar{D}_{GRS}}{\bar{D}_{GDS}} = \frac{27{,}000 \text{ km}}{10{,}000 \text{ km}} = 2.7$$

The ratio of the diameters of Jupiter and Neptune is:

$$\frac{\bar{D}_{Jup}}{\bar{D}_{Nep}} = \frac{142{,}800 \text{ km}}{49{,}500 \text{ km}} = 2.9$$

These results support the notion that the Great Dark Spot is a scaled-down version of the Great Red Spot.

CHAPTER

17

INTERPLANETARY VAGABONDS

Chapter Summary

This chapter discusses the characteristics of the smaller bodies in the solar system. The use of Bode's law to establish the sizes of planetary orbits is presented in the context of the discovery of the asteroids. Meteors, meteoroids and meteorites are defined and classified. The value of meteorites in determining the age and possible origin of the solar system is noted. The structure and orbital properties of comets are reviewed.

17-1 Bode's law led to the discovery of asteroids between the orbits of Mars and Jupiter

Bode's law is described and its significance is discussed. The original publication of this relation by Titius is noted. The discovery of the minor planet Ceres by Piazzi is related and the value of contributions of Gauss in celestial mechanics are discussed in the context of the verification process.

17-2 Many asteroids orbit the Sun between the orbits of Mars and Jupiter

Subsequent discoveries of lesser minor planets are reviewed. The value of the application of photography in searches for minor planets is noted. The numbers and sizes of asteroids are discussed. The asteroid belt is defined and asteroids having unusual orbital characteristics are identified.

17-3 Jupiter's gravity affects the structure of the asteroid belt and captures asteroids along its orbit

The gravitational influence of Jupiter on the asteroids to form the Kirkwood gaps is explained. The Trojan asteroids are discussed and related to two of the five Lagrangian points.

17-4 Asteroids occasionally collide with each other and with the inner planets

The characteristics of Amor and Apollo asteroids are defined. The information which can be derived from the light variations of asteroids is reviewed. The fragmentation of larger bodies to produce Hirayama families is described. Impact craters and meteoroids are identified. The theory that a meteorite may have been responsible for the extinction of the dinosaurs is examined.

Box 17-1 Lagrangian points and the restricted three-body problem

The definitions of all five Lagrangian points are discussed and explained. Stable and unstable Lagrangian points are distinguished. Examples of satellites at the stable Lagrangian points in the Saturnian system are noted. The suitability of Lagrangian points for space colonization is examined.

17-5 Meteorites are classified as stones, stony irons, or irons, depending on their composition

Distinctions are made between meteoroids, meteors and meteorites. The classification of meteorites is described and the relative numbers of each class are discussed. Widmanstätten patterns are explained and their value noted.

17-6 Some meteorites retain traces of the early solar system

The origins of the different types of meteorites from the fracturing of a larger parent asteroid are discussed. The nature of carbonaceous chondrites and their roles in evaluating the original composition of the solar nebula, in establishing the age of the solar system, and in providing evidence about the formation of the solar system are reviewed.

17-7 A comet is a dusty chunk of ice that becomes partly vaporized as it passes near the Sun

The general nature of cometary material is presented. The specific structural elements are defined and described. The two types of comet tails are distinguished. The short period comets are compared with those found in the Oort cloud.

17-8 Cometary dust and debris rain down on the Earth during meteor showers

The destruction of comets by the gravitational attraction of the Sun is described and Sun-grazing comets are noted. The nature of meteor showers are reviewed and the origin of the particles causing such showers is examined.

Box 17-2 Meteor showers

A table is provided which displays the range in dates at maximum, the coordinates of the radiant, the hourly rate and velocity for the most dramatic meteor showers. The conditions for optimum viewing are explained.

Teaching Hints and Strategies

[Refer to the Teaching Hints and Strategies section of Chapter 7 of this Instructor's Guide for general strategies applicable to solar system objects.]

Bode's law (section 17-1) is an interesting relationship which can serve to focus attention on the verification process which is an important element in modern science. The discovery of Uranus and of asteroids between the orbits of Mars and Jupiter were both consistent with Bode's law. However, the subsequent discoveries of Neptune and Pluto have clearly shown that the relationship is not valid beyond the orbit of Uranus. Newton's quest to understand the underlying meaning of Kepler's laws led to the revision of mechanics and resulted in much more general replacements and full understanding of those laws. The attempt to find the underlying cause of Bode's law has failed to produce a similar result. A comparison of these two cases will provide students some insight into the verification process in science.

The *asteroids* (section 17-2) are more descriptively referred to as the minor planets. Ask your students to explain which name is better and to defend their choices. Be sure to note that the number of asteroids of any given mass increases as the mass decreases. Thus smaller asteroids are much more numerous than larger ones. It should be made clear that the combined mass of all of the asteroids is smaller than the masses of the major satellites. Thus the asteroids probably do not represent the remnants of a substantial, fractured planet. The origin of meteoroids is presented at the end of section 17-4 in the context of collisions involving asteroids.

The *Kirkwood gaps* (section 17-3) in the asteroid belt are similar to the Cassini and Encke divisions in Saturn's rings, with Jupiter being the primary controlling agent. You might have students use Kepler's third law to compute the exact orbit sizes for asteroids in the Kirkwood gaps when given the period ratios and the period of Jupiter. Compare these values to those shown in Figure 17-3.

Collisions of asteroids (section 17-4) may be responsible both for building larger objects by accretion in the early stages of solar system evolution, and for fragmenting large objects into smaller bodies in more recent times. Classes of asteroids having orbits which carry them into the inner solar system, such as Apollo and Aten asteroids, are of more immediate interest to inhabitants of Earth as they pass near us and permit much more careful observation and have much greater potential for accidental or planned interactions. It is important to recognize that asteroids are generally made of materials similar to those found in the terrestrial planets and do not change dramatically like comets do if they should happen to come close to the Sun.

When discussing the *Lagrange points* (Box 17-1) be sure to make a clear distinction between the stable and the unstable points. Remind students that as the three objects orbit the center of mass of the two massive objects, their relative positions remain fixed.

Students often fail to recognize the distinctions between meteors, meteorites and meteoroids unless special care is taken to identify them. When the *classification of meteorites* (section 17-5) in terms of their composition is reviewed, be sure to point out the difference between the relative numbers of each class 1) when found accidentally on the ground and 2) when found as a result of observed falls. The first category is reflected by museum collections. The latter category provides a much more representative sample of what is in orbit around the Sun. This is an excellent example of observational selection effects which frequently impact observational astronomy. Such selection effects are present when conducting surveys for stars (more luminous stars are seen to much greater distances) and for galaxies (same as for stars and also spirals are much more conspicuous than ellipticals).

In addition to the *ages of meteorites* (section 17-6) they provide evidence of the destruction of more massive objects in the past by virtue of their Widmanstätten patterns. The value of meteorites in establishing the age of the solar system should also be noted. You might want to mention the contrast in the physical properties of the asteroids found in the inner and outer regions of the asteroid belt.

The major structural elements of *comets* (section 17-7) should be noted with special emphasis on the fact that the vast sizes of the coma, hydrogen envelope and tail do not imply large masses for comets. The dirty snowball model should be noted with emphasis on the changes that accompany the heating of volatile materials in a very low pressure environment. The process of sublimation, as demonstrated by dry ice, can be a useful conceptual tool in describing the transition which occurs as comets approach the Sun. Their very low density should be stressed. The analysis of the chemical composition of the gases in comets is made possible by the emission lines produced as a result of this low density environment. The similarity of the chemical composition of comets with that of the Jovian planets should be noted as well. Be sure to point out the difference between the spectrum reflected by dust particles and that emitted by the gas atoms as examples of spectroscopic analysis. The orbital characteristics of comets are very different from those of the asteroids and should be compared (eccentricities and orientation of orbital planes). The very large eccentricities of comet orbits can be easily computed using the information in Box 4-3. The distance of the Oort cloud relative to the orbit of Mercury can serve as rough examples of aphelion and perihelion distances respectively. The application of Kepler's second law to orbits of comets is interesting. Emphasize the very great differences between perihelion and aphelion distances and use the conservation of angular momentum as defined in Box 23-1 to translate these values into relative velocities. The transient nature of comets should be noted and their role in the generation of meteor showers is significant for casual observers of the night sky. The collision of comets and meteoroids with the Earth is of interest to students. The differences between impacts of these two classes of objects arise due to the different compositions involved. Cometary materials are volatile in nature and tend to vaporize upon impact with the Earth's atmosphere. The Tunguska event of 1908 showed little evidence of an impact with a solid object. The Barringer crater in New Mexico, however, indicates the result of an impact with a solid object made of refractory materials, such as those found in asteroids.

Meteor showers (section 17-8 and Box 17-2) have become popular as many television meteorologists will note when major meteor showers are about to take place. Students usually lose some of their interest when told that the best observing times are in the early morning hours. Recommend that they make their observations on their way home from a party. Be sure to note the detrimental impact of a bright moon on the detection of meteors.

Answers to Chapter 17 Computational Questions

14. We can use Kepler's third law to find the planet's orbital period as follows:

$$P^2 = a^3$$
$$P = (a^3)^{1/2}$$
$$= [\,(77.2)^3\,]^{1/2} \text{ yrs}$$
$$= (4.601 \times 10^5)^{1/2} \text{ yrs}$$
$$= 6.78 \times 10^2 \text{ yrs} = 678 \text{ yrs}$$

The planet's angular velocity is:

$$\omega = \frac{360°}{678 \text{ yrs}} = 0.53°/\text{yr}$$

Since astronomers can measure the proper motions of stars with an accuracy of a small fraction of an arc second per year, this is an easily detectable motion.

15. Both members of the double asteroid are at the same distance from the

Sun (i.e., source of illumination) and at the same distance from the Earth (i.e., location of observer). Thus these distances need not enter our calculation for the ratio of the brightnesses of the two members (b_1/b_2).

Let d_1/d_2 and A_1/A_2 be the ratios of the diameters and albedos of the two members respectively. Then

$$\frac{b_1}{b_2} = \left(\frac{d_1}{d_2}\right)^2 \left(\frac{A_1}{A_2}\right)$$

$$16 = \left(\frac{d_1}{d_2}\right)^2 \quad (1)$$

$$\frac{d_1}{d_2} = 4 \quad \text{and} \quad d_1 = 4d_2$$

If the larger asteroid has $d_1 = 200$ km then $d_2 = \frac{1}{4}d_1 = 50$ km

18. In general for an ellipse, the major axis ($2a$) is equal to the sum of the perihelion (d_{pe}) and aphelion (d_{ap}) distances as shown below:

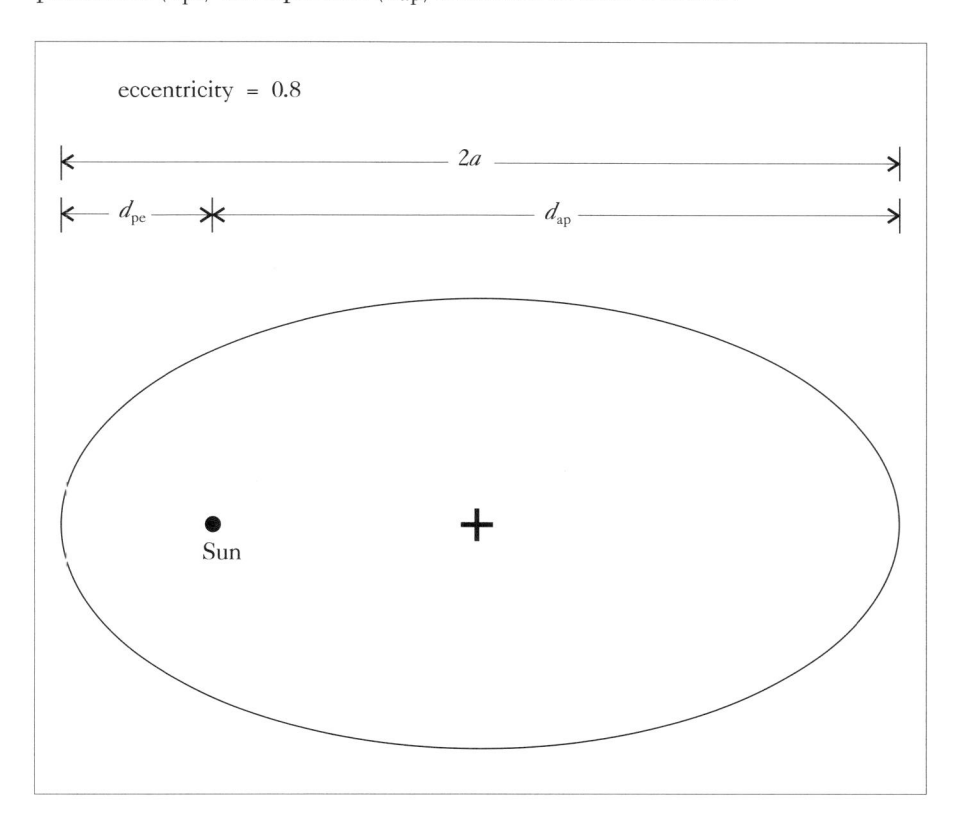

eccentricity = 0.8

```
                eccentricity = 0.98
|←————————————————— 2a —————————————————→|
|←*————————————————— d_ap ——————————————→|
 d_pe
```

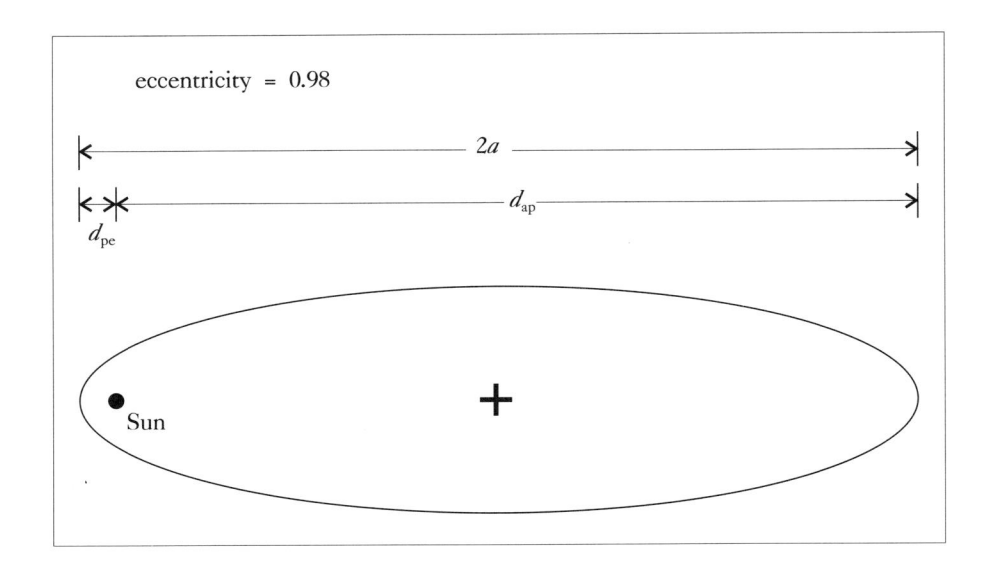

(ellipse with Sun at left focus and + at center)

For a highly elongated cometary orbit, like the one shown in the figure above, the perihelion distance is very much smaller than the aphelion distance. Therefore the aphelion distance is essentially equal to twice the semimajor axis (a), or

$$a = \frac{1}{2}d_{ap}$$

From Kepler's third law we have:

$$P = a^{3/2} = (a^3)^{1/2}$$

Comet a: $a = \frac{1}{2}(100 \text{ AU}) = 50 \text{ AU}$

$$P = 50^{3/2} = 354 \text{ yrs yrs}$$

$$\text{lifetime} = 100P = 35{,}400 \text{ yrs}$$

Comet b: $a = \frac{1}{2}(1000 \text{ AU}) = 500 \text{ AU}$

$$P = 500^{3/2} \text{ yrs} = 11{,}200 \text{ yrs}$$

$$\text{lifetime} = 100P = 1.12 \text{ million years}$$

Comet c: $a = \frac{1}{2}(10{,}000 \text{ AU}) = 5{,}000 \text{ AU}$

$$P = 5000^{3/2} \text{ yrs} = 3.54 \times 10^5 \text{ yrs}$$

$$\text{lifetime} = 100P = 35.4 \text{ million years}$$

Comet d: $a = \frac{1}{2}(100{,}000 \text{ AU}) = 50{,}000 \text{ AU}$

$$P = (50{,}000)^{3/2} \text{ yrs} = 1.12 \times 10^7 \text{ yrs}$$
$$\text{lifetime} = 100P = 1.12 \text{ billion years}$$

19. On pages 297 and 298 of the text we read that 95% of meteorites are stones and 4% are irons. We therefore assume:

95% of parent asteroid is rocky mantle

4% of parent asteroid is iron core

thus

$$V_{\text{core}} = \frac{4}{95} V_{\text{mantle}}$$

Making use of the formula for the volume of a sphere we get:

$$\frac{4}{3}\pi R_{\text{core}}^3 = \frac{4}{95}\left[\frac{4}{3}\pi R_{\text{total}}^3 - \frac{4}{3}\pi R_{\text{core}}^3\right]$$

$$R_{\text{core}}^3 = \frac{4}{95}\left[R_{\text{total}}^3 - R_{\text{core}}^3\right]$$

$$\frac{95}{4} R_{\text{core}}^3 = R_{\text{total}}^3 - R_{\text{core}}^3$$

$$R_{\text{total}}^3 = R_{\text{core}}^3\left(\frac{95}{4} + 1\right) = \frac{99}{4} R_{\text{core}}^3$$

$$= 25 R_{\text{core}}^3$$

Thus $R_{\text{total}} = R_{\text{core}}(25)^{1/3} = 2.91\, R_{\text{core}}$

or alternatively:

$$R_{\text{core}} = 0.34 R_{\text{total}}$$

So the radius of the iron core is about a third that of the entire asteroid.

For the Earth, the radius of the iron core is 7000 km/12,700 km or about 0.55 that of the entire planet.

Chapter

18

Our Star

Chapter Summary

This chapter describes the physical properties of the solar interior and atmosphere. The fusion of hydrogen to helium is explained. A summary of the process of computing stellar models includes a discussion of energy transport mechanisms. The solar neutrino problem is examined. The layers of the solar atmosphere are listed and described as are some of the conspicuous features observed in the solar atmosphere. Solar seismology is examined and a short list of satellites used for solar observations is presented.

18-1 The photosphere is the lowest of three main layers in the Sun's atmosphere

The physical conditions in the photosphere are described. Granulation is discussed and its convective origins are revealed. Limb darkening is explained.

Box 18-1 Sun data

The physical properties of the Sun are listed. Apparent and absolute magnitudes are provided. The orbital motion of the Sun in the Galaxy is outlined.

18-2 The chromosphere is located between the photosphere and the Sun's outermost atmosphere

The physical conditions in the chromosphere are described. The emission lines commonly detected in the chromosphere are noted. The nature of spicules and supergranules are discussed.

18-3 The corona is the outermost layer of the Sun's atmosphere

The physical conditions in the corona are described. Coronal streamers, transients and holes are discussed. The spectral features which indicate very high temperatures are identified.

18-4 The number of sunspots varies with an 11-year period

Sunspots are described. The umbra and penumbra of sunspots are defined. The relative darkness of sunspots is explained in terms of the Stefan–Boltzmann law. The sunspots cycle is described and the distribution of sunspots in solar latitude with time is displayed.

18-5 Sunspots are one of many phenomena produced by the Sun's magnetic field and a 22-year solar cycle

The Zeeman effect is introduced and the magnetic field of the Sun is discussed. Interactions between magnetic fields and plasmas are examined. Prominences and solar flares are characterized. The strong magnetic fields associated with sunspots are explained in terms of differential rotation. The Maunder minimum and its related effects on the Earth's climate are described.

Box 18-2 The Zeeman effect

The splitting of energy levels in the presence of magnetic fields is illustrated and discussed in the context of atomic structure.

18-6 The Sun's energy is produced by thermonuclear reactions in the core of the Sun

The history of attempts to identify the age of the Sun is reviewed briefly. The conversion of mass to energy is explained and the fusion of hydrogen into helium is described with calculations of typical energies released. Nuclear fusion reactions are discussed. The strong and weak nuclear forces are introduced. Luminosity and flux are defined.

Box 18-3 The proton–proton chain and the CNO cycle

The individual steps of the proton–proton and CNO cycles are described and their importance in low- and high-mass stars is examined.

18-7 A theoretical model of the Sun shows how energy gets from the Sun's center to its surface

Hydrostatic and thermal equilibrium conditions are defined. Energy transport by convection, conduction and radiation is discussed. The equations of stellar structure are noted and the characteristics of a stellar model are reviewed. The data for a theoretical solar model are displayed in tabular and graphic forms.

18-8 The mystery of the missing neutrinos inspires speculation about the Sun's interior

The characteristics of neutrinos are reviewed and attempts to detect solar neutrinos are described. The differences between predicted and observed detection rates are noted and potential explanations are discussed.

18-9 Solar seismology and new satellites are among the latest tools for solar research

Vibrations of the solar surface are described and the value of solar seismology in studies of the solar interior is noted. Some of the achievements of satellite studies of the Sun are reviewed and future missions are identified.

Essay by Arthur B. C. Walker, Jr.
The Solar Corona

Essay Summary

The history of the study of the Sun's corona is reviewed. The strange coronal spectral lines are identified as coming from highly ionized iron. Mechanisms for heating the corona to the observed high temperatures are proposed. Recent x-ray observations are described.

The question asked in the essay is: Which of the three mechanisms heats the corona? It might be useful to have students outline the steps that brought us to the point where we could ask that question, beginning with the first observations of the corona during eclipses. Students could then see how our present knowledge of the corona has descended from a series of observations, attempted explanations, new questions asked and answered, and improvements in technology. Students could then propose what the next step should be and how our model of the corona would change as a result of possible outcomes of that step.

Teaching Hints and Strategies

The Sun is a unique star for astronomers. It is the only star for which direct evidence of an age is available. The ages of rocks on Earth and the Moon and of meteorites place a definite limit on the age of the Sun. (The Sun is also the only star for which we can routinely resolve and examine individual surface features.)

When introducing *solar data* (Box 18-1) be certain to emphasize the relative values rather than the absolute values. Relative values are much more meaningful.

Before discussing the layers of the solar atmosphere, take a few minutes to explain that the solar atmosphere is continuous in nature and that the temperature falls as altitude increases in the *photosphere* (section 18-1), but rises through the chromosphere and corona. The density decreases with increasing distance from the center of the Sun, reaching very low values in the corona. Point out that limb darkening observations permit the calculation of a detailed model of the physical conditions in the solar atmosphere and note that these observations are not available for any other star. Granulation is an excellent example of the Stefan–Boltzmann law. Be sure to explain that the dark regions are only dark relative to the hotter surroundings. Also remind students that the absorption spectrum from stars is produced in this layer. Studying the absorption spectra of stars generally provides information about their photospheres.

The most dramatic photographs of the Sun are often the filtergrams or spectroheliograms which show the radiation from the *chromosphere* (section 18-2). Take a few minutes to explain how the light from the photosphere can be eliminated using filters or a dispersing element to permit the recording of the chromosphere. Draw an absorption line with an emission core to indicate the radiation recorded when the rest of the spectrum is eliminated. You might also note that information about stellar chromospheres can be obtained by applying similar techniques to other stars, although imaging is not possible.

The dramatic large-scale changes in the shape and extent of the *corona* (section 18-3) should be noted. The very high temperatures in the corona tend to cause students to develop incorrect ideas concerning its thermal energy content. Explain that the density is very low which reduces the energy content to smaller values in spite of the very high temperatures. The recent discovery of coronal holes and their link to the solar wind are a fine example of the application of space observatories to advance our understanding of the solar system in other spectral regions. The value of coronagraphs on the Earth's surface can be compared to the value of coronagraphs in space. The scattering of light by the Earth's atmosphere reduces the effectiveness of a coronagraph on the Earth's surface.

The *sunspot cycle* (section 18-4) should be clearly distinguished from the solar cycle as this confuses many students. The dark regions in sunspots are explained using the laws of blackbody radiation just as granulation was. The principal difference is that sunspots persist for much longer periods than granulation cells due to the supporting forces supplied by magnetic fields. The influence of the solar wind on the Earth and on other planets is also worth noting. The diagrams showing the impact of solar wind on planetary magnetospheres in Figures 8-19, 13-14 and the figure in Box 11-2 should be noted.

Make certain that the students understand that the *Zeeman effect* (Box 18-2) is manifested by spectral line splitting while the Doppler effect produces line broadening or changes in position. The effect of weak magnetic fields can be confused with line broadening in some cases.

George Ellery Hale almost single-handedly originated and promoted astrophysics. He felt that there should be a close connection between the laboratory sciences and observational astronomy and this partnership would allow us to more fully understand the stars. The Zeeman effect is an excellent example of this. Babcock's theory (section 18-5) can be illustrated using a series of chalkboard drawings similar to Figure 18-18. Start with magnetic field lines like those in planetary magnetospheres. Use arrows to keep track of the field directions distinguishing between magnetic north and south poles as places where the arrows point into and out of the surface. As the magnetic field lines wind up due to differential rotation, there is a tendency for them to penetrate through the surface because magnetic field lines in the same direction repel one another. Meanwhile the bundles of magnetic field lines in the two hemispheres attract because they run in opposite directions. This causes the latitude shift.

The concept of conservation of *energy* (section 18-6) should be examined again and its implications for the eventual demise of the Sun noted. The connection between conservation of energy and maximum possible age can be enhanced by reviewing the history of attempts to establish the age of the Sun and potential energy sources which might satisfy the solar luminosity for that period of time. An interesting example for mass conversion was provided by George Abell. He noted that the conversion of 1 gram of hydrogen into helium resulted in enough energy to raise a 1000 ton battleship 40 miles above the surface of the Earth. Point out to the students that one gram is about the mass of a paper clip. After the conversion almost all of the mass remains. Only 1% is converted to energy. The concept of mass conversion is frequently confused with mass loss in stellar evolution. Make it clear that mass to energy conversion results only in a very small reduction of the mass of a star in its entire evolutionary cycle.

Many students mistakenly assume that the *CNO cycle* (Box 18-3) converts carbon into oxygen. Stress that the original carbon is returned and the net result is just another process for the conversion of hydrogen into helium. You can explain why higher temperatures are required for the CNO cycle by reviewing the repulsive nature of the like electrical charges in atomic nuclei. The nuclear force only binds the particles together when an enormous electrical force is overcome by the high kinetic energies of nuclei having high temperatures. The larger numbers of protons in the nuclei involved in the CNO cycle generate more repulsion and require larger kinetic energy which means higher temperature.

When introducing the subject of *theoretical stellar models* (section 18-7), it is interesting to note that we probably know more about the physical conditions in the interiors of remote stars than we know about conditions in the interiors of planets and satellites in the solar system. The reason for this is that stars are generally gaseous throughout and gases are more easily described mathematically and physically than liquids and solids. The equilibrium conditions described are based on direct observations of the Sun and stars. You might note that pulsating variable stars are an exception here. The equations which describe stellar structure are just mathematical representations of these conditions and the conservation of mass and conservation of energy laws. Common examples of radiation, convection and conduction should be presented. You might point out that the convective zone is produced by the blocking of radiation flow in that region of the solar interior where the temperature drops sufficiently low so that hydrogen and helium can exist in their neutral states. This blocking of photons results in the mass motions which are characteristic of convection and which manifest themselves as granulation in the solar atmosphere.

The *mystery of the solar neutrinos* (section 18-8) can be discussed in the context of the detection of neutrinos from the supernova 1987A described in Chapter 22. The detection serves as an excellent example of serendipity in science. You might also review this section after discussing stellar evolution to reinforce some of the problems which remain unsolved in an otherwise rather successful theory.

Solar seismology (section 18-9) can be introduced by reviewing related techniques in the Earth's interior and the interior of the Moon. This process may permit the gathering of observational data about the structure of the solar interior which could be used to test theoretical models of the Sun.

Answers to Chapter 18 Computational Questions

18. From Box 18-1 on page 312 we have

$$M_\odot = 1.99 \times 10^{30} \text{ kg}$$

$$R_\odot = 6.96 \times 10^5 \text{ km}$$

$$= 6.96 \times 10^8 \text{ m}$$

The Sun's average density ($\bar{\rho}$) is therefore

$$\bar{\rho} = \frac{M_\odot}{\frac{4}{3}\pi R_\odot^3}$$

$$= \frac{1.99 \times 10^{30}}{\frac{4}{3}(3.1416)(6.96 \times 10^8)^3} \text{ kg/m}^3$$

$$= 1410 \text{ kg/m}^3$$

From the data given in Appendix 3 we see that the Sun's average density is quite similar to that of Jupiter, Uranus, and Neptune.

21. The problem involves the straightforward use of Einstein's equation $E = mc^2$.

(a) $E = mc^2 = (2 \times 10^{-26} \text{ kg})(3 \times 10^8 \text{ m/s})^2$

134 CHAPTER 18

$$= 2 \times 9 \times 10^{-26+16} \text{ J}$$
$$= 1.8 \times 10^{-9} \text{ J}$$

(b) $E = mc^2 = (1 \text{ kg})(3 \times 10^8 \text{ m/s})^2$
$$= 9 \times 10^{16} \text{ J}$$

(c) $E = mc^2 = (6 \times 10^{24} \text{ kg})(3 \times 10^8 \text{ m/s})^2$
$$= 54 \times 10^{24+16} \text{ J}$$
$$= 5.4 \times 10^{41} \text{ J}$$

22. In Box 18-1 on page 312 we see that the luminosity of the Sun is

$$L_\odot = 3.90 \times 10^{26} \text{ W}$$

(Remember that 1 watt = 1 joule/second)

Using the answers from Question 21, we calculate:

(a) To produce the energy contained in the mass of a carbon atom, the Sun must shine for

$$\frac{E}{L_\odot} = \frac{1.8 \times 10^{-9} \text{ J}}{3.9 \times 10^{26} \text{ J/s}}$$
$$= \left(\frac{1.8}{3.9}\right) \times 10^{-9-26} \text{ s}$$
$$= 4.6 \times 10^{-36} \text{ s}$$

(b) To produce the energy contained in a 1 kilogram mass, the Sun must shine for

$$\frac{E}{L_\odot} = \frac{9 \times 10^{16} \text{ J}}{3.9 \times 10^{26} \text{ J/s}}$$
$$= \left(\frac{9}{3.9}\right) \times 10^{16-26} \text{ s}$$
$$= 2.3 \times 10^{-10} \text{ s}$$

(c) To produce the energy contained in the mass of the Earth, the Sun must shine for

$$\frac{E}{L_\odot} = \frac{5.4 \times 10^{41} \text{ J}}{3.9 \times 10^{26} \text{ J/s}}$$

$$= \left(\frac{5.4}{3.9}\right) \times 10^{41-26} \text{ s}$$

$$= 1.38 \times 10^{15} \text{ s} \left(\frac{1 \text{ yr}}{3.16 \times 10^7 \text{ s}}\right)$$

$$= 4.4 \times 10^7 \text{ yrs}$$

$$= 44 \text{ million years}$$

23. The Sun's luminosity $L_\odot$ is 3.9×10^{26} J/s. Therefore the luminosity of Sirius is

$$40 L_\odot = 40 \times 3.9 \times 10^{26} \text{ J/s}$$

$$= 1.56 \times 10^{28} \text{ J/s}$$

Each second Sirius produced 1.56×10^{28} joules of energy. Since $E = mc^2$, this energy output entails the conversion each second of

$$m = \frac{E}{c^2}$$

$$= \frac{1.56 \times 10^{28} \text{ J}}{(3 \times 10^8 \text{ m/s})^2}$$

$$= \left(\frac{1.56}{9}\right) \times 10^{28-16} \text{ kg}$$

$$= 1.73 \times 10^{11} \text{ kg}$$

This mass, which is completely converted into energy each second, is only 0.7% of the mass of hydrogen involved in the reaction. Thus the total amount of hydrogen burned each second is

$$\frac{1}{0.007}(1.73 \times 10^{11}) \text{ kg} = 2.5 \times 10^{13} \text{ kg}$$

24. The Sun's luminosity is

$$L_\odot = 3.90 \times 10^{26} \text{ J/s}$$

This entails a mass-to-energy conversion rate of

$$\frac{L_\odot}{c^2} = \frac{3.90 \times 10^{26} \text{ J/s}}{(3 \times 10^8 \text{ m/s})^2}$$

$$= \left(\frac{3.90}{9}\right) \times 10^{26-16} \text{ kg/s}$$

$$= 4.33 \times 10^9 \text{ kg/s}$$

$$5 \text{ billion years} = 5 \times 10^9 \text{ yr} \left(\frac{3.16 \times 10^7 \text{ s}}{1 \text{ yr}}\right)$$

$$= 1.58 \times 10^{17} \text{ seconds}$$

Therefore the mass converted into energy in 5 billion years is

$$(4.33 \times 10^9 \text{ kg/s})(1.58 \times 10^{17} \text{ s})$$

$$= 6.84 \times 10^{26} \text{ kg}$$

When hydrogen is converted into helium, only 0.7 percent of the mass is transformed into energy. Therefore, the amount of hydrogen used up in the Sun's energy production over the next five billion years is:

$$\frac{6.84 \times 10^{26} \text{ kg}}{0.007} = 9.77 \times 10^{28} \text{ kg}$$

The mass of the Sun is 1.99×10^{30} kg, and so the amount of hydrogen used up is:

$$\frac{9.77 \times 10^{28}}{1.99 \times 10^{30}} = 4.9\% \text{ of the Sun's mass}$$

The present composition of the Sun is:

hydrogen = 0.74 $M_\odot$ = 0.74 $(1.99 \times 10^{30}$ kg$)$ = 1.47×10^{30} kg

helium = 0.25 $M_\odot$ = 0.25 $(1.99 \times 10^{30}$ kg$)$ = 4.98×10^{29} kg

heavy elements = 0.01 $M_\odot$

In 5 billion years the composition of the Sun will be:

hydrogen = 1.47×10^{30} kg $- 9.77 \times 10^{28}$ kg

$$= (1.47 - 0.098) \times 10^{30} \text{ kg}$$

$$= 1.37 \times 10^{30} \text{ kg}$$

Therefore, in five billion years, the fraction of the Sun's mass that will be hydrogen is

$$\frac{1.37 \times 10^{30} \text{ kg}}{1.99 \times 10^{30} \text{ kg}} = \frac{1.37}{1.99} = 0.688 \approx 69\%$$

Similarly, the helium content of the Sun will be:

helium = 4.98×10^{29} kg + 9.77×10^{28} kg − $0.007(9.77 \times 10^{28}$ kg)
$= 4.98 \times 10^{29} + (9.77 − 0.07) \times 10^{28}$
$= (4.98 + 0.97) \times 10^{29}$ kg
$= 5.95 \times 10^{29}$ kg

So the helium abundance is

$$\frac{5.95 \times 10^{29} \text{ kg}}{1.99 \times 10^{30} \text{ kg}} = 0.299 \approx 30\%$$

25. In the text we find that

$$T_{photosphere} = 5800 \text{ K}$$
$$T_{chromosphere} = 50{,}000 \text{ K}$$
$$T_{corona} = 1.5 \times 10^6 \text{ K}$$

Wien's law states:

$$\lambda_{max} = \frac{2.9 \times 10^{-3} \text{ mK}}{T}$$

(a) photosphere:

$$\lambda_{max} = \frac{2.9 \times 10^{-3}}{5800} \text{ m} = 5 \times 10^{-7} \text{ m}$$
$$= 500 \text{ nm} = \text{visible light}$$

(b) chromosphere:

$$\lambda_{max} = \frac{2.9 \times 10^{-3}}{5 \times 10^4} \text{ m} = 5.8 \times 10^{-8} \text{ m}$$
$$= 58 \text{ nm} = \text{ultraviolet}$$

(c) corona:

$$\lambda_{max} = \frac{2.9 \times 10^{-3}}{1.5 \times 10^6} = 1.93 \times 10^{-9} \text{ m}$$
$$= 1.93 \text{ nm} = \text{X rays}$$

These calculations tell us that the photosphere can be studied with the ground-based telescopes, but the chromosphere and corona are best observed from spacecraft and satellites.

26. In the text we read that the temperature of the photosphere is 5800 K. The Stefan–Boltzmann law (see page 86) relates energy flux (F) to temperature (T):

$$F = \sigma T^4$$

Let the subscripts p and s refer to the photosphere and sunspot, respectively. Thus

$$\frac{F_s}{F_p} = \frac{\sigma T_s^4}{\sigma T_p^4} = \left(\frac{T_s}{T_p}\right)^4$$

$$= \left(\frac{5800 - 1500}{5800}\right)^4$$

$$= \left(\frac{43}{58}\right)^4 = (0.741)^4$$

$$= 0.302$$

Therefore each square meter of a sunspot is only about 30% as bright as a square meter of undisturbed photosphere.

CHAPTER

19

THE NATURE OF STARS

Chapter Summary

This chapter reviews the methods used by astronomers to establish the basic physical properties of stars. The determination of the distances to stars is described, followed by methods for determining the motions of stars. Magnitude systems are explained. The connection between absolute magnitudes and luminosities is revealed. Techniques of photometry and spectroscopy are presented. These observational techniques are applied to the detection and analysis of binary stars. The H–R diagram and mass-luminosity relation are discussed.

19-1 Careful measurements of the positions and motions of stars reveal stellar distances and space velocities

The general concept of parallax is explored and the relationship between parallax angle and distance is developed. Typical parallaxes for a few nearby stars are discussed. Proper motions, tangential velocities, radial velocities and space velocities are defined and explained.

Box 19-1 The moving cluster method

The Pleiades and Hyades clusters are introduced and the equations for determining the cluster distance from velocity data are presented.

19-2 A star's luminosity can be determined from its apparent magnitude and distance

The modern definition of the apparent magnitude system is presented. The absolute magnitude of a star is defined. Brightness and luminosity are explained and the inverse-square law of light is examined and illustrated.

Box 19-2 Magnitude and brightness

The mathematical conversion of magnitude differences into brightness ratios is related and several examples are given. The use of apparent magnitudes and distances to compute absolute magnitudes is explained.

Box 19-3 Bolometric magnitude and luminosity

The concepts of bolometric magnitudes and bolometric corrections are explained. Expressions are presented for computation of stellar luminosity in solar units from absolute bolometric magnitudes of the star and the Sun. A few examples are solved.

19-3 Dim stars are more common than bright ones

Luminosity function is defined and the luminosity function in the vicinity of the Sun is presented. The relative scarcity of high luminosity stars is noted.

19-4 A star's color reveals its surface temperature

The application of multicolor photoelectric photometry to the measurement of stellar colors is discussed. Color indices are described and general properties of the UBV photometric system are reviewed. Apparent magnitudes and color indices of a few bright stars are listed. The relationship between B-V color and temperature is displayed graphically.

19-5 Stars are classified according to the appearance of their spectra in a way that reveals their surface temperatures

The history of spectral classification is summarized. The role of the Bohr model of the atom in explaining spectral lines is reviewed. The basic criteria for one-dimensional classification of stellar spectra are presented and the order of the spectral sequence is explained. The subdivision of the original temperature classes is pointed out and the roles of excitation and ionization in the variations of line strengths are noted.

Box 19-4 Equivalent widths and line strengths

The use of CCD's in recording spectra is mentioned and several examples are displayed. The notation for neutral and ionized atoms is defined. The concept of equivalent widths is explained and the behavior of line strength with temperature for several line-producing species is shown.

19-6 Hertzsprung–Russell diagrams demonstrate that there are different kinds of stars

The main regions of the H–R diagram are described. Relative sizes of stars in these regions are revealed from qualitative applications of the Stefan–Boltzmann law to stars having the same color.

Box 19-5 Stellar radii

Relationships involving luminosity, temperature and radius are derived and sample sizes relative to the Sun are provided for stars. The lines representing stars of equal size in the H–R diagram are shown.

19-7 Luminosity class is a useful way of denoting a star's brightness

The MKK system is defined. The use of spectral class and luminosity class to find the absolute magnitude via the H–R diagram is explained.

19-8 Binary stars provide crucial information about stellar masses

Double stars, binary stars and visual binary stars are distinguished. The application of Newton's derived form of Kepler's third law to determine the system mass is presented. The concept of the center of mass is reviewed and the mass–luminosity relation for main-sequence stars is described and illustrated. The significance of the main-sequence in the H–R diagram as a mass sequence is discussed.

19-9 Binary systems that cannot be examined visually can still be detected and analyzed

Spectrum binaries, spectroscopic binaries and eclipsing binaries are described. The Doppler effect is reviewed and radial velocity variations with time are displayed graphically. The connection between relative speeds and relative masses in a binary system is discussed.

19-10 Light curves of eclipsing binaries provide detailed information about the two stars

The geometry of eclipsing binary stars and the resulting light curves are illustrated and explained. Some of the details, such as tidal distortions, hot-spot reflection and stellar radius determinations are discussed.

Teaching Hints and Strategies

The material in this chapter is generally more difficult for the students to master than previous material. It is somewhat more quantitative and requires that they apply the concepts presented in Chapters 4 and 5. The emphasis is on how the results are obtained as well as the description of the results. The material presents an opportunity to discuss the use of relationships (equations) to translate observed quantities into basic information about the physical properties of stars. It is helpful to point out to students that the ability of astronomers to extract the basic data from passive observations of faint sources of light is confusing at first but, once understood, provides great insight into the fundamental nature of modern astrophysics.

The use of the word parallax (section 19-1) in astronomy tends to be confusing to some students. It can refer to its general definition, to half the angle through which a star appears to move each year as a result of the Earth's orbital motion, or to the distance of a star. Be sure you alert the students to these related but somewhat different meanings. You can illustrate parallax by having students hold their pencils up in front of their faces, close one eye and then alternate eyes. The inverse relation to distance can be shown by holding the pencil close to their faces and then at full arm's length and noticing the difference in apparent movement.

The total or space velocity of a star requires both the radial and tangential parts. These are measured by entirely different methods. The Doppler effect measurement sounds to the student conceptually more diffecult but is simpler to perform. Therefore there are tens of thousands of stars whose radial velocities are known. Proper motion and tangential velocity are simpler to visualize but extremely difficult measurements to perform leading to a relatively few stars with accurate tangential velocities. Also, the same set of stars for which proper motion measurements are possible have parallax motions superimposed and this further complicates analysis.

When developing the modern definition of the *apparent magnitude system*, it is important to stress the backward numbering system used for magnitudes and to demonstrate that it is a logarithmic system. The reversal of the numbering derives from the decision of Hipparchus to label the brightest stars as first magnitude. Remind students to review section 3-6. What initially seemed to be a logical choice has resulted in a system in which the numbers get smaller as the quantity being measured gets larger. Therefore, magnitude is actually a measure of dimness rather than brightness. The origin of the logarithmic nature of the system is related to the response of the human eye. It proves to be very useful in that an enormous range of brightnesses is represented by a small range of magnitudes. It should also be noted that the zero point is arbitrarily defined and does not indicate zero brightness. Many students are familiar with the Richter scale for measuring earthquakes. Point out the similarity to the stellar magnitude scale. The motivation for using both is to compress a very wide range of measurement into a scale using only a small range of numbers. Emphasize that the magnitude system based on the amount of light reaching the Earth from objects at very different distances does not describe the intrinsic properties of the stars. Clearly identify the need to determine absolute magnitudes in order to compare intrinsic brightnesses.

Emphasize that *absolute magnitude and luminosity* (Box 19-3) are just different ways of expressing a similar quantity. The application of the inverse square law of light to convert apparent magnitudes to absolute magnitudes will reinforce these concepts. Compute the absolute magnitude of a star with an apparent magnitude of +5 and a distance of 1 pc. Repeat this calculation using an observed distance of 100 pc. The factor of 10 change in distance results in a factor of 100 change in brightness or a magnitude difference of 5 magnitudes in both cases. Use logic rather than equations to stress the process without getting involved in number crunching. Follow the same procedure with the Sun. The results permit a comparison of the intrinsic brightnesses of the stars. Ask how the Sun would appear in comparison to other stars if it were moved to this modest distance, as stellar distances go. This provides the first evidence that the Sun may not be as luminous as it seems to us. Also note that the star having an absolute magnitude of zero does not imply zero luminosity, but a luminosity 100 times greater than that of the Sun. The absolute magnitude system, like the apparent magnitude system, has an arbitrary zero point.

Although the Sun is near the middle of the range of luminosity possible, there are far fewer stars brighter than the Sun compared to those fainter than it. The assumption is that the luminosity function (section 19-3) for the Sun's vicinity is typical. The only region of space where the census of stars is complete is a small volume centered on the Sun. As one imagines larger and larger volumes, stars at the faint end are lost because they are too faint even for telescopes. By the time the volume is large enough to include the bright appearing nighttime stars, an enormous number of faint stars have dropped below the threshold of visibility. Therefore the visible nighttime stars are not typical. This point needs to be explained carefully.

When discussing the use of *colors* (section 19-4), as measured by photometry, to determine the *surface temperatures* of stars, refer students to the discussion of blackbody radiation in Chapter 5. Describe how observations of a blackbody in the laboratory could be used to calibrate color indices with specific temperatures. Be sure to point out that the color indices are differences between magnitudes and, therefore, represent ratios of brightness in these two colors. Also indicate that these color indices have arbitrarily defined zero points like the apparent and absolute magnitude systems. Remind students that the temperatures in stellar interiors rise dramatically as one moves from the surface to the core.

Be sure to alert students to the fact that the difference in magnitudes is equal to the exponent of the base of the *magnitude* system when computing the *brightness* ratio. Also note that the equation used in the examples in Box 19-3 has three unknowns. If any two of these variables are known, then the third can be computed. The methods of spectroscopic and photometric parallaxes are based on determining d when given M and m. You might discuss the distance modulus (m-M) in the solution for distance.

The major advantage of *spectral classification* (section 19-5) in temperature degterminations is that a spectral type can be assigned easily based on visual inspection and requires no accurate measurements as color indices do. In addition, spectral types are not altered by the presence of interstellar material or by atmospheric absorption. The principal disadvantage of spectral classification is the spreading out of the starlight to form the spectrum. This necessitates extending exposure times and makes the acquisition of data much more tedious than for photometry. Use of objective prisms can make this process more efficient and has resulted in the vast majority of spectral types to date. It is helpful to explain the decreasing strength of the Balmer lines with both increasing and decreasing temperature as one moves away from the A stars as resulting from insufficient excitation in the latter case and increasing ionization in the former as shown in the figure in Box 19-4. Also stress that the presence of the lines of an atom or ion indicates that it is present, but the absence of its lines do not necessarily indicate that it is absent.

The development of two-dimensional spectral classification from the *H–R diagram* (section 19-6) can be an effective method of reviewing the application of the Stefan–Boltzmann law to estimate the sizes of stars. The main sequence reflects the increasing luminosity of blackbodies with increasing temperature. As the temperature ratio for red and blue stars is about a factor of 10, the luminosity ratio should only be about a factor of 10,000 for similar sized objects. The greater range of luminosities indicates a size increase up the main sequence as well. Comparison of luminosities of stars having a common temperature type leads naturally to the size variations for dwarf, giant and supergiant stars and the concept of luminosity classes. The spectral variations which accompany these size differences are evidence of pressure differences and surface gravity differences for these different stars. Remind students that the temperature increases to the left in the H–R diagram because color indices or spectral types were originally plotted. The value of the H–R diagram should be discussed in the contexts of spectroscopic parallaxes and stellar evolution. Trigonometric parallaxes are useful only for stars which are closer than about 100 pc. Point out that the determination of the two-dimensional spectral type for a star can be used with the H–R diagram for nearby stars to estimate the star's absolute magnitude. This absolute magnitude is used with its apparent magnitude to compute its distance. This is how the nearby structure of our Galaxy is determined. The differences in basic stellar properties lead naturally to the question of their interrelationships. Are red giant stars and white dwarf stars inherently different types of stars or do they represent different stages in the life cycle of a single type of star? The relative numbers of stars in the various regions of the H–R diagram provide information about the relative times spent by stars in those stages.

You should point out that the *bolometric magnitudes* (Box 19-3) for stars are difficult to observe. Photometric detectors are not uniformly sensitive to all wavelengths of the electromagnetic spectrum. Direct measurements of long wavelength radiation are more accurate from the surface of the Earth than for short wavelength radiation. Thus, bolometric corrections for red stars were more accurate than for blue stars. Early bolometric corrections for blue stars were computed rather than observed. Fortunately, blue stars have much more simple stellar atmospheres than do red stars. Discuss how space-based observatories have changed this situation.

As an exercise in the determination of *stellar radii* (Box 19-5) you could have students compute the sizes for M0 giant and supergiant stars compared to an M0 dwarf star. Do the same for B0 giant and supergiant stars compared to a B0 dwarf. Have them compare the two ranges in size.

Be sure to note that the mass–luminosity relation determined from *binary star* (section 19-8) data is well established for main-sequence stars only. Emphasize that it indicates that the main sequence is a mass sequence. It is also of interest that it is based on mass determinations for about fifty visual binary stars for the lower mass end and about fifty eclipsing binaries for the massive stars. This diagram is also of value in the interpretation of stellar evolution. Since nearly 95% of all stars are main-sequence stars, their masses can be estimated using this relationship even when they are not members of binary star systems. Their absolute magnitudes can be computed directly from known distances or indirectly using the H–R diagram and two-dimensional spectral types.

The use of radial velocity variations in *spectroscopic binary systems* (section 19-9) implies the application of Newton's second law of motion. Changes in velocity constitute an acceleration which is evidence of a force. The study of spectroscopic binary stars is a major tool in the search for unseen companions such as black holes and planets. Although it is not emphasized in the text, you might want to define the characteristics of astrometric binary stars at this time. The nonlinear proper motions of such stars also indicate forces exerted by unseen objects and have been used for several decades to search for low-mass companions to stars. The Doppler effect is rather difficult for most students to grasp and generally requires careful explanation in class. The use of modern radial velocity spectrometers in the last decade has revolutionized the search techniques for low-mass objects such as planets and brown dwarfs. Such techniques are just beginning to detect such objects. Remind students that orbital periods of the Jovian planets are typical periods to be observed. It takes a few years for such periods to be detected when a new technology becomes available.

When discussing *eclipsing binary stars* (section 19-10), it might be worth noting that intrinsic and eclipsing variable stars are both manifested by a periodic variation in their light and radial velocity curves, but can usually be unambiguously distinguished from each other. Some of the numerous properties of stars which can be extracted from light curve analysis should be listed.

Answers to Chapter 19 Computational Questions

3. From the inverse-square law (see Figure 19-6 and the text on page 342), we know that the apparent brightness (b) of the light source is inversely proportional to its distance (d) from an observer:

$$b \propto \frac{1}{d^2}$$

We can therefore write the proportion

$$\frac{b_{\text{Jupiter}}}{b_{\text{Earth}}} = \frac{\left(\frac{1}{5.2}\right)^2}{(1)^2}$$

$$= \frac{1}{(5.2)^2} = \frac{1}{27.04} = 0.037$$

Therefore, when viewed from Jupiter, the Sun is only 3.7% as bright as it appears when viewed from Earth.

4. First convert the star's distance (d) from AU to parsecs. Since 1 pc equals 206,265 AU, we have

$$d = 10^6 \text{ AU} \left(\frac{1 \text{ pc}}{206,265 \text{ AU}}\right)$$

$$= \frac{10^6}{0.2063 \times 10^6} \text{ pc} = 4.85 \text{ pc}$$

The star's parallax (p) is

$$p = \frac{1}{d} = \frac{1}{4.85} \text{ arc sec}$$

$$= 0.206 \text{ arc sec}$$

$$\approx 0.2 \text{ arc sec}$$

5. We must use the equation given on page 339 that relates tangential velocity (v_t), proper motion (μ), and distance (d):

$$v_t = 4.74 \, \mu d$$

Thus

$$d = \frac{v_t}{4.74 \, \mu}$$

$$= \frac{50 \text{ km/s}}{4.74 \, (0.005 \text{ arc sec/yr})}$$

$$= \frac{50}{0.0237} \text{ pc}$$

$$= 2110 \text{ pc}$$

The star is about 2100 pc from Earth.

The proper motion of Barnard's star is 10.3″/yr.

$$v_t = 4.74\ \mu d = 4.74(10.3''/\text{yr})(2110 \text{ pc})$$

$$= 1.03 \times 10^5 \text{ km/sec} \approx \frac{1}{3} c$$

The star would need a tangential velocity nearly one-third the speed of light in order to exhibit a proper motion equal to that of Barnard's star.

6. We must use the equation given on page 339 that relates space velocity (v), radial velocity (v_r), and tangential velocity (v_t):

$$v = (v_r^2 + v_t^2)^{1/2}$$

Squaring both sides of this equation gives:

$$v^2 = v_r^2 + v_t^2$$

So

$$v_t^2 = v^2 - v_r^2$$

$$= (65 \text{ km/s})^2 - (35 \text{ km/s})^2$$

$$= (4225 - 1225) \text{ km}^2/\text{s}^2$$

$$= 3000 \text{ km}^2/\text{s}^2$$

$$v_t = 55 \text{ km/s}$$

The star has a tangential velocity of 55 km/s.

15. Inspecting the graph in Figure 19-19 on page 357, we see that a main-sequence star with a luminosity of 10^4 $L_\odot$ has a mass of about 10 $M_\odot$. Similarly, a 0.1 $M_\odot$ star has a luminosity of about 10^{-3} $L_\odot$.

16. The image brightness is proportional to area of the telescope's mirror, which is proportional to the square of the mirror's diameter (D):

$$\text{image brightness} \propto D^2$$

A 60-inch mirror can produce star images that are $(60/6)^2 = 10^2$ or 100 times brighter than a 6-inch mirror can. Thus a 60-inch mirror detects stars that are 100 times fainter. Since a factor of 100 in brightness corresponds to a difference of 5 magnitudes, the 60-inch telescope can detect stars with a magnitude of

148 CHAPTER 19

$$12 + 5 = 17$$

17. This problem involves a simple application of the equation on page 338 that relates parallax (p) to distance (d):

$$d = \frac{1}{p}$$

$$= \frac{1}{0.232}$$

$$= 4.31 \text{ pc}$$

18. We must first use the equation given on page 339 that relates tangential velocity (v_t), proper motion (μ), and distance (d).

$$v_t = 4.74 \, \mu d$$

From the previous problem, we know that the distance to Van Maanen's star is $d = 4.31$ pc. The star's tangential velocity therefore is:

$$v_t = 4.74 (2.95''/\text{yr})(4.31 \text{ pc})$$

$$= 60.3 \text{ km/s}$$

We can now calculate the star's space velocity (v):

$$v^2 = v_t^2 + v_r^2$$

$$= (60.3 \text{ km/s})^2 + (54 \text{ km/s})^2$$

$$= 6550 \text{ km}^2/\text{s}^2$$

$$v = 80.9 \text{ km/s}$$

$$\approx 81 \text{ km/s}$$

19. (a) We must use the formula for the Doppler shift given on page 339:

$$\frac{\lambda - \lambda_0}{\lambda_0} = \frac{v_r}{c}$$

$$\lambda = 656.50 \text{ nm}, \quad \lambda_0 = 656.28 \text{ nm}$$

and so

$$\frac{v_r}{c} = \frac{656.50 - 656.28}{656.28}$$

$$= \frac{0.22}{656.28}$$

$$= 3.35 \times 10^{-4}$$

$$v_r = (3.35 \times 10^{-4})c$$
$$= (3.35 \times 10^{-4})(3 \times 10^5 \text{ km/s})$$
$$= 101 \text{ km/s}$$

Since v_r is positive, the star is moving away from Earth.

19. (b) Let $\Delta \lambda = \lambda - \lambda_0$

$$\frac{\Delta \lambda}{\lambda_0} = \frac{v_r}{c}$$

For H β, $\lambda_0 = 486.13$ nm, and so

$$\Delta \lambda = \lambda_0 \left(\frac{v_r}{c}\right)$$
$$= (486.13 \text{ nm})(3.35 \times 10^{-4})$$
$$= 0.16 \text{ nm}$$

Thus H β in the star's spectrum has a wavelength of:

$$\lambda = \lambda_0 + \Delta \lambda$$
$$= (486.13 + 0.16) \text{ nm}$$
$$= 486.29 \text{ nm}$$

20. Recall from Box 5-2 on page 86 that the solar constant is the average flux of solar energy arriving at Earth and is equal to 1370 W/m². This is a measure of the apparent brightness of the Sun, as seen from a distance of 1 AU.

In general, the inverse-square law (see page 342) relates apparent brightness (b), luminosity (L) and distance (r) by:

$$b = \frac{L}{4 \pi r^2} \propto \frac{1}{r^2}$$

When $r = 1$ AU, $b = 1370$ W/m², so we can use the following proportion to calculate r if $b = 1$ W/m²:

$$\frac{1370 \text{ W/m}^2}{1 \text{ W/m}^2} = \frac{r^2}{(1 \text{ AU})^2}$$

Thus $r = \sqrt{1370}$ AU

$\qquad = 37$ AU

22. We must use the equation given of page 344 that relates absolute magnitude (M), apparent magnitude (m), and distance (d):

$$M = m - 5 \log (d/10)$$

Since $M = 0.0$ and $m = 16.0$, we get

$$0.0 = 16.0 - 5 \log (d/10)$$
$$5 \log (d/10) = 16.0$$
$$\log (d/10) = \frac{16.0}{5} = 3.2$$

Raising 10 to the power of both sides of this equation and remembering that

$$10^{\log x} = x$$

we get

$$\frac{d}{10} = 10^{3.2} = 1585 \text{ pc}$$
$$d = 15{,}850 \text{ pc}$$

The star cluster is about 16 kiloparsecs away.

23. We can use the equation of page 344 that relates apparent magnitude (m), absolute magnitude (M), and distance (d):

$$m - M = 5 \log d - 5$$
$$0.37 - M = 5 \log (3.51) - 5$$
$$M = 0.37 + 5 - 5 \log (3.51)$$
$$= 5.37 - 5(0.545)$$
$$= 2.64$$

Therefore Procyon has an absolute magnitude of +2.64.

In Box 19-2 on page 344, we saw that a difference in magnitude is related to the logarithm of the ratio of brightnesses:

$$M_2 - M_1 = 2.5 \log \left(\frac{b_1}{b_2}\right)$$

let star 1 be Procyon, so $M_1 = 2.64$

let star 2 be the Sun, so $M_2 = 4.84$

so we have

$$4.84 - 2.64 = 2.5 \log \left(\frac{b_1}{b_2}\right)$$

$$\log\left(\frac{b_1}{b_2}\right) = \frac{4.84 - 2.64}{2.5}$$

$$= 0.88$$

Thus $\frac{b_1}{b_2} = 10^{0.88} = 7.59$

Procyon is about 7.6 times more luminous than the Sun.

24. To be visible to the naked eye, a star must have an apparent magnitude of at least 6.0. Let r_1 be the distance to Barnard's star for it to have an apparent magnitude of $M_1 = +6.00$. From pages 342 and 344 we have the relations

$$m_2 - m_1 = 2.5 \log\left(\frac{b_1}{b_2}\right)$$

and $b = \frac{L}{4\pi r^2}$

Thus $m_2 - m_1 = 2.5 \log\left(\frac{L_1/4\pi r_1^2}{L_2/4\pi r_2^2}\right)$

Let the subscript 2 designate the observed apparent magnitude, distance, and luminosity of Barnard's star:

$$m_2 = +9.54 \text{ and } r_2 = 1.81 \text{ pc}$$

Since the luminosity of Barnard's star does not depend on its distance from Earth $L_1 = L_2$, and we have:

$$m_2 - m_1 = 2.5 \log\left(\frac{r_2^2}{r_1^2}\right)$$

$$9.54 - 6.00 = 2.5 \times 2 \times \log\left(\frac{r_2}{r_1}\right)$$

$$= 5 \log\left(\frac{r_2}{r_1}\right)$$

Thus $\log\left(\frac{r_2}{r_1}\right) = \frac{9.54 - 6.00}{5}$

$$= 0.708$$

and so $\frac{r_2}{r_1} = 10^{0.708} = 5.11$

$$r_1 = \frac{r_2}{5.11} = \frac{1.81 \text{ pc}}{5.11} = 0.35 \text{ pc}$$

Therefore Barnard's star could be seen with the naked eye if it were nearer than 0.35 pc from Earth.

25. Let ω be the angular velocity of the star (measured in radians per second), v_t be the tangential velocity of the star, and r be the distance to the star. As shown in the diagram below, we have:

$$v_t = \omega r$$

We want v_t measured in km/s. We must convert r from km to pc, and we must convert ω from radians/second to arc sec per year. These conversions are done as follows:

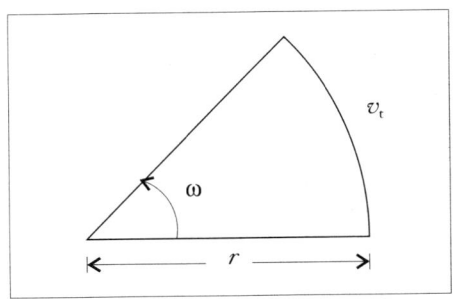

$$r(\text{km}) = r(\text{pc}) \left(\frac{3.0856 \times 10^{13} \text{ km}}{1 \text{ pc}} \right)$$

$$= r(\text{pc}) \times 3.0856 \times 10^{13}$$

Also

$$\omega \left(\frac{\text{rad}}{\text{s}} \right) = \omega \left(\frac{\text{arc sec}}{\text{yr}} \right) \left(\frac{1 \text{ rad}}{206,265 \text{ arc sec}} \right) \left(\frac{1 \text{ yr}}{3.1557 \times 10^7 \text{ s}} \right)$$

$$= \omega \left(\frac{\text{arc sec}}{\text{yr}} \right) \times 1.5363 \times 10^{-13}$$

Substituting for $r(\text{km})$ and $\omega(\text{rad/s})$, we get:

$$v_t = \omega(\text{rad/s}) \, r(\text{km})$$

$$= \omega(\text{arc sec/yr}) \times 1.5363 \times 10^{-13} \times r(\text{pc}) \times 3.0856 \times 10^{+13}$$

$$= 4.74 \omega(\text{arc sec/yr}) r(\text{pc})$$

But by definition $\mu = \omega(\text{arc sec/yr})$ and so

$$v_t = 4.74 \, \mu r$$

26. Let the subscripts 1 and 2 indicate the stars. From the equation of page 343 we have:

$$m_1 - M_1 = 5 \log r_1 - 5$$

and

$$m_2 - M_2 = 5 \log r_2 - 5$$

Since the stars have the same apparent magnitude, $m_1 = m_2$, and so:

$$M_2 - M_1 = 5 \log r_1 - 5 \log r_2$$

$$= 5(\log r_1 - \log r_2)$$

$$= 5 \log\left(\frac{r_1}{r_2}\right)$$

$$= 5 \log 10 = 5$$

Thus the absolute magnitudes of the two stars differ by 5 magnitudes. Since a difference in 5 magnitudes corresponds to a factor of 100 in brightness, we conclude that the distant star is 100 times more luminous than the nearer star.

27. Let the subscripts b and a indicate values before and after the outburst. For instance,

$$\rho_b = \text{density before outburst}$$

$$\rho_a = \text{density after outburst}$$

In general, for a sphere $r = \dfrac{M}{\frac{4}{3}\pi R^3}$ and so

$$\frac{\rho_a}{\rho_b} = \frac{\dfrac{M_a}{\frac{4}{3}\pi R_a^3}}{\dfrac{M_b}{\frac{4}{3}\pi R_b^3}} = \frac{M_a R_b^3}{M_b R_a^3}$$

Assuming that the mass of the star does not change as a result of the outburst, $M_a = M_b$, which gives us $\dfrac{\rho_a}{\rho_b} = \dfrac{R_b^3}{R_a^3} = 8$

Thus we find

$$\frac{R_a}{R_b} = \sqrt[3]{8} = 2$$

From Box 19-5 on page 352 we see that

$$\frac{L_a}{L_b} = \left(\frac{R_a}{R_b}\right)^2 \left(\frac{T_a}{T_b}\right)^4$$

We are told that $T_a = 2T_b$, and so

$$\frac{L_a}{L_b} = (2)^2 (2)^4 = 4 \times 16 = 64$$

After the outburst, the star's radius is twice as large as before and its luminosity is 64 times greater than before.

154   CHAPTER 19

28. The appropriate form of Kepler's third law (see Box 4-3 on page 69) is:

$$P^2 = \left[\frac{4\pi^2}{G(M_1 + M_2)}\right] a^3$$

As noted on page 355, if a is measured in AU, P is measured in years, and the masses M_1 and M_2 are in solar masses, then this equation can be written:

$$M_1 + M_2 = \frac{a^3}{P^2}$$

The small-angle formula tells us:

$$a = \frac{\alpha r}{206{,}265}$$

where a is measured in parsecs. If we convert to AU, then the small-angle formula becomes

$$a(\text{AU}) = \alpha r(\text{pc})$$

We are told that $\alpha = 4.5$ arc sec, so

$$a(\text{AU}) = 4.5 r(\text{pc})$$

We are also told that the parallax (p) is 0.2 arc sec. Therefore the distance to the binary is

$$r(\text{pc}) = \frac{1}{p} = \frac{1}{0.2} = 5 \text{ parsecs}$$

Consequently, the semimajor axis of the orbit in AUs is

$$a = 4.5 \times 5 = 22.5 \text{ AU}$$

Turning to Kepler's third law, with $P = 87.7$ years, we get:

$$M_1 + M_2 = \frac{a^3}{P^2} = \frac{(22.5)^3}{(87.7)^2} = \frac{1.139 \times 10^4}{7.691 \times 10^3}$$

$$= 1.48 \, M_\odot$$

Thus the sum of the masses of the two stars is $1.48 \, M_\odot$.

29. We use the form of Kepler's third law given on page 355:

$$M_1 + M_2 = \frac{a^3}{P^2}$$

$$= \frac{(2)^3}{(4)^2} = \frac{8}{16} = \frac{1}{2}$$

But we are told that $M_1/M_2 = 3.0$ and so

$$M_1 = 3M_2$$

thus

$$3M_2 + M_2 = \frac{1}{2}$$

$$4M_2 = \frac{1}{2}$$

$$M_2 = \frac{1}{8} M_\odot$$

and

$$M_1 = \frac{3}{8} M_\odot$$

The masses of the individual stars in this system are therefore $\frac{1}{8} M_\odot$ and $\frac{3}{8} M_\odot$.

30. We can make use of the equation given in Box 19-5 on page 352:

$$\frac{L_*}{L_\odot} = \left(\frac{R_*}{R_\odot}\right)^2 \left(\frac{T_*}{T_\odot}\right)^4$$

We are told $\frac{L_*}{L_\odot} = 64{,}000$ and $\frac{T_*}{T_\odot} = 3$, and so

$$\left(\frac{R_*}{R_\odot}\right)^2 = \frac{(L_*/L_\odot)}{(T_*/T_\odot)^4} = \frac{64{,}000}{(3)^4} = 790$$

$$\frac{R_*}{R_\odot} = \sqrt{790} = 28$$

Therefore Rigel's diameter is about 28 times larger than the Sun's diameter.

31. A spectral type of G8 corresponds to a surface temperature of about 4700 K.

$$M_* - M_\odot = -2.5 \log\left(\frac{L_*}{L_\odot}\right)$$

We are told that $M_* = 0.7$ and we know that $M_\odot = 4.84$, so:

156   CHAPTER 19

or . . .

$$0.7 - 4.84 = -2.5 \log\left(\frac{L_*}{L_\odot}\right)$$

$$\log\left(\frac{L_*}{L_\odot}\right) = \frac{4.14}{2.5} = 1.66$$

$$\frac{L_*}{L_\odot} = 10^{1.66} = 45.7$$

Now we can use the equation

$$\frac{L_*}{L_\odot} = \left(\frac{R_*}{R_\odot}\right)^2 \left(\frac{T_*}{T_\odot}\right)^4$$

$$\left(\frac{R_*}{R_\odot}\right)^2 = \frac{L_*/L_\odot}{(T_*/T_\odot)^4} = \frac{45.7}{(4700/5800)^4} = \frac{45.7}{0.431} = 106$$

And finally

$$\frac{R_*}{R_\odot} = \sqrt{106} = 10.3$$

Summarizing our results for Capella, we have:

$$T_* = 4700 \text{ K}$$

$$L_* = 45.7 \, L_\odot$$

$$R_* = 10.3 \, R_\odot$$

32. The absolute magnitude of a A1V star is about +1.6, and so

$$M = +1.6$$

We are told that the apparent magnitude is

$$m = 1.58$$

The appropriate equation (from page 343) is:

$$m - M = 5 \log d - 5$$

$$1.58 - 1.6 = 5 \log d - 5$$

$$5 \log d = -0.02 + 5 = 4.98$$

$$\log d = \frac{4.98}{5} = 0.996$$

$$d = 10^{0.996} = 9.9 \text{ pc}$$

Thus the distance to Castor is about 10 parsecs.

CHAPTER 20

THE BIRTH OF STARS

Chapter Summary

This chapter describes interstellar material, objects associated with star formation, and pre–main-sequence phases of stellar evolution. Observational evidence of interstellar material such as interstellar extinction, interstellar reddening and reflection nebulae are discussed. Objects detected in space which are generally associated with the star formation process are described. Predictions of the theory of pre–main-sequence evolution are displayed in the H–R diagram. Cocoon nebulae, T Tauri stars, emission nebulae and reflection nebulae are reviewed. Open or galactic clusters and stellar associations are mentioned briefly. Giant molecular clouds are described and star formation processes are identified.

20-1 Interstellar gas and dust pervade the Galaxy

A brief reveiw of the nature of the interstellar medium is provided. Dark clouds, reflection nebulae, interstellar extinction and interstellar reddening are described.

20-2 Protostars form in cold, dark nebulae and then evolve to become young main-sequence stars

Dark nebulae, Barnard objects and Bok globules are discussed. The physical conditions in Bok globules are described and the formation of protostars is traced. The theoretical calculations of Henyey and Hayashi are noted and reviewed and evolutionary tracks are displayed on an H–R diagram. Physical descriptions of the expected observations associated with these stages of evolution are provided.

20-3 A vigorous ejection of matter often accompanies the birth of a star

The mass motions which are associated with T Tauri stars, Herbig–Haro objects and jets of material are described.

20-4 Young star clusters are found in H II regions

Emission and reflection nebulae are discussed and their connections to young stellar associations and star clusters are examined. General properties of open star clusters and stellar associations are noted.

Box 20-1 Famous H II regions

This table lists the distance, diameter, mass and density for each of five of the brightest emission nebulae.

20-5 Star birth can begin in giant molecular clouds

The general properties of the interstellar medium and giant molecular clouds are described. The mapping of spiral arms using molecular emission is discussed. The pattern of star formation in giant molecular clouds is examined.

Box 20-2 Interstellar molecules

Some of the interstellar molecules detected to date are grouped according to the number of constituent atoms. The frequent appearance of carbon atoms is noted and discussed in the context of organic compounds.

20-6 Processes that compress the interstellar medium can trigger star birth

Supernova explosions and supernova remnants are described in the context of star formation. The hypothesis that such an explosion was responsible for the formation of the Sun is noted again.

Teaching Hints and Strategies

The impact of *interstellar material* (section 20-1) on attempts to map out the structure of our stellar system provides a very interesting chapter in the history of cosmology. You might discuss with your class the results of Herschel and of Kapteyn which seemed to imply that the Sun was centrally located in our stellar system. This result seems to imply that the Sun occupies a special place in the universe. The discovery of interstellar material in the 1930's which finally revealed the source of our mistaken interpretation, our incorrect assumption that there is nothing between the stars to diminish the intensity of starlight, finally confirmed once and for all that the Sun did not occupy a special place in our stellar system.

When introducing students to *dark nebulae and Bok globules* (section 20-2), explain that our assumption that the solar system formed from a large, cold, dark cloud is not just speculation, but is based on observed objects like the ones described here. Remind students that the formation of the solar system was primarily the formation of a star (99.9% of the entire mass of the solar system is in the Sun). Observations of Bok globules can support calculations of their future status. Many will collapse. Conservation of energy will ensure that the gravitational potential energy they now possess will be converted to the thermal energy and radiation which is characteristic of stars. Emphasize that observations of objects of such low temperature are best accomplished by radio and infrared telescopes. Visible radiation is very weak and is readily absorbed by the dust grains in the objects. The heating in the centers of collapsing globules leads to much hotter dense clumps or protostars (section 20-2). At some point the stars will become sufficiently hot that the surrounding cocoon nebula will be destroyed by the higher energy photons in the radiation field of the hotter objects. The protostar will eventually become visible and the temperature and luminosity will be sufficiently high so that the point representing the developing star will appear on the H–R diagram. Describe the changes which take place in the cloud in terms of the basic laws of nature. The collapse is assured by Newton's laws of motion due to the presence of unbalanced forces. Heating and emission are due to conservation of energy and to the laws of thermodynamics. Higher temperatures alter the spectrum of radiation. The substantial increase in high energy photons in the radiation field eventually causes breakdown of the dust grains and molecules and alters the excitation and ionization states of the matter.

The *ejection of matter* (section 20-3) which accompanies star birth can be addressed in the context of mass loss in the early solar nebula. Remind students that the volatile materials in the vicinity of the terrestrial planets was ejected early in the evolution of the solar system. It is estimated that the solar system may have had as much as three solar masses initially.

The connection between *young star clusters and H II regions* (section 20-4) provides direct visible evidence of the link between massive, young stars and gas and dust clouds. Point out that older star clusters contain much less gas and dust. Explain that a lack of O and B stars to ionize the hydrogen could preclude visible emission, but that no evidence of cold neutral hydrogen gas, such as 21 cm emission, or of dust grains, such as reddening or general obscuration, is detected. Also remind students that dark lanes and spots on emission nebulae are usually dark nebulae and not regions lacking hydrogen gas.

When discussing *interstellar molecules* (Box 20-2) you might have the students determine the frequency of each chemical element in such structures. This will reinforce the cosmic abundances described in Chapter 7. Many students tend to forget that many of the complex atoms are not very abundant while the simple atoms are much more common. You might point out that all atoms are not equally capable of forming molecules.

Giant molecular clouds (section 20-5) serve as very effective spiral-arm tracers. The very spectacular spiral arms seen in galaxies beyond the Milky Way are generated by light from H II regions. These regions are usually associated with giant molecular cloud complexes. Similar clouds can be detected by radio telescopes at great distances in the Galaxy while the visible light from H II regions cannot. The mapping of spiral structure of the Galaxy is thus facilitated by studies of the distribution of such objects. The distribution of 21 cm radiation of H I regions tends to be different from that of the H II regions (see Chapter 25).

The role of *supernova explosions* (section 20-6) in triggering star formation is a rather interesting link between star death and star birth. Remind students that the entire evolutionary cycle of a massive star which will produce a supernova is orders of magnitude shorter than the life cycle of the Sun. Many supernova cycles could be concluded in much less than one percent of the main-sequence lifetime of the Sun.

Answers to Chapter 20 Computational Questions

14. First of all, let's calculate the number of hydrogen atoms in one solar mass:

$$1 \, M_\odot = 1.989 \times 10^{30} \, \text{kg}$$

$$1 \, \text{H atom} = 1.67 \times 10^{-27} \, \text{kg}$$

Thus 1 $M_\odot$ contains $\dfrac{1.989 \times 10^{30}}{1.67 \times 10^{-27}} = 1.19 \times 10^{57}$ atoms

A Bok globule of 100 $M_\odot$ contains 100 times as many atoms, or 1.19×10^{59} atoms.

The volume of a sphere of radius r is $\dfrac{4}{3}\pi r^3$, and so the volume of the Bok globule is:

$$V = \dfrac{4}{3}\pi(9.46 \times 10^{15} \, \text{m})^3$$

$$= 3.546 \times 10^{48} \, \text{m}^3$$

where we have made use of the fact that

$$1 \, \text{ly} = 9.46 \times 10^{15} \, \text{m}$$

The density of atoms therefore is:

$$\rho = \dfrac{1.19 \times 10^{59} \, \text{atoms}}{3.546 \times 10^{48} \, \text{m}^3} = 3.36 \times 10^{10} \, \text{atoms/m}^3$$

But 1 m = 100 cm, and so 1 m^3 = 10^6 cm^3 which gives us

$$\rho = 3.36 \times 10^4 \text{ atoms/cm}^3$$
$$= 33{,}600 \text{ atoms/cm}^3$$

Comparing this result with the densities listed in Box 20-1 on page 373, we see that the density of atoms in a Bok globule is roughly 100 times greater than the density of atoms in an H II region.

15. The total absorption is:

$$2 \text{ magnitudes/kpc} \times 1.5 \text{ kpc} = 3 \text{ magnitudes}$$

We can use equation from Box 19-2 on page 344

$$m_2 - m_1 = 2.5 \log\left(\frac{b_1}{b_2}\right)$$

and take $m_2 - m_1 = -3$

thus

$$\log\left(\frac{b_1}{b_2}\right) = -\frac{3.0}{2.5} = -1.2$$

$$\frac{b_1}{b_2} = 10^{-1.2} = 0.063$$

Approximately 6.3 percent of the photons leaving the star cluster actually make it to Earth.

16. We make use of the equation given in Box 19-4 on page 352:

$$\left(\frac{L_*}{L_\odot}\right) = \left(\frac{R_*}{R_\odot}\right)^2 \left(\frac{T_*}{T_\odot}\right)^4$$

where the subscript * indicates values for the protosun.

We are told $\frac{L_*}{L_\odot} = 1000$ and also

$$\frac{T_*}{T_\odot} = \frac{1000}{5800} = 0.1724$$

thus

$$\left(\frac{R_*}{R_\odot}\right)^2 = \frac{(L_*/L_\odot)}{(T_*/T_\odot)^4} = \frac{1000}{(0.1724)^4}$$

$$= 1.13 \times 10^6$$

CHAPTER 20

$$\frac{R_*}{R_\odot} = (1.13 \times 10^6)^{1/2} = 1.06 \times 10^3 = 1060$$

$$\approx 1000$$

The radius of the protosun was about 1000 times larger than that of the present Sun.

18. Since the molecular weight of ethyl alcohol is 46 and the mass of a hydrogen atom is 1.67×10^{-27} kg, the mass of a single ethyl alcohol molecule is:

$$46 \times 1.67 \times 10^{-27} \text{ kg} = 7.68 \times 10^{-26} \text{ kg}$$

The number of molecules we would need for 10 g = 0.01 kg is

$$0.01 \text{ kg} \left(\frac{1 \text{ molecule}}{7.68 \times 10^{-26} \text{ kg}} \right)$$

$$= 1.30 \times 10^{23} \text{ molecules}$$

If there is 1 molecule per 10^8 m^3, the required volume is:

$$1.30 \times 10^{23} \text{ molecules} \left(\frac{10^8 \text{ m}^3}{1 \text{ molecule}} \right)$$

$$= 1.3 \times 10^{31} \text{ m}^3$$

Since 1 AU = 1.496×10^{11} m, 1 AU^3 = 3.35×10^{33} m^3, and so the required volume can be written as:

$$1.3 \times 10^{31} \text{ m}^3 \left(\frac{1 \text{ AU}^3}{3.35 \times 10^{33} \text{ m}^3} \right) = 3.9 \times 10^{-3} \text{ AU}^3$$

or about $\frac{1}{256}$ of a cubic AU.

CHAPTER 21

STELLAR MATURITY AND OLD AGE

Chapter Summary

The main-sequence and post–main–sequence stages in stellar evolution are presented in the context of nuclear fusion reactions. Properties of open and globular star clusters are expanded and the concept of three stellar populations is developed. Mass loss by single stars, mass transfer in binary systems and pulsational instability are related to the late phases of stellar evolution.

21-1 When core hydrogen burning ceases, a main-sequence star becomes a red giant

The current abundances of the elements hydrogen and helium as a function of radius in the solar interior are contrasted with the same quantities 4.5 billion years ago. Hydrogen shell burning is described. The structure of a red giant is compared to that of the Sun.

Box 21-1 Main-sequence lifetimes

The relationship between mass and main-sequence lifetimes is derived and applied to a star with four times the mass of the Sun.

21-2 Helium burning begins at the center of a red giant

The triple alpha process is described. The nature of degenerate gases and perfect gases are discussed and core helium burning and the helium flash are examined.

21-3 Evolutionary tracks on the H–R diagram reveal the ages of star clusters

Evolutionary tracks from the main sequence to the red giant branch are shown on an H–R diagram for stars of several different masses. Theoretical H–R diagrams as a function of time are shown for a hypothetical cluster based on computer simulations. The turnoff point as an age indicator is explained. Horizontal branch stars in globular clusters are discussed. Metallicity differences between globular and open star clusters and stellar populations are examined in the context of cluster age determinations and stellar evolution.

21-4 Supergiants and red giants typically show mass loss

The method for detecting mass loss spectroscopically is described and typical mass loss rates reported. Circumstellar envelopes are mentioned and properties of Wolf–Rayet stars are reviewed.

21-5 Mass transfer in close binary systems can produce unusual double stars

Mass transfer in a binary system is described with references to Roche lobes and the inner Lagrangian point. The classification of binary systems as detached, semi-detached, contact, or overcontact systems is described. Examples of three types of systems are illustrated with characteristic light curves. The formation of accretion disks is examined.

21-6 Many mature stars pulsate

The instability strip in the H–R diagram is noted. Cepheid variables are described and the period–luminosity relation is revealed. The evolutionary status of Cepheid variables is explained and RR Lyrae stars are contrasted with Cepheid variables.

Teaching Hints and Strategies

Stellar evolution can be one of the most exciting topics in modern astrophysics. It is helpful for students to place this topic in its proper context. Evolution means slow change. Remind them that in biological evolution, slow change takes place over several generations. In stellar evolution the slow change refers to the description of the life cycles of individual stars. To provide some insight into stellar evolution, pose an analogous problem for a biologist. You want to be able to describe the life cycles of all of the animals on the continent of Africa. The observational data is a collection of still photographs taken from an airplane as it flew over. What might you conclude about the evolution of small and large elephants, large and small lions, tigers, leopards and cheetahs and about caterpillars and moths? This is similar in time scale to an astronomer trying to describe the life cycles of objects having lifetimes of billions of years based on observations taken over a 100 year period. (One second is to one year as 100 years is to 30 billion years.) The obvious problem for observational astronomers is the long time scale for evolution. However, for a theoretician, rapid changes are a major problem. Thus theoretical calculations are valid only in cases where there is very slow or no change while observations of change are of value only when change is reasonably fast. Theory and observation are thus complimentary.

An examination of the interplay of theory and observation is very useful when discussing the *termination of core hydrogen burning and subsequent evolutionary changes* (section 21-1). Explain how theoretical stellar models can be shown to reproduce the main sequence if they are assumed to have different masses but the same chemical composition. Point out that this is called the zero-age main sequence because it is the beginning of hydrogen burning, or nuclear fusion processes in general, which are the defining characteristic of a star. Ask what changes will take place in the model of such a star. Our understanding of nuclear reactions allows us to predict just how much hydrogen is converted to helium each second to account for the energy radiated by the star. Over a period of thousands of years the chemical composition in the core changes in a precise way. A new model is computed using the slightly altered core composition. This process is repeated until changes become evident in the overall structure and at the surface. By using a sequence of static models a slowly evolving star can be traced. The theory predicts that the core will collapse and get hotter while the outer layers will expand and cool to form a red giant.

This is an excellent time to demonstrate the *main–sequence lifetimes* (Box 21-1) of stars of different mass. You could have students apply the equations derived in Box 21-1 or develop the following simplified comparison. Use round numbers to facilitate the calculations. A B0 star with an absolute magnitude of −5 and 15 solar masses and an M0 star with an absolute magnitude of +10 and 0.5 solar masses are convenient. Their total energies relative to the Sun are roughly in proportion to their mass ratio. By dividing total energy by luminosity the main-sequence lifetime can be computed. Make the calculation relative to the Sun to simplify. Stress the concepts rather than the arithmetic. The result is that more massive stars have more potential nuclear energy, but use it up very rapidly and, therefore, evolve sooner. The analogy to the mileage of cars is useful. The capacity of the gas tank is not the only factor in determining how far one can travel on a full tank. Very good or very bad mileage can more than compensate for different tank capacities.

When introducing *core helium burning* (section 21-2) you might note that the number of nuclear fusion cycles available to a star and its rate of energy consumption are controlled by its initial mass. Also point out the role of convection in transporting the nuclei generated in the core to outer layers. The concept of degeneracy is presented and the instability associated with the helium flash in the core is discussed. It is important to note that this ends the continuous series of stellar models generated by computers for low-mass stars.

In high-mass stars the degeneracy is not a factor and the models remain stable to much more advanced stages.

Star clusters (section 21-3) can be described as the kindest gift of nature in our unraveling of stellar evolution. They represent a convenient cosmic laboratory. The stars in a given cluster have the same chemical composition and age but different masses. The stars in open clusters have about the same chemical compositions but different ages. The stars in globular clusters have about the same age but different chemical compositions than open clusters. Thus, nature has provided us with a series of objects to permit a test of our theory. The main sequence in the H–R diagram of a cluster can be described as a candle which burns down with time. Ages for clusters are assigned based on such observations. The red giant limit for evolutionary models of low-mass stars is transcended by the age ranges of clusters. The globular clusters are older than the oldest open clusters. Globular clusters have large numbers of horizontal branch stars which must represent the next phase in the life cycles of low-mass stars. The different properties of the population I and II stars are consistent with the evolutionary scenario and lead one to describe the evolution of the Milky Way in much the same way as the solar system.

Again, be certain that students don't confuse *mass loss* (section 21-4) with the mass defect in nuclear fusion. A simple calculation of the surface gravity the Sun will have when it becomes a red giant compared to its current value will help illustrate the relative magnitudes of the forces involved. The masses of horizontal branch stars seem to be much smaller than the masses of the main-sequence stars from which they are thought to have evolved. This is of interest in the context of mass loss. The enrichment of the interstellar medium is also important in understanding the different chemical compositions of star clusters and the aging process in the Galaxy.

When discussing light curves of *close eclipsing binary stars* (section 21-5) it is useful to show examples of published light curves to illustrate the differences between the individual observations and the computed fit to the data. It is interesting to discuss the problems associated with observations of binary systems having periods close to one synodic month or close to one day.

Pulsating variable stars (section 21-6) should be identified as representatives of stages in the evolutionary cycles of most stars rather than a unique property of special objects. Pulsational instability is probably related to mass loss in general and possibly to formation of planetary nebulae. You might have students compare the light curves shown in Figure 21-15 with those in Chapter 19 for eclipsing binaries. The light curves of semi-detached and contact eclipsing binaries, like those shown in Figure 21-13, might be displayed here to show that all eclipsing binaries do not have nice neat light curves and that the distinction between the light curves of eclipsing and pulsating variable stars may not be as straightforward as it seems.

Answers to Chapter 21 Computational Questions

4. (a) From Table 21-1 on page 386 we see that the lifetime of a 25 $M_\odot$ star is about 3 million years. Since the Sun's main-sequence lifetime is 10 billion years, the 25 $M_\odot$ star lasts only

$$\frac{3 \times 10^6}{10 \times 10^9} = 3 \times 10^{6-9-1} = 3 \times 10^{-4}$$

as long as the Sun does.

(b) From Table 21-1 we see that a 3 $M_\odot$ star has a lifetime 5×10^8 years. In Box 21-1 we see that main-sequence lifetime is inversely proportional to $M^{2.5}$:

$$T \propto \frac{1}{M^{2.5}}$$

We therefore write the following proportion:

$$\frac{T}{5 \times 10^8 \text{ yr}} = \frac{(3 M_\odot)^{2.5}}{(5 M_\odot)^{2.5}}$$

Thus the main-sequence lifetime of a 5 $M_\odot$ star is:

$$T = 5 \times 10^8 \left(\frac{3}{5}\right)^{2.5} \text{ yr}$$

$$= 5 \times 10^8 \times 0.279 \text{ yr}$$

$$= 1.4 \times 10^8 \text{ yrs}$$

Comparing this lifetime with that of the Sun:

$$\frac{1.4 \times 10^8}{10 \times 10^9} = 1.4 \times 10^{8-10} = 1.4 \times 10^{-2}$$

So the star's lifetime is only 1.4×10^{-2} times that of the Sun.

15. In "Tips and Tools" we read that the Sun consumes 6×10^{11} kg of hydrogen each second. Since its birth 5 billion years ago, the Sun has consumed

$$6 \times 10^{11} \text{ kg/s} \times 5 \times 10^9 \text{ yr} \left(\frac{3.16 \times 10^7 \text{ s}}{1 \text{ yr}}\right)$$

$$= 9.48 \times 10^{28} \text{ kg of hydrogen}$$

The Sun's present composition is 74% hydrogen

$$= 0.74 \times M_\odot = 0.74 \times 1.99 \times 10^{30} \text{ kg}$$
$$= 1.47 \times 10^{30} \text{ kg}$$

Thus the Sun's initial hydrogen content was

$$1.47 \times 10^{30} \text{ kg} + 0.09 \times 10^{30} \text{ kg} = 1.56 \times 10^{30} \text{ kg}$$

Over the next 5 billion years, the Sun will devour another 9.48×10^{28} kg of hydrogen, bringing the total amount consumed to:

$$2 \times 9.48 \times 10^{28} \text{ kg} \approx 1.9 \times 10^{29} \text{ kg}$$

The fraction of the Sun's initial hydrogen content that will be converted into helium over its entire 10-billion-year lifetime thus is:

$$\frac{1.9 \times 10^{29} \text{ kg}}{1.56 \times 10^{30} \text{ kg}} \approx 0.12 = 12 \text{ percent}$$

16. In Box 21-1 on page 386 we saw that the lifetime (T) of a star is inversely proportional to its mass (M) to the 2.5 power:

$$T \propto \frac{1}{M^{2.5}}$$

We can therefore write the following proportion between a star (designated by the subscript *) and the Sun:

$$\frac{T_\odot}{T_*} = \left(\frac{M_*}{M_\odot}\right)^{5/2}$$

Thus

$$\frac{M_*}{M_\odot} = \left(\frac{T_\odot}{T_*}\right)^{2/5}$$

Take $T_\odot = 10 \times 10^9$ yr and $T_* = 2.8 \times 10^9$ yr.

$$\frac{M_*}{M_\odot} = \left(\frac{10 \times 10^9}{2.8 \times 10^9}\right)^{2/5} = \left(\frac{10}{2.8}\right)^{2/5}$$
$$= (3.57)^{2/5} = 1.66$$

The star can have a mass as large as 1.66 $M_\odot$.

17. The Sun's mass is 74 percent hydrogen:

$$0.74 \, M_\odot = 0.74 \times 1.99 \times 10^{30} \, \text{kg}$$
$$= 1.47 \times 10^{30} \, \text{kg}$$

Over the next 5 billion years, the Sun will consume

$$6 \times 10^{11} \, \text{kg/s} \times 5 \times 10^9 \, \text{yr} \left(\frac{3.16 \times 10^7 \, \text{s}}{1 \, \text{yr}} \right)$$
$$= 9.48 \times 10^{28} \, \text{kg of hydrogen}$$

Thus, when it becomes a red giant, the Sun's supply of hydrogen will be:

$$1.47 \times 10^{30} \, \text{kg} - 9.48 \times 10^{28} \, \text{kg} = 1.38 \times 10^{30} \, \text{kg}.$$

Since the Sun's luminosity will be 100 times greater than it is today, the Sun will consume hydrogen at a rate of

$$100 \times 6 \times 10^{11} \, \text{kg/s} = 6 \times 10^{13} \, \text{kg/s}$$

Thus the hydrogen will last for

$$\frac{1.38 \times 10^{30} \, \text{kg}}{6 \times 10^{13} \, \text{kg/s}} = 2.3 \times 10^{16} \, \text{s} \left(\frac{1 \, \text{yr}}{3.16 \times 10^7 \, \text{s}} \right) = 7.3 \times 10^8 \, \text{yr}$$
$$= 730 \, \text{million years}$$

19. The average apparent magnitude is:

$$\frac{8.97 + 9.95}{2} = 9.46$$

The absolute magnitude (M) of an RR Lyrae variable is 0.0. Since interstellar extinction dims the star by 0.5 magnitude, the actual apparent magnitude is:

$$m = 9.46 - 0.5 = 8.96$$

We can now use the equation given on page 375:

$$m - M = 5 \log d - 5$$
$$8.96 - 0 = 5 \log d - 5$$
$$5 \log d = 8.96 + 5 = 13.96$$
$$\log d = \frac{13.96}{5} = 2.79$$
$$d = 10^{2.79} = 617 \, \text{pc}$$
$$\approx 620 \, \text{parsecs}$$

20. The average apparent magnitude of Polaris is

$$\frac{1.92 + 2.02}{2} = 1.97$$

From the graph in Figure 21-16 on page 397, we estimate

$$L \approx 2 \times 10^3 \, L_\odot$$

We can compute Polaris's absolute magnitude (M) by comparing Polaris with the Sun, whose absolute magnitude is +4.8 as follows:

$$M - 4.8 = 2.5 \log \left(\frac{L_\odot}{2 \times 10^3 \, L_\odot} \right)$$

$$M = 4.8 + 2.5 \log \left(\frac{1}{2 \times 10^3} \right)$$

$$= 4.8 + 2.5 \log (1) - 2.5 \log (2000)$$

$$= 4.8 + 0 - 8.2 = -3.4$$

To find Polaris's distance, we can now use the equation

$$m - M = 5 \log d - 5$$

$$1.97 + 3.4 = 5 \log d - 5$$

$$5 \log d = 1.97 + 3.4 + 5 = 10.37$$

$$\log d = \frac{10.37}{5} = 2.07$$

$$d = 10^{2.07} = 117 \text{ parsecs}$$

In view of inaccuracies in reading the graph in Figure 21-16, we shall conclude that the distance to Polaris is about 100 pc. Incidentally, as Polaris moves along its evolutionary track toward lower temperatures on the H–R diagram, it has left the instability strip and its pulsations are rapidly decaying. The amplitude of its oscillations had been observed to be fairly constant for some 70 years. Then, around 1960, a rapid decline in Polaris's pulsation amplitude began. Polaris may stop pulsating altogether by 1995!

CHAPTER 22

THE DEATHS OF STARS

Chapter Summary

This chapter introduces the end stages of stellar evolution. Shell helium burning is described as are the periods of instability and subsequent ejection of the outer layers of the star. The characteristics of planetary nebulae and white dwarfs are reviewed. The advanced fusion cycles of high-mass stars are noted and the processes leading to a supernova explosion are explained. The mechanisms for the various types of supernovae are explained and the properties of supernova remnants are examined.

22-1 Aging low-mass stars dredge up the ashes of thermonuclear burning from deep within their interiors

Depletion of helium in the cores of red giant stars, the subsequent helium shell burning, and evolution onto the AGB are described. The convectional processes leading to the production of carbon stars is explained.

Box 22-1 Famous planetary nebulae

The distances, apparent magnitudes of central stars and angular diameters are listed for seven bright planetary nebulae.

22-2 The burned-out core of a low-mass star cools and contracts until it becomes a white dwarf

The transition of the central cores of low-mass stars into white dwarfs is described. The mass-radius relation is noted. Electron degeneracy and the Chandrasekhar limit are discussed. The detection of white dwarf stars in space is reviewed.

Box 22-2 The s and r processes

Details of the s process and r process of neutron capture are revealed. Specific isotopes associated with each of these processes are identified.

22-3 High-mass stars create heavy elements in their cores

Carbon, oxygen and neon burning reactions are listed and reference is made to silicon burning. Neutron capture is mentioned and the s and r processes are noted.

22-4 High-mass stars die violently by blowing themselves apart in supernova explosions

The processes of photodisintegration and neutronization are examined. The approximate time scales for the various nuclear reaction cycles are listed. Neutron stars and black holes are identified as the likely end phases of the evolutionary cycles of massive stars. A supercomputer simulation of the core of a massive star during supernova is presented.

22-5 In 1987 a nearby supernova gave us a close-up look at the death of a massive star

Details are provided about the supernova discovered in the Large Magellanic Cloud in February, 1987. The unusual nature of this supernova is revealed. The detection of neutrinos from this object is also noted.

22-6 White dwarfs in close binary systems can also become supernovae

Type I and Type II supernovae are discussed. Type II supernovae are associated with explosions of massive single stars. Type I supernovae are associated with binary systems containing white dwarfs. Light curves for both types of supernovae are displayed.

22-7 A supernova remnant can be detected at many wavelengths for many years

Supernova remnants are described and the observational information in visible, radio and X-ray wavelength regions is discussed. The rate of supernova explosions in galaxies is examined.

22-8 Neutrinos and gravity waves emanate from supernovae

The experiments which were designed to detect proton decay are described in the context of neutrino detection. Cerenkov radiation is defined and the limits on neutrino mass computed from supernova 1987A are related. The relationship of gravity waves to supernovae is examined and attempts to detect gravity waves are discussed.

Teaching Hints and Strategies

The "dredging-up" described in section 22-1 refers to stars which are leaving the main sequence. Earlier, as main-sequence stars, there is very little, if any, vertical mixing. This is why the spectra of main-sequence stars shows the original composition even though large changes are occurring in the core.

When one describes the instabilities which accompany the *advanced evolutionary stages of low-mass stars* (section 22-2), it is important that the students realize that the advanced stages of stellar evolution are controlled by the mass of the star as it enters those stages. The mass-loss mechanisms described in sections 21-5, 21-7 and 22-2 reduce the mass of the original star considerably so that the mass of the star on the main sequence can be several times greater than the mass near the end of its lifetime. This becomes an interesting aspect of the search for black holes. Just because stars on the main sequence have up to 60 or 80 solar masses does not mean that such massive objects must exist in the latest phases of stellar evolution.

Physical properties of *planetary nebulae* can be revealed by having students use the information provided in Box 22-1 to compute the true sizes of planetary nebulae. Assume spherical symmetry and a typical mass ejection of 0.2 to 0.5 solar masses to estimate the average density. Compare this with the density of Earth's atmosphere.

It is wise when introducing *white dwarf stars* (section 22-3) to prepare students for the conceptual basis for stellar death. Review the conditions for hydrostatic equilibrium and the cycle of collapse followed by nuclear fusion which provides the energy radiated by stars. The fact that a star can be prevented from further collapse at a given stage, in the context of conservation of energy and the blackbody radiation laws, leads to the conclusion that such a star will cool continuously as it radiates. Review the support mechanisms for main-sequence and giant stars (gas pressure, radiation pressure, degenerate-electron pressure). Note that, for low-mass stars, degenerate-electron pressure supports white dwarf stars and precludes further collapse and more exotic nuclear fusion cycles. The mass of the star controls the gravitational force which would overcome the internal pressure. If more than 1.4 solar masses is present, gravitational forces will cause further collapse. Less mass results in a white dwarf. Repeat this comparison for neutron stars. It is also important to establish an observational basis for the exotic objects to be described. The physical properties of white dwarfs were established observationally by H. N. Russell early in this century when he determined absolute magnitudes for these stars. Their enormous densities were implied by the small sizes and normal masses. Their theoretical explanation was not provided until decades later. On the other hand, neutron stars and black holes were predicted to exist long before they were detected. It is helpful for students to be exposed to this theme to reinforce the interaction between observation and theory.

Stress that *advanced nuclear fusion cycles* (section 22-4) and subsequent disintegration of the normal atomic nuclei can only occur for massive stars. Explain why the energy available from nuclear fission and nuclear fusion cycles both end when the element iron is reached. Also, the energy available from reactions decreases as this limit is approached. This can be described using stability curves for nuclei or curves showing the binding energy per nucleon. Such a discussion can help to illustrate the relative energies available from fusion processes and fission processes. The accelerating rate of nuclear fuel depletion in the advanced stages of evolution are best understood in this context. Also stress the fact that nuclei with more than 26 protons in the nucleus involve absorption of energy from the star and can not release net energy by the fusion process. As only hydrogen and helium are produced in the Big Bang, the other elements must be accounted for. Elements up to iron (26 protons) can be generated in normal exothermic fusion reactions. Elements with more protons than iron are only produced in large numbers in the explosive processes in supernovae. It can be interesting to consider the changes in the solar system which might occur if the Sun could become a supernova. (Be sure to point out that it can't.)

To reinforce the concept of isotopes when discussing *s and r processes* (Box 22-2), it is helpful to show students a table of the nuclides of the elements. Such a table presents the broad range of nuclides of the elements in an organized display. This table is for nuclear physicists what the periodic table of the elements is for the chemist. Make an enlargement of a small section to illustrate the information provided.

The final stages of the evolution of a high mass star (section 22-5) occur very quickly. Point out that the relative time scales for the different events are even more extreme since the main-sequence stage is so long. It is instructive to liken the lifetime of a star to one year and calculate which days of that year and which times during those days the various events occur.

The significance of *Supernova 1987A* (section 22-6) lies in the fact that it is the nearest supernova observed in modern times and that it lies in a stellar system whose distance is well known. The conversion of observed physical quantities into true values will be much more reliable due to the high precision of the distance. It is somewhat unfortunate that it is not a typical Type II supernova. It originated due to the explosion of a blue supergiant while it is thought that most Type II supernovae are exploding red supergiant stars.

Supernova classification (section 22-7) began as a means of organizing the similarities and differences in the appearances of the light curves and the spectra of different events. The information gathered has been analyzed to reveal the underlying physical differences between Type I and Type II supernovae. Remind students that novae are similar to Type I supernovae in that they involve mass transfer in a binary star system. However, the processes in the case of novae take place on a much smaller scale. Much less material is trans-

The Deaths of Stars 175

ferred in a nova outburst and the white dwarf star is not seriously affected. For a Type I supernova the white dwarf star is quite likely destroyed by the event.

Remind students that the actual explosion of the progenitor to SN1993J (Box 22-3) occurred 10 million years ago and the light signaling the event arrived here this year. All the light produced in M81 in the last 10 million years is on the way toward us. If the supernova rate in M81 is one every 50 years (section 22-7), then there are photons from 200,000 more supernovae already enroute from just this one galaxy.

Students should be reminded of the role supernovae and *supernova remnants* (section 22-8) play in star formation as well as star death. Refer to section 20-6 which deals with supernova-induced solar system formation and the section in Chapter 25 dealing with self-sustaining star formation in some spiral galaxies.

Although *neutrinos and gravity waves* (section 22-9) should both be emitted by supernovae, neutrinos have been detected but gravity waves have not. The search for gravity waves has been pursued for several years with little success. The detection of gravity waves reported by Weber has never been reproduced by other scientists with more sensitive instruments.

Answers to Chapter 22 Computational Questions

6. As we saw on page 405, the radius (R) of a white dwarf is inversely proportional to its mass (M) to the 1/3 power:

$$R \propto \frac{1}{M^{1/3}}$$

We can therefore write the following proportion for two white dwarfs:

$$\frac{R_1}{R_2} = \left(\frac{M_2}{M_1}\right)^{1/3}$$

The Chandrasekhar limit is 1.4 $M_\odot$, and so

$$\frac{R_1}{1.7 \times 10^4} = \left(\frac{1.0}{1.4}\right)^{1/3}$$

$$R_1 = 1.7 \times 10^4 \, (0.714)^{1/3} \text{ km}$$

$$= 1.52 \times 10^4 \text{ km}$$

Thus a white dwarf with a mass equal to the Chandrasekhar limit has a radius of about 15,200 km.

14. In Box 19-5 on page 352 we learned

$$\frac{L_*}{L_\odot} = \left(\frac{R_*}{R_\odot}\right)^2 \left(\frac{T_*}{T_\odot}\right)^4$$

We are told $\frac{L_*}{L_\odot} = 1000$ and we know

$$\frac{T_*}{T_\odot} = \frac{100{,}000}{5800} = 17.24$$

Thus

$$1000 = \left(\frac{R_*}{R_\odot}\right)^2 (17.24)^4$$

$$\left(\frac{R_*}{R_\odot}\right)^2 = \frac{1000}{(17.24)^4}$$

$$\frac{R_*}{R_\odot} = \frac{\sqrt{1000}}{(17.24)^2} = 0.106$$

$$R_* = 0.106\, R_\odot$$

16. In Box 22-1 on page 404 we saw that the angular diameter of the Ring Nebula is

$$\alpha = 1 \text{ arc min} = 60 \text{ arc sec}$$

and its distance from Earth is

$$d = 4000 \text{ ly} \left(\frac{1 \text{ pc}}{3.26 \text{ ly}}\right) = 1230 \text{ pc}$$

We can use the small-angle formula to calculate its linear diameter (D):

$$D = \frac{\alpha d}{206{,}265}$$

$$= \frac{60 \times 1230}{206{,}265} = 0.36 \text{ pc} \left(\frac{3.09 \times 10^{13} \text{ km}}{1 \text{ pc}}\right)$$

$$= 1.1 \times 10^{13} \text{ km}$$

So the radius of the Ring Nebula is:

$$R = \frac{1}{2}D = 0.55 \times 10^{13} \text{ km}$$

Since the expansion velocity is 20 km/s, we can now compute the time (T) that has elapsed since the outburst as follows:

$$T = \frac{5.5 \times 10^{12} \text{ km}}{20 \text{ km/s}}$$

$$= 2.75 \times 10^{11} \text{ s} \left(\frac{1 \text{ yr}}{3.16 \times 10^7 \text{ s}}\right)$$

$$= 0.87 \times 10^{11-7} \text{ yr} = 8700 \text{ yr}$$

$$\approx 9000 \text{ years}$$

The outburst occurred nearly 9000 years ago.

17. The white dwarf's mass is

$$M = 1 M_\odot = 1.989 \times 10^{30} \text{ kg}$$

and its radius (see Appendix 2) is:

$$R = R_{\text{Earth}} = \frac{1}{2}(12{,}756 \text{ km})$$

$$= 6.378 \times 10^6 \text{ m}$$

Thus the average density ($\bar{\rho}$) is:

$$\bar{\rho} = \frac{M}{\frac{4}{3}\pi R^3}$$

$$= \frac{1.989 \times 10^{30} \text{ kg}}{\frac{4}{3}\pi(6.378 \times 10^6 \text{ m})^3}$$

$$= \frac{1.989 \times 10^{30}}{1.087 \times 10^{21}} \text{ kg/m}^3$$

$$= 1.83 \times 10^9 \text{ kg/m}^3$$

In Box 7-2 on Page 135 we find the following formula for escape velocity:

$$v_{\text{esc}} = \left(\frac{2GM}{R}\right)^{1/2}$$

$$= \left(\frac{2 \times 6.67 \times 10^{-11} \times 1.989 \times 10^{30}}{6.378 \times 10^6}\right)^{1/2}$$

$$= \left(\frac{26.53}{6.378} \times 10^{13}\right)^{1/2} = (41.60 \times 10^{12})^{1/2}$$

$$= 6.45 \times 10^6 \text{ m/s} \left(\frac{1 \text{ km}}{10^3 \text{ m}}\right)$$

$$= 6450 \text{ km/s}$$

In-falling, gas strikes the star's surface at a speed of 6450 km/s.

19. In Appendix 5 we find the distance to Betelgeuse:

$$d = 310 \text{ ly} = 95 \text{ pc}$$

Assume that Betelgeuse will become a Type II supernova with a peak absolute magnitude of

$$M \approx -17$$

We can use the equation on page 343 to compute the apparent magnitude (m):

$$m - M = 5 \log d - 5$$
$$m = M + 5 \log d - 5$$
$$= -17 + 5 \log (95) - 5$$
$$= -22 + 5 \times 1.98 = -12$$

This is approximately the same brightness as the full moon (see Figure 19-5 on page 342).

20. Since its spectrum showed hydrogen lines, the supernova was of Type II. We can therefore assume that the peak absolute magnitude was

$$M \approx -17$$

Since $m = 12$ we can again turn to the equation

$$m - M = 5 \log d - 5$$
$$5 \log d = m - M + 5$$
$$= 12 + 17 + 5 = 34$$
$$\log d = \frac{34}{5} = 6.8$$
$$d = 10^{6.8} = 6.3 \times 10^6 \text{ pc}$$
$$= 6.3 \text{ Mpc}$$

The galaxy M100 is approximately 6 Mpc from Earth.

CHAPTER

23

NEUTRON STARS

Chapter Summary

This chapter covers the physical properties of neutron stars and their connection with pulsars and bursters. Conservation of angular momentum is defined and discussed. The properties of novae are explained.

23-1 The discovery of pulsars stimulated interest in neutron stars

The predictions of the existence of neutron stars in the 1930's is reviewed and neutron degeneracy is examined. The physical properties of neutron stars are outlined. The discovery of pulsars is related and their relationship to supernova remnants is established.

23-2 Pulsars are rapidly rotating neutron stars with intense magnetic fields

The transformation from normal stars to neutron stars is described qualitatively. The characteristics of magnetic fields in neutron stars are discussed. The Crab Nebula pulsar and the Vela pulsar are examined in some detail. The plans for the search of the SN1987A remnant for the expected pulsar are outlined.

Box 23-1 The conservation of angular momentum

Angular velocity and angular momentum are defined. The conservation of angular momentum of a solid body is discussed in the context of the collapse of a star.

23-3 Pulsars gradually slow down as they radiate energy into space

The slowing of the Crab pulsar is related to the luminosity of the nebula. Synchrotron radiation is defined. "Glitches" in rotation period are attributed to starquakes in the pulsar crusts.

23-4 Superfluidity and superconductivity are among the strange properties of neutron stars

Superfluidity and superconductivity are defined and are shown to be properties of the interiors of neutron stars. A model of the structure of a neutron star is presented.

23-5 The fastest pulsars were probably created by mass transfer in close binary systems

The discovery of millisecond pulsars is mentioned. The nature of millisecond pulsars as descendants of slower pulsars that have acquired mass from binary companions is explained.

23-6 Pulsating X-ray sources are neutron stars in close binary systems

The early development of X-ray astronomy is reviewed. X-ray pulsars are discussed in the context of binary star systems and accretion disks.

23-7 High-speed jets of matter can be ejected from an accreting neutron star

The peculiar properties of SS433 are analyzed in the context of gas jets ejected from a compact object at the center of a precessing accretion disk in a binary star system.

23-8 Explosive thermonuclear processes on white dwarfs and neutron stars produce novae and bursters

Novae are contrasted with supernovae. The fusion processes on the surface of a white dwarf star are discussed. The properties of X-ray bursters are described and explained in the context of mass transfer in a binary system containing a neutron star.

Teaching Hints and Strategies

The connection between pulsars (sections 23-1 and 23-2) and neutron stars can be used to extend the discussion of the observational verification process for theoretical calculations noted in section 22-3—Teaching Hints for white dwarf stars. Neutron stars were hypothesized in the 1930's. The predictions of the existence of such objects generated little interest because the chances of detecting visible radiation from their surfaces was thought to be very remote. Their properties were given much more extensive consideration in the 1960's when explanations for bright X-ray sources were being sought. The detection of pulsars in the radio spectral region provided just the right data at the right time to show that the existence of neutron stars was the only reasonable explanation consistent with all of the observed properties. Subsequent studies all seem to be consistent with the interpretation of pulsars as being rotating neutron stars.

You might want to review the logical process of elimination which led to this conclusion.

The *conservation of angular momentum* (Box 23-1) can be related to orbital motion discussed in section 4-4 and Box 4-4 and to the evolution of the solar nebula described in section 7-7.

The rate of slowing of the Crab pulsar (section 23-3) and the measured energy output of the nebula were the clues that finally identified pulsars as rotating neutron stars. The luminosity of the nebula was too great for a 900-year-old supernova remnant. Therefore there had to be a source inside the nebula. By assuming a rotating object slowing by the measured amount and equating the energy loss to the observed luminosity, the mass and size of the pulsar could be calculated. The results matched the theoretical predictions for neutron stars.

Superfluidity and superconductivity (section 23-4) represent some of the bizarre physical conditions which can be studied by using the celestial laboratories provided by nature. Conditions expected in the interiors of neutron stars can never be reproduced on Earth. However, theoretical models of the interiors of neutron stars which involve these physical conditions can be tested against observations.

Pulsars that speed up again due to mass transfer (section 23-5) gain spin angular momentum from the loss of orbital angular momentum of their companions. The total angular momentum of the system is conserved.

Students often confuse *X-ray pulsars* (section 23-6) with radio pulsars. Both types of pulsars involve rotating neutron stars with strong magnetic fields. Explain that radio pulsars are usually single stars which accelerate charged particles near the poles. These accelerated charged particles emit the low energy radio radiation. The X-ray pulsars are members of binary systems. The transfer of materials from the giant star to the neutron star is channeled by the magnetic field to the poles. The high temperatures of this in-falling material produce the high energy X rays. You might want to review some of the assumptions made in the computation of stellar models in the context of close binary systems. For example, the spherical symmetry is removed by the presence of tidal and centrifugal forces. The close proximity of the stars also causes thermal interaction and asymmetry. These problems, and many others, make the computation of models for binary stars virtually impossible. The vast range of possible interaction parameters make the number of potential models very large. The observed models proposed to explain the observations can also be used to illustrate the role of mass transfer in such systems. The association of a neutron star of low mass in a system with a high-mass primary presents an apparent problem if one assumes that they formed at the same time and the more massive star evolves at the faster rate. The obvious answer is mass transfer which complicates theoretical models even more.

Jets of material (section 23-7) have been discovered in stellar systems

182 CHAPTER 23

ranging in size from binary star systems to quasars and giant elliptical galaxies. These jets are often explained using models involving accretion disks surrounding neutron stars or black holes. The precession of the accretion disk which produces the 164 day periodicity of SS 433 can be related to the Earth's precession.

Novae and bursters (23-8) are discussed together due to the fact that both involve mass transfer onto an evolved stellar object. You might review the relative sizes of white dwarfs and neutron stars and compare these to the duration of the increases in radiation for visible light from the nova event and of the X-ray burst as shown in Figures 23-18 and 23-19. Discuss how the light-travel time for an object of given size will limit the duration of the event. This same reasoning is used in Chapter 27 to evaluate the size of quasars. The distinction between novae and supernovae can be reinforced by noting the relative masses ejected in both cases. A typical nova results in the ejection of a small fraction of a solar mass which is much less mass than a typical planetary nebula contains. Supernovae generally eject several solar masses. Remind students that novae may repeat frequently due to the relatively mild process while a supernova is a one-time event.

Answers to Chapter 23 Computational Questions

13. Convert the distance from parsecs to light-years:

$$2000 \text{ pc} \left(\frac{3.26 \text{ ly}}{1 \text{ pc}} \right) = 6520 \text{ ly}$$

Since the supernova was observed in 1054 AD, the light from the explosion left the doomed star 6520 years earlier, or in 5466 BC.

16. The pulsar's signal is Doppler shifted by an amount

$$\Delta \lambda = \lambda_0 \left(\frac{v}{c} \right)$$

If P_0 is the pulsar's unshifted period and ΔP is the change in the period, we can write

$$\Delta P = P_0 \left(\frac{v}{c} \right)$$

Since v = 30 km/s, we get

$$\Delta P = \frac{30 \text{ km/s}}{300{,}000 \text{ km/s}} P_0$$

$$= 10^{-4} P_0$$

Neutron Stars

The effect is greatest on the ecliptic, because then the largest component of the Earth's orbital motion is directed toward or away from the pulsar.

17. For a sphere of mass M and radius R the average density ($\bar{\rho}$) is given by:

$$\bar{\rho} = \frac{M}{\frac{4}{3}\pi R^3}$$

For a neutron:

$$\bar{\rho} = \frac{1.7 \times 10^{-27}}{\frac{4}{3}\pi (10^{-15})^3} = \frac{1.7}{4.19} \times 10^{-27+45}$$

$$= 4.1 \times 10^{17} \text{ kg/m}^3$$

For a neutron star, use the parameters given in Figure 23-9 on page 457:

$$M = 1.3 \, M_{\odot} = 1.3 \times 1.989 \times 10^{30} \text{ kg}$$

$$= 2.59 \times 10^{30} \text{ kg}$$

$$R = 16 \text{ km} = 1.6 \times 10^4 \text{ m}$$

and so

$$\bar{\rho} = \frac{2.59 \times 10^{30}}{\frac{4}{3}\pi (1.6 \times 10^4)^3} = \frac{2.59}{17.16} \times 10^{30-12}$$

$$= 1.5 \times 10^{17} \text{ kg/m}^3$$

19. The average angular diameter of the Crab Nebula is 5 arc minutes = 300 arc sec.

The time that has elapsed since the explosion is 1994 – 1054 = 940 years.

The expansion rate therefore is:

$$\frac{300 \text{ arc sec}}{940 \text{ yr}} = 0.32 \text{ arc sec/yr}$$

To observe a change of 1 arc sec, we must wait for

$$\frac{1 \text{ arc sec}}{0.32 \text{ arc sec/yr}} = 3.1 \text{ years}$$

20. In general, we can relate tangential velocity (v_t), proper motion (μ), and distance (d) by the equation given on page 339:

$$v_t = 4.74 \, \mu d$$

Assume $v_r = v_t$. Since $v_r = 1200$ km/s we have

$$1200 \text{ km/s} = 4.74 \, \mu d$$

If the overall expansion rate of the Crab Nebula is 0.23 arc sec/yr, then the rate at which the nebula's outer layers are receding from the center is $\frac{1}{2}(0.23)$ arc sec/yr = μ, and we get

$$d = \frac{1200}{4.74 \left(\frac{0.23}{2}\right)} = 2200 \text{ pc}$$

This answer agrees with the generally accepted distance of about 2 kpc, to within an accuracy of 10 percent.

The mean angular radius of the Crab Nebula is

$$\frac{1}{2}\left(\frac{4+6}{2}\right) = 2.5 \text{ arc min}$$

$$= 150 \text{ arc sec}$$

Since this radius is today observed to increase at a rate of $\frac{1}{2}(0.23$ arc sec/yr), the total expansion time is

$$\frac{150 \text{ arc sec}}{\frac{1}{2}(0.23 \text{ arc sec/yr})} = 1300 \text{ yrs}$$

So, this calculation suggests the expansion began 1300 years ago, around 700 AD. This is obviously too far in the past. The presently observed expansion rate is somewhat lower than the rate the nebula had immediately following the explosion.

CHAPTER 24

BLACK HOLES

Chapter Summary

This chapter describes the structural elements and characteristics of black holes. The special and general theories of relativity are reviewed. The structural elements of black holes are defined and the three observable characteristics are discussed. The status of the observational search for black holes in binary systems is presented.

24-1 The general theory of relativity describes gravity in terms of the geometry of space and time

The basic assumptions about space and time which form the bases for Newtonian mechanics and special relativity are compared. The general theory of relativity and the principle of equivalence are discussed.

Box 24-1 Some comments on special relativity

The relative values of length, time and mass for observers moving relative to each other are discussed. Lorentz transformations are used to convert proper values of length, time and mass into relative values. The limiting nature of the speed of light is explained in the context of relative mass.

24-2 A black hole is a simple object that has only a "center" and a "surface"

The concepts of the event horizon, Schwarzschild radius and singularity are developed in the context of the collapse of a massive star. The unpredictable nature of naked singularities is noted.

24-3 The structure of a black hole can be completely described with only three numbers

Mass, electrical charge and angular momentum are identified as the only measurable attributes of a black hole. The nature of the ergosphere is described.

Box 24-2 Gravitational radiation

The characteristics of gravitational radiation are described. Problems associated with the detection of gravitational radiation are examined and likely sources are discussed. Evidence of the emission of gravitational radiation by the binary pulsar is reviewed.

24-4 A black hole distorts the images of background stars and galaxies

The potential value of gravitational lensing for detecting black holes is described. The detection of gravitational lensing produced by other galaxies is discussed and Einstein rings are reviewed.

24-5 Black holes have been discovered in binary star systems

Evidence that black holes having several solar masses have been found in binary systems is reviewed. The characteristics of Cygnus X-1 are described in detail. The nature of LMC X-3 is noted briefly. Several other flickering X-ray sources are mentioned.

Box 24-3 Wormholes and time machines

The geometries of an Einstein–Rosen bridge and a wormhole are shown. The illogic of using wormholes for time travel is explained in terms of causality violation.

Teaching Hints and Strategies

The brief *review of the theories of relativity* (section 24-1) which were introduced and discussed in section 4-8. As relativity is usually difficult for even the best students to grasp, it would be wise to cover this material again. Emphasize that relativity was developed because it was needed. Newtonian physics did not predict the results of observations and experiments correctly. Remind students that the predictions of relativity are not in conflict with intuition (as they seem at first) because no one has ever experienced the high speeds and strong gravities involved.

When introducing the *event horizon and singularity* (section 24-2), emphasize that the radius of the black hole does not define the limit of the mass distribution, as the radius does in the case of most other objects, but the region of space from which light cannot escape. The equation for the value of the Schwarzschild radius can be derived, using simple algebra, from the classical definitions of kinetic and potential energy if you have defined these terms at an earlier point in the course. Just set the kinetic energy equal to the potential energy to establish the escape velocity defined in Box 7-2 and substitute the speed of light for the escape velocity and solve for R. Any amount of mass, if sufficiently compressed, can become a black hole. Have your students calculate the Schwartzchild radius for various objects including their own bodies. Students find these results interesting. (See problems 2 and 14).

You might continue the identification of support mechanisms for stars to the extreme case of no stabilizing outward force for more than three solar masses. Remember to point out that there is an assumption being made here. The assumption is that there is no pressure in nature greater than that exerted by degenerate neutrons. As our knowledge of degenerate electrons and neutrons is only about fifty years old, it might be wise to ask students to consider whether this assumption might be revised in the future.

The Lorentz transformations in *special relativity* (Box 24-1) are relatively easy to explain and provide insight into the relative nature of basic measured quantities. Be certain to point out that no inertial reference frame is special in any way. Observers moving relative to each other both observe the same time dilation and length contraction as the other. There is no way to determine if one or both observers is moving. Only the relative motion is important. In general relativity this is not the case. One encounters time dilation due to the presence of a gravitational field. This time dilation is not due to motion but to the strength of the gravitational field. Different observers get the same results.

The observable properties of black holes, mass, electrical charge and angular momentum (section 24-3), describe the object completely. The public perception of black holes tends to be that of overpowering attraction. Remind students that the strange properties of black holes are valid only in the immediate vicinity of the object. The forces are very strong only when distances are very small. The compact nature of a black hole permits very close approach to the center. You might find it helpful to compute in class the relative forces of gravity on the surfaces of the Sun, a white dwarf star and a neutron star to review the effect of the inverse square nature of gravitation. Remind students that the surfaces of these objects are progressively smaller which permits closer approach to the center. While electrically charged black holes are theoretically possible, it is unlikely that such an object would arise in nature due to the strong tendency for the quantities of positive and negative electrical charge to be equal in most situations. Rotating black holes, on the other hand, are almost ensured. Any small rotation of the parent material will be magnified by the conservation of angular momentum during the collapse to form the object as discussed in Box 23-1. If you haven't already done so, demonstrate this using a rotating stool and weights. The rapid rotation periods of pulsars (neutron stars) clearly indicate the consequences of rotation and the conservation of angular momentum.

Gravitational lensing (section 24-4) has been detected in several celestial objects. Most cases involve very remote galaxies or quasars. The luminous arc and Einstein ring shown in Figures 24-9 and 24-10 are explained by some astronomers as examples of gravitational lensing. Lensing of quasars is taken as evidence of their enormous distances sence they must be beyond the lensing galaxy which itself is so remote as to be almost undetectable.

Evidence of *gravitational radiation* (Box 24-2) has been provided by detailed studies of the binary pulsar. Only one such object has been discovered to date. The changes in period seem to be consistent with the emission of gravity waves as predicted by general relativity.

Cygnus X-1 is certainly the best studied object which contains a *stellar-mass black hole candidate* (section 24-5). Point out the roles played by X-ray astronomy in detecting compact objects and in monitoring the activity of the "unseen" object. The X rays are probably produced by an accretion disk surrounding the object. The system in the Large Magellanic Cloud (LMC X-3) and A0620-00 are also very strong candidates, although they have not been studied as long. The rapid evolutionary cycles of massive stars were noted in earlier chapters and the chemical evolution of the interstellar medium was described. This might be a logical time to ask how many black holes might exist in space. The black hole candidates which we see are members of binary systems which are very young. Black holes which form as single stars would not be detectable by such techniques and in older binary systems the second star would have evolved long ago. In such a system there would be no mass transfer to sustain an accretion disk to the present time. The chemical compositions, spatial motions and galactic distributions of the stellar populations imply a process of chemical enrichment in the galaxy which must have been fueled by supernovae in the past. A substantial population of black holes may be present as part of the residue of such a process. It is also interesting to examine the relative numbers of main-sequence stars in this context. The space density of stars on the main sequence rises as you go down the main sequence and reaches a peak in the late M stars. This is due, at least in part, to the disparate lifetimes of these stars. While an O star completely evolves in less than 10 million years, an M star will remain on the main sequence for tens of billions of years. Virtually every M dwarf star ever formed (assuming an age of the universe of 10-20 billion years) is still on the main sequence. Only those O stars formed in the past few million years are seen as main-sequence stars at the present time.

Many students are familiar with the terms "black hole" and "wormhole" (Box 24-3) because of the Star Trek and Deep Space Nine TV programs. Point out where science ends and fiction begins. Black holes belong on the science side and wormholes on the fiction side (for the time being).

Answers to Chapter 24 Computational Questions

2. On page 443 we saw that the Schwarzschild radius is given by:

$$R_{Sch} = \frac{2GM}{c^2}$$

In Box 13-1 on page 249 we find the mass of Jupiter:

$$M = 1.90 \times 10^{27} \text{ kg}$$

$$R_{Sch} = \frac{2 \times 6.67 \times 10^{-11} \times 1.90 \times 10^{27}}{(3 \times 10^8)^2} \text{ m}$$

$$= \frac{25.3 \times 10^{16}}{9 \times 10^{16}} \text{ m}$$

$$= 2.8 \text{ m}$$

12. We must first calculate the Schwarzschild radius of a 10 $M_\odot$ black hole:

$$M = 10 \text{ } M_\odot = 1.989 \times 10^{31} \text{ kg} \approx 2 \times 10^{31} \text{ kg}$$

$$R_{Sch} = \frac{2GM}{c^2}$$

$$= \frac{2 \times 6.67 \times 10^{-11} \times 2 \times 10^{31}}{(3 \times 10^8)^2} \text{ m}$$

$$= 2.96 \times 10^4 \text{ m} = 29.6 \text{ km}$$

The required density (ρ) is then:

$$\rho = \frac{M}{\frac{4}{3}\pi R^3} = \frac{2 \times 10^{31} \text{ kg}}{\frac{4}{3}\pi (2.96 \times 10^4 \text{ m})^3}$$

$$= \frac{2 \times 10^{31}}{108.6 \times 10^{12}} \approx \frac{2.0}{1.1} \times 10^{31-14}$$

$$= 1.8 \times 10^{17} \text{ kg/m}^3$$

13. The average density is given by

$$\bar{\rho} = \frac{M}{\frac{4}{3}\pi R^3}$$

but

$$R = R_{Sch} = \frac{2GM}{c^2} \text{ and so}$$

$$\bar{\rho} = \frac{M}{\frac{4}{3}\pi \left(\frac{2GM}{c^2}\right)^3} = \frac{3Mc^6}{4\pi(8G^3M^3)}$$

$$= \frac{3c^6}{32\pi G^3 M^2} \propto \frac{1}{M^2}$$

Thus

$$\bar{\rho} \propto \frac{1}{M^2}$$

Now we can write

$$M^2 = \frac{3c^6}{32\pi G^3 \bar{\rho}}$$

Substituting in the values of c and G and using $\bar{\rho} = 1000 \text{ kg/m}^3$ we get:

$$M^2 = \frac{3(3 \times 10^8)^6}{32\pi(6.67 \times 10^{-11})^3 (1000)} \text{ kg}^2$$

$$= \frac{3 \times 729 \times 10^{48}}{(100.5)(296.7 \times 10^{-33}) \times 10^3} \text{ kg}^2$$

$$= \frac{2187 \times 10^{48}}{29{,}820 \times 10^{-30}} = \frac{21.87 \times 10^{50}}{2.98 \times 10^{-26}} \text{ kg}^2$$

$$= 7.34 \times 10^{76} \text{ kg}^2$$

And finally

$$M = \left(7.34 \times 10^{76}\right)^{1/2} = 2.7 \times 10^{38} \text{ kg}$$

but

$$1 \, M_\odot \approx 2 \times 10^{30} \text{ kg and so}$$

$$M = 1.4 \times 10^8 \, M_\odot$$

$$= 140 \text{ million solar masses}$$

14. In general

$$R_{Sch} = \frac{2GM}{c^2}$$

For ease of computation, we first calculate the value of the constants:

$$\frac{2G}{c^2} = \frac{2(6.67 \times 10^{-11})}{(3 \times 10^8)^2} = \frac{13.34 \times 10^{-11}}{9 \times 10^{16}}$$

$$= 1.48 \times 10^{-27} \text{ m/kg}$$

(a) For Earth, $M = 5.976 \times 10^{24}$ kg

$$R_{Sch} = (1.48 \times 10^{-27})(5.976 \times 10^{24})$$

$$= 8.89 \times 10^{-3} \text{ m} \approx 9 \text{ mm}$$

$$\bar{\rho} = \frac{M}{\frac{4}{3}\pi R^3} = \frac{5.976 \times 10^{24}}{\frac{4}{3}\pi(8.89 \times 10^{-3})^3}$$

$$= \frac{5.976 \times 10^{24}}{2943 \times 10^{-9}} = \frac{5.976}{2.943} \times 10^{24+6}$$

$$\approx 2 \times 10^{30} \text{ kg/m}^3$$

(b) For the Sun, $M = 1\ M_\odot = 1.989 \times 10^{30}$ kg

$$R_{Sch} = (1.48 \times 10^{-27})(1.989 \times 10^{30})$$

$$= 2.94 \times 10^3 \text{ m} \approx 3 \text{ km}$$

$$\bar{\rho} = \frac{M}{\frac{4}{3}\pi R^3} = \frac{1.989 \times 10^{30}}{\frac{4\pi}{3}(2.94 \times 10^3)^3}$$

$$= \frac{1.989 \times 10^{30}}{106.4 \times 10^9} = \frac{1.989}{1.064} \times 10^{30-11}$$

$$= 1.87 \times 10^{19} \text{ kg/m}^3 \approx 2 \times 10^{19} \text{ kg/m}^3$$

(c) For the Galaxy,

$$M = 1.1 \times 10^{11}\ M_\odot = (1.1 \times 10^{11})(1.989 \times 10^{30} \text{ kg})$$

$$= 2.19 \times 10^{41} \text{ kg}$$

$$R_{Sch} = (1.48 \times 10^{-27})(2.19 \times 10^{41})$$

$$= 3.24 \times 10^{14} \text{ m} \left(\frac{1 \text{ AU}}{1.496 \times 10^{11} \text{ m}} \right)$$

$$= 2.17 \times 10^3 \text{ AU} \approx 2200 \text{ AU}$$

$$\bar{\rho} = \frac{M}{\frac{4}{3}\pi R^3} = \frac{2.19 \times 10^{41} \text{ kg}}{\frac{4}{3}\pi (3.24 \times 10^{14} \text{ m})^3}$$

$$= \frac{2.19 \times 10^{41} \text{ kg}}{142.5 \times 10^{42} \text{ m}^3}$$

$$= \frac{2.19}{1.425} \times 10^{41-44} \text{ kg/m}^3$$

$$= 1.54 \times 10^{-3} \text{ kg/m}^3 \left(\frac{10^3 \text{ g}}{1 \text{ kg}} \right)$$

$$\approx 2 \text{ g/m}^3$$

15. This problem requires that we use the equation for the dilation of time given in Box 24-1 on page 440:

$$T = \frac{T_o}{\left(1 - \frac{v^2}{c^2}\right)^{1/2}}$$

$$10 = \frac{2}{\left(1 - \frac{v^2}{c^2}\right)^{1/2}}$$

$$\left(1 - \frac{v^2}{c^2}\right)^{1/2} = \frac{1}{5}$$

$$1 - \frac{v^2}{c^2} = \frac{1}{25}$$

$$\frac{v^2}{c^2} = 1 - \frac{1}{25} = \frac{24}{25} = 0.96$$

$$\frac{v}{c} = \sqrt{0.96} = 0.98$$

$$v = 0.98c$$

16. We must use the equation for the Fitzgerald–Lorentz contraction given in Box 24-1 on page 441:

$$L = L_0 \left(1 - \frac{v^2}{c^2}\right)^{1/2}$$

$$\frac{L}{L_0} = 0.01 = \left(1 - \frac{v^2}{c^2}\right)^{1/2}$$

$$1 - \frac{v^2}{c^2} = 10^{-4}$$

$$\frac{v^2}{c^2} = 1 - 10^{-4} = 0.9999$$

$$\frac{v}{c} = 0.99995$$

$$v = 0.99995c$$

17. We must use the Lorentz transformation for mass given in Box 24-1 on page 441:

$$M = \frac{M_0}{\left(1 - \frac{v^2}{c^2}\right)^{1/2}}$$

$$\frac{M}{M_0} = 2 = \frac{1}{\left(1 - \frac{v^2}{c^2}\right)^{1/2}}$$

$$\frac{1}{1 - \frac{v^2}{c^2}} = 4$$

$$1 - v^2/c^2 = \frac{1}{4}$$

$$v^2/c^2 = 1 - \frac{1}{4} = \frac{3}{4}$$

$$\frac{v}{c} = \frac{\sqrt{3}}{2} = 0.87$$

$$v = 0.87c$$

194 CHAPTER 24

18. We can use Kepler's third law:

$$M_1 + M_2 = \frac{a^3}{P^2}$$

but we must express P in years and a in AU so that the star's masses are measured in solar masses.

$$P = 7.5 \text{ hrs} \left(\frac{1 \text{ d}}{24 \text{ hr}}\right)\left(\frac{1 \text{ yr}}{365.24 \text{ d}}\right)$$

$$= 8.84 \times 10^{-4} \text{ yr}$$

$$a = 2.8 \, R_\odot \left(\frac{6.96 \times 10^8 \text{ m}}{1 R_A}\right)\left(\frac{1 \text{ AU}}{1.496 \times 10^{11} \text{ m}}\right)$$

$$= 1.30 \times 10^{-2} \text{ AU}$$

$$\frac{a^3}{P^2} = \frac{(1.30 \times 10^{-2})^3}{(8.84 \times 10^{-4})^2} = \frac{2.20 \times 10^{-6}}{78.1 \times 10^{-8}}$$

$$= 2.8 \, M_\odot$$

This is a reasonable answer because neutron stars have masses less than 3 $M_\odot$.

19. The orbital period of A0620-00 is

$$P = 0.32 \text{ day} \left(\frac{1 \text{ yr}}{365.24 \text{ day}}\right) = 8.76 \times 10^{-4} \text{ yr}$$

also

$$P = 0.32 \text{ day} \left(\frac{24 \text{ hr}}{1 \text{ day}}\right)\left(\frac{3600 \text{ s}}{1 \text{ hr}}\right) = 2.76 \times 10^4 \text{ s}$$

Assuming the orbit is circular, the period (P) times the velocity (v) equals the orbit's circumference ($2\pi a$):

$$P v = 2 \pi a$$

Thus

$$a = \frac{P v}{2 \pi} = \frac{(2.76 \times 10^4 \text{ s})(457 \text{ km/s})}{6.283}$$

$$= 2.01 \times 10^6 \text{ km} \left(\frac{1 \text{ AU}}{1.496 \times 10^{11} \text{ m}}\right)\left(\frac{10^3 \text{ m}}{1 \text{ km}}\right)$$

$$= 1.34 \times 10^{-2} \text{ AU}$$

Now we can use Kepler's third law:

$$M_1 + M_2 = \frac{a^3}{P^2} = \frac{(1.34 \times 10^{-2})^3}{(8.76 \times 10^{-4})^2}$$

$$= \frac{2.41 \times 10^{-6}}{76.7 \times 10^{-8}} = 3.1 \text{ M}_\odot$$

Assuming that the mass of the K5V star is negligible, $M_2 \approx 0$, and so

$$M_1 \approx 3.1 \text{ M}_\odot$$

CHAPTER 25

OUR GALAXY

Chapter Summary

This chapter describes the principal structural elements of the Milky Way. The determinations of the size and shape of the Milky Way are reviewed and the structural elements are identified. The use of radio astronomy to map the spiral arms and establish the rotation curve of the Galaxy are explained. Self-sustaining and spiral density wave star formation mechanisms are examined. The nucleus of the Galaxy is discussed in the context of observations in various spectral regions and its possible relationship to active galaxies is considered.

25-1 Interstellar dust hides much of the Milky Way Galaxy from view

The development of our modern view of the structure of the Galaxy is related. The problems associated with interstellar extinction are discussed. The appearance of the Milky Way in the IR is shown.

25-2 The Sun is located in the disk of our Galaxy, about 25,000 light-years from the galactic center

Henrietta Leavitt's discovery of the P–L relation for Cepheids and RR Lyrae stars is reviewed. Shapley's application of variable star distances to map the globular cluster distribution is explained. The galactic nucleus, disk, central bulge and halo are identified.

25-3 The spiral structure of our Galaxy has been plotted from radio and optical observations of star-forming regions

The physical explanation for 21 cm radiation is provided and the mapping process for distant spiral features is introduced. Maps made from radio mapping of H I regions are compared with photographs which emphasize optical emission from H II regions.

Box 25-1 The Local Bubble and the Geminga Pulsar

The probability that the Sun lies inside a huge cavity in the interstellar medium dubbed the Local Bubble is discussed. The discovery of the Geminga Pulsar is described and the possible identification of this pulsar with the supernova which generated the bubble is made.

25-4 Moving at half a million miles per hour, the Sun takes about 200 million years to complete one orbit of our Galaxy

The rotation of the Galaxy is described and the orbital speed and period of rotation of the Sun are noted. The concept of differential rotation is reviewed and the galactic rotation curve is presented and discussed. It is suggested that substantial amounts of dark matter may remain undetected.

Box 25-2 Estimating the mass inside the Sun's orbit

Kepler's third law is used to calculate the mass of the Galaxy interior to the Sun's orbit.

25-5 Self-sustaining star formation can produce spiral arms

Spiral galaxies are classified as flocculent spirals or grand-design spirals. The former are explained as having resulted primarily from stochastic, self-propagating star formation distorted by differential rotation.

25-6 Spiral arms are caused by density waves that sweep around a galaxy

Density waves and kinematic waves are described in some detail. The shock action of the density wave is noted and its role in star formation is revealed. Enrichment of the heavy element abundances of the interstellar medium is discussed and the subsequent formation of metal rich stars is examined.

25-7 Infrared and radio observations are used to probe the galactic nucleus, whose nature is poorly understood

The appearance of the night sky at the galactic center is described. The limitations due to interstellar absorption of observations made in visible light are noted again. Optical, infrared and radio images and maps of the galactic center are compared and contrasted. Filaments and spiral patterns of gas and dust are shown and examined in the context of magnetic fields and a massive object. The hypothesis of a supermassive black hole in the galactic nucleus is presented as are other interpretations of the observations.

Teaching Hints and Strategies

The "Kapteyn Universe" described in section 25-1 supported the estimates by certain biblical authorities that the universe is only a few thousand years old. In that short time, light from stars only up to a few thousand light-years distance could have reached us. This was in stark contrast to evidence from the science of geology that claimed that the Earth must be many millions of years old. Now, all the true sciences are in much better agreement.

The transformation to the *modern model of the Galaxy* (section 25-2) with the Sun far removed from the center from a Sun-centered view which took place in the first quarter of this century should be given special emphasis. The roles of variable stars and globular clusters and the assumption of no appreciable interstellar extinction should be clearly outlined. An observing session at a dark sight or a planetarium presentation can be most effective in conveying the symmetries and asymmetries of the distribution of objects in our stellar system. Star charts in *Sky and Telescope* and *Astronomy* can be useful in helping students to identify these objects. In the sky the fainter stars, which contribute the faint glow of the Milky Way, show strong concentration to the plane of the Galaxy. This concentration for the brightest stars is not as conspicuous. Point out the supergiant stars such as Rigel, Betelgeuse, Antares, and Deneb. Show how they cluster along the plane of the Galaxy as do the brightest Messier emission nebulae. Remind students that emission nebulae are illuminated by the very youngest star clusters and point out how these clusters clump along the Milky Way to hint at the locations of the spiral arms in our Galaxy. Describe the concentration of galaxies toward the galactic poles and the concentration of the globular clusters toward Sagittarius. If you are in a planetarium you can project slides of these objects as you point out their positions to help students associate these objects with galactic structure. You can also discuss the concept of stellar populations and describe the evolution of the Galaxy as implied by the distribution of these objects and our understanding of stellar evolution.

The *spiral arms* (section 25-3) can be identified as a highly compressed subdivision of the galactic disk to permit a better perception of the location of such population I objects as H I regions and massive O and B main-sequence stars. The limited horizon for visual observations should be described completely at this point if it has not been done before. The concentration of dust into a very thin central layer of the galactic disk is responsible for the general obscuration and reddening in the central plane of the Milky Way. Remind students that this was responsible for the adoption of Sun-centered models of our Galaxy up until the 1920s. The ability of the longer wavelength radiation to penetrate these cool dust clouds makes infrared and radio telescopes much better suited to the study of the large-scale structure of the spiral arms. You might explain the changing color of the Sun with altitude to provide an analogous phenomenon which they have observed. The color of the daytime sky is also related to the wavelength dependence of scattering. You can note that the only other tracer which is commonly used is the carbon monoxide molecule. Be sure to explain that assumptions have to be made about the rotation of the Galaxy and about the peculiar motions of the source clouds in order to construct a map based on the observations. Similar types of observations can be made for spiral galaxies other than the Milky Way.

The determination of the *galactic rotation curve* (section 25-4) presents some serious complications which have resulted in the alteration of our models of the Galaxy in recent years. It has been noted in recent years from observations of rotation curves for other galaxies that nearly all spiral galaxies seem to have very large quantities of unseen matter or dark matter. This is related to what used to be referred to as the "problem of the missing mass." The mass seems to be there, but is not seen, and so is more recently referred to as unseen or dark matter. The rotation curves of other spiral galaxies seem to be flat at large distances from the center which implies much larger amounts of matter than is indicated by the radiation detected. This might be of interest in discussions of the numbers of black holes which are present in the Galaxy. Evidence has been presented in recent years which suggests that the fraction of massive stars formed from a cloud of gas and dust may not be constant in time. More of the more massive stars may have been formed early in the history of the Galaxy than at present. The assumptions that the star formation rate is well known and that the initial mass function is constant are currently being vigorously debated. These assumptions determine the number of stellar remnants to be expected at the present time in our Galaxy and in other spiral galaxies. The application of Kepler's third law, as modified by Newton, to establish the mass of the Galaxy interior to the orbit of the Sun about the galactic center (Box 25-2) should be noted to illustrate the broad applicability of the basic laws of nature.

It may also be of interest when discussing *self propagating star formation* (section 25-5) to compute the number of revolutions the Sun has completed in the lifetime of the Galaxy. Point out that differential rotation presents a real problem in maintaining spiral patterns of long-lived objects. However, the objects which are spiral arm tracers are generally all very young objects such as massive O and B main-sequence stars and supergiants which have short lifetimes compared to the revolution times of galactic orbits. This precludes the destruction of the spiral patterns due to differential rotation.

When discussing *spiral density waves* (section 25-6) it should be remembered that students frequently have difficulty in separating the properties of a wave from the properties of the particles which maintain the wave pattern. A simple string tied to a fixed object provides an analogy to the density waves. Discuss the properties of the wave pattern and distinguish them from the properties of the particles in the string itself. Explain that the spiral patterns in the galaxy are not fixed with respect to the motions of objects. The pattern is a wave disturbance which passes through the material in the Galaxy. As it passes through dark clouds and giant molecular cloud complexes it compresses them to initiate star formation processes. The very youngest stars, O and B main-sequence stars, illuminate their surrounding clouds of interstellar material and produce H II regions which are the most visible spiral tracers in galaxies. After the density wave has passed, the massive stars destroy the gas and dust clouds from which they formed, rapidly evolve and die, leaving behind little visual evidence of their passage when viewed from vast distances. While the density wave theory provides a possible explanation for the existence of spiral patterns, the next problem is to explain where the density wave comes from. You might reflect here on the neverending series of questions which characterizes modern science. Rather than being a body of knowledge to be memorized, it is the process of looking ever deeper into the processes of nature. The results of most experiments seem to suggest more questions than were originally answered. This is the fun of scientific research.

The use of *infrared and radio astronomy techniques to study the nucleus of the Galaxy* (section 25-7) should be noted as should the vast improvement in the resolving powers which have been achieved in recent years due to new telescopes, space telescopes and interferometric techniques. It is rather interesting to compute the linear size of an object at the distance of the galactic nucleus which subtends an angle of one arc second to illustrate the difficulty in establishing the fine structural detail in the nucleus. Be certain also to distinguish between the nucleus and the nuclear bulge.

Answers to Chapter 25 Computational Questions

8. The orbital velocity is

$$v = 400 \text{ km/s} \left(\frac{10^3 \text{ m}}{1 \text{ km}} \right) = 4 \times 10^5 \text{ m/s}$$

The distance from the galaxy's center is:

$$r = 20{,}000 \text{ pc} \left(\frac{3.09 \times 10^{16} \text{m}}{1 \text{pc}} \right) = 6.18 \times 10^{20} \text{ m}$$

(a) The orbital period (P) of the gas cloud equals the circumference of its orbit ($2\pi r$) divided by its orbital velocity:

$$P = \frac{2\pi r}{v} = \frac{2\pi \times 6.18 \times 10^{20} \text{ m}}{4 \times 10^5 \text{ m/s}}$$

$$= 9.71 \times 10^{15} \text{ s} \left(\frac{1 \text{ yr}}{3.156 \times 10^7 \text{ s}} \right)$$

$$= 3.08 \times 10^8 \text{ yr}$$

$$\approx 300 \text{ million years.}$$

(b) We can now use the form of Kepler's third law given in Box 25-2:

$$M = \frac{rv^2}{G}$$

$$= \frac{(6.18 \times 10^{20})(4 \times 10^5)^2}{6.67 \times 10^{-11}} \text{ kg}$$

$$= 14.8 \times 10^{41} \text{ kg} \left(\frac{1 \, M_\odot}{1.99 \times 10^{30} \text{ kg}}\right)$$

$$= 7.4 \times 10^{11} \, M_\odot$$

9. The age of the solar system is 5×10^9 yr. Our orbital period is 2×10^8 yr. Therefore, the number of trips around the center of our Galaxy is:

$$\frac{5 \times 10^9 \text{ yr}}{2 \times 10^8 \text{ yr/revolution}} = 25 \text{ revolutions}$$

13. As we saw in Box 25-2:

$$M = \frac{rv^2}{G}$$

where M is the mass interior to a distance r from the galactic center, and v is the orbital velocity at the distance r.

An error Δr in distance gives an error ΔM in mass:

$$M + \Delta M = \left(\frac{v^2}{G}\right)(r + \Delta r)$$

$$= \frac{v^2 r}{G} + \frac{v^2}{G}\Delta r$$

$$= M + \left(\frac{v^2}{G}\right)\Delta r$$

and so

$$\Delta M = \left(\frac{v^2}{G}\right)\Delta r$$

Thus a 10% error in distance produces a 10% error in mass.

An error Δv in velocity gives an error ΔM in mass:

$$M + \Delta M = \left(\frac{r}{G}\right)(v + \Delta v)^2$$

$$= \left(\frac{r}{G}\right)(v^2 + 2v\,\Delta v + (\Delta v)^2)$$

Since Δv is assumed to be small (e.g., 10%) $(\Delta v)^2$ is negligible (e.g,. 10 % × 10% = 1%) and so

$$M + \Delta M = \frac{rv^2}{G} + \left(\frac{2rv}{G}\right)\Delta v$$

$$= M + \left(\frac{2rv}{G}\right)\Delta v$$

thus

$$\Delta M = 2\left(\frac{rv}{G}\right)\Delta v$$

Consequently, a 10% error in velocity produces a 20% error in mass.

15. We can use the form of Kepler's third law given in Box 25-2:

$$M = \frac{rv^2}{G}$$

where

$$r = 60{,}000 \text{ ly}\left(\frac{9.46 \times 10^{15} \text{ m}}{1 \text{ ly}}\right)$$

$$= 5.68 \times 10^{20} \text{ m}$$

and

$$v = 300 \text{ km/s} = 3 \times 10^5 \text{ m/s}$$

and so we get

$$M = \frac{(5.68 \times 10^{20})(3 \times 10^5)^2}{6.67 \times 10^{-11}} \text{ kg}$$

$$= 7.66 \times 10^{41} \text{ kg}\left(\frac{1 \text{ M}_\odot}{1.99 \times 10^{30} \text{ kg}}\right)$$

$$= 3.9 \times 10^{11} \text{ M}_\odot$$

$$\approx 400 \text{ billion solar masses}$$

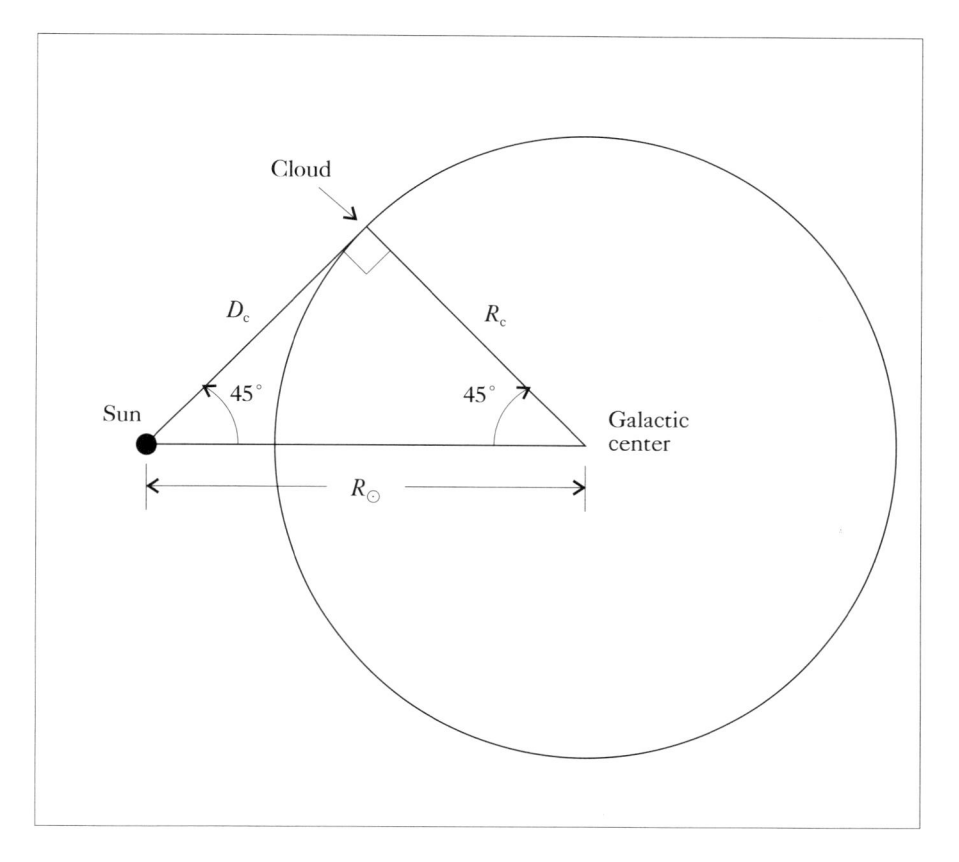

16. It will be helpful to draw a diagram as shown below:

$R_\odot$ is the distance between the Sun and the galactic center.

R_c is the distance between the cloud and the galactic center.

D_c is the distance between the Sun and the cloud.

Since the triangle formed by the Sun, the galactic center and the cloud is an isosceles right triangle, we can write:

$$\frac{R_c}{R_\odot} = \sin 45°$$

or

$$R_c = R_\odot \sin 45° = 0.707 R_\odot$$

And also

$$D_c = R_c = R_\odot \sin 45° = 0.707 R_\odot$$

Since $R_\odot = 25{,}000$ ly

$$D_c = R_c = (0.707)(25{,}000 \text{ ly})$$
$$= 17{,}700 \text{ ly}$$
$$\approx 18{,}000 \text{ light-years}$$

17. The volume of the Galaxy is:

$$V_G = \pi \left(\frac{D}{2}\right)^2 \times T$$

where D = diameter of Galaxy = 25,000 parsecs
and T = thickness of Galaxy = 600 parsecs

$$V_G = \pi (12{,}500 \text{ pc})^2 (600 \text{ pc})$$
$$= 2.945 \times 10^{11} \text{ pc}^3$$

The volume of a sphere of radius 300 pc is

$$V_s = \frac{4}{3}\pi R^3 = \frac{4}{3}\pi (300 \text{ pc})^3$$
$$= 1.131 \times 10^8 \text{ pc}^3$$

The number (N) of radius 300 pc spheres in the Galaxy is:

$$N = \frac{V_G}{V_s} = \frac{2.945 \times 10^{11}}{1.131 \times 10^8} = 2.60 \times 10^3 = 2600$$

If the distribution of supernova is totally random, then one in 2600 explosions will be within 300 pc of the Sun.

The time for 2600 explosions to occur at a rate of three per century is

$$\frac{2600}{3/100 \text{ yr}} = 8.67 \times 10^4 \text{ yrs}$$
$$\approx 90{,}000 \text{ years}$$

Thus we expect a supernova within 300 pc of the Sun about once every 90,000 years.

CHAPTER 26

GALAXIES

Chapter Summary

This chapter reviews the origin of our expanding universe theories and the Hubble classification system for galaxies. Clusters of galaxies are described and the dark-matter problem for clusters of galaxies is presented. The results of galaxy interactions are explored. The Hubble law is explained and the uncertainty of the value of the Hubble constant is evaluated. The process of determining the distances of external galaxies is outlined.

26-1 Hubble proved that the spiral nebulae are far beyond the Milky Way

Edwin Hubble's discovery of Cepheids in M31 is reviewed. The modern Cepheid distance to this galaxy is given.

Box 26-1 Cepheid Variables as indicators of distance

The HST program to establish the distance to IC 4182 using Cepheids is reviewed and the result presented. The importance of this galaxy as a calibration point for Type Ia supernovae is established.

26-2 Hubble devised a system for classifying galaxies according to their appearance

Hubble's classification scheme for galaxies is outlined in some detail. Distinctions are made between dwarf and giant elliptical galaxies. The correlations of stellar motions with elliptical class is examined. The tuning-fork diagram is illustrated and possible evolutionary connections between types are suggested.

26-3 The Hubble law states that the redshifts of remote galaxies are proportional to their distances from Earth

The history of the discovery of the Hubble law is reviewed briefly. The interpretation of the Hubble law as evidence of an expanding universe is presented. The formula for the Hubble law is provided and is displayed graphically. Some of the problems which are encountered in calibrating the Hubble law are described. The current uncertainty in the value to be adopted for the Hubble constant is presented.

Box 26-2 The Hubble law as a distance indicator

A detailed mathematical description of the Hubble law is presented. Examples are provided of calculations of distances to galaxies using the Hubble law from the observations of the wavelengths of spectral lines.

26-4 Galaxies are grouped in clusters that are members of superclusters

The distribution of galaxies in clusters is noted. Clusters of galaxies are classified as rich or poor and as regular or irregular. Larger scale superclustering and large scale voids are described. Maps made by radio telescopes and optical telescopes which show the distributions of clusters and superclusters of galaxies are shown and discussed. Large-scale structures such as the Great Wall are noted.

Box 26-3 The Local Group

This table lists Hubble type, distance and diameter for the galaxies which are members of our local group.

26-5 Most of the matter in the universe has yet to be discovered

The dark-matter problem is described. The discovery of hot gas between galaxies by X-ray telescopes is reviewed. Evidence of the existence of extended halos around galaxies is explored.

26-6 Colliding galaxies produce starbursts, spiral arms, and other spectacular phenomena

Interactions between galaxies which produce starbursts, galaxy mergers and galactic cannibalism are described. Tidal stripping of materials from galaxies is examined.

26-7 Spiral galaxies were more common in the past than they are today

Galactic evolution is discussed and the range in ages of galaxies with increasing distance is discussed. The changing colors and luminosities of galaxies due to stellar evolution are described. Elliptical and spiral galaxies are contrasted in terms of their content and evolution.

Essay by Alan Dressler
The Great Attractor

Essay Summary

The three-dimensional distribution of galaxies is discussed. The dark-matter problem is briefly reviewed. A distinction is made between the distribution of mass and the distribution of galaxies. The determination and analysis of galaxy motions is identified as the means for determining the large-scale distribution of mass in the universe and the location and mass of the Great Attractor.

This essay is closely related to material presented in sections 26-5, 29-5 and 29-7. It might be interesting at this point to discuss the need for more large telescopes and the role of more efficient radiation detectors in projects such as this one. The time involved in obtaining a spectrogram of a galaxy is related to the aperture of the telescope and to the efficiency of the detector. The more remote galaxies are very faint to begin with. When you spread the light out to obtain the spectrogram necessary to determine the radial velocity, the light levels are very low, which results in very long exposure times. The results of this investigation reflect an enormous amount of telescope time on the largest telescopes on Earth. Without large telescopes and efficient detectors, such a project could never be considered. The concept of determining mass distributions without seeing the material involved is similar to determining the mass distribution in the Galaxy by determining the motions of the component stars and nebulae as described in section 25-4.

Teaching Hints and Strategies

The *work by Hubble* in determining *distances of galaxies* (section 26-1) and in establishing a simple morphological classification system (section 26-2) was very important in resolving the question of the nature of spiral nebulae and in establishing the validity of the island universe theories. A review of the evidence regarding the nature of spiral nebulae (galaxies) as it existed in the early 1920's was part of the Shapley–Curtis debate in 1920 and can be very informative. (See *Astronomy of the 20th Century*; Otto Struve and Velta Zebergs; Macmillan, New York, 1962.) The discovery of periodic variable stars in M31 by Hubble resolved the controversy. You might point out the advantage of variable stars over spectroscopic parallaxes in the determinations of distances to remote objects. The period of a variable star is established from photometric measurements of its combined light. The recording of a spectrum involves spreading the light out and requires much more observing time. The Hubble classification system presents an opportunity to discuss the value of classification in advancing our understanding of nature. Classification of stellar spectra has led to an understanding of the basic controlling influences of temperature, density and chemical composition. This has been expanded through the use of the H–R diagram to include an understanding of the processes of stellar evolution. The classification process in galaxies is more complicated. The tuning-fork diagram of Hubble is often interpreted in the context of galaxy evolution even though Hubble never indicated such an application. More recent systems to classify galaxies have adopted different criteria for classification. Our current understanding of the relationships between various galaxy types is still quite unclear.

The major astronomical figures in the early-middle part of the century are fascinating. Biographical tidbits interspersed in your lectures will bring to life the subject of this chapter. Especially interesting are Shapley, Hubble, Humason, Baade, Zwickey and Sandage. These are the titans of extragalactic astronomy.

Hubble's law (section 26-3) and the origin of the expanding universe theories are described. It is common to see astronomy articles in the newspapers announcing the astounding discovery that the universe is twice as big or only half as big as it was thought to be. Try to impress on your students that the estimates of the size and age of the universe are dependent upon the calibration of the Hubble constant and that there is a long history of uncertainty associated with its determination. This is a nice time to reinforce the concept of error bars or uncertainty associated with all values in science. The interesting aspect of this controversy is that there are two groups who feel that their values of the Hubble constant are correct and that the difference between their values is much larger than the combined uncertainties which they adopt. Note that the larger value for the Hubble Constant results in an age for the universe that is less than the ages of globular clusters (section 21-3). Also, the larger value of H_0 implies closer distances to all galaxies beyond the Local Group. Closer distances, in turn, imply smaller physical dimensions, If $H_0 = 90$, then the Milky Way and M31 would virtually be the largest spiral galaxies in the universe—this is unlikely!

When discussing the application of the *Hubble law as a distance indicator* (Box 26-2), it is wise to review the process by which the Hubble constant is determined. Explain the use of different types of stars, nebulae and star clusters (section 26-4) to estimate the distances to galaxies and discuss the determination of radial velocities, being sure to note the large uncertainties in both of these quantities. You might want to plot a graph on an overhead projector using data for a few galaxies and show the error bars. You might even use the different distances adopted by the two groups of astronomers and plot in two colors to illustrate the nature of their differences. Be sure to emphasize the limitations imposed by distance when using the standard candles. In nearby galaxies we can see periodic variable stars, supernovae, globular clusters, novae and giant H II regions. As distances increase the fainter objects can no longer be seen. It is interesting to plot an extended H–R diagram showing the absolute magnitudes for the standard candles commonly used. Choose a limiting apparent magnitude and compute the distances to which each standard candle could be seen. Note these distances on the right side of the diagram. Make a transparency and move a mask up from the bottom to indicate just what types of objects can be seen at given distances.

You might compare the classification systems for *clusters of galaxies* (section 26-5) with that used for star clusters. Be sure to caution the students to be careful not to confuse clusters of galaxies with star clusters.

Compare what you might observe on a photograph of *the Local Group* (Box 26-3) taken from far away with what we know to be present. You might also review the selection effects which plague our census of galaxies. The spirals are much easier to recognize in visual scanning of photographs than are irregulars and ellipticals. The number of ellipticals in a volume of space is usually greater than the number of spirals. The dwarf ellipticals are also much more difficult to see at large distances.

The discussion of the *dark-matter problem* (section 26-6) is presented formally in this section. While this name for the problem is used to describe it in the current literature, you might alert students that many older articles refer to the dark matter as the "missing mass." The logic behind the name change is that the matter is almost certainly there and should not be labeled as missing. The question is why we do not see radiation to indicate its presence. The first evidence of vast quantities of dark matter was the fact that the gravitational forces in clusters of galaxies, as derived from the motions of the galaxy members, are much larger than was predicted based on the luminosity of the cluster members and our understanding of the mass–luminosity relation. This discrepancy is about a factor of ten. A related problem has been noted in the Milky Way in the vicinity of the Sun based on studies of motions of stars perpendicular to the plane of the Milky Way. The difference here is only about a factor of two and is not too much larger than the uncertainties in the calculations and observations. The extended halos implied by the rotation curves in spiral galaxies like the Milky Way are additional evidence of the unseen matter. You might review the types of objects which might be responsible for the gravitational force without producing radiation. The references at the end of this chapter in the text review the objects which have been proposed to account for the forces and some of the reasons for rejecting many of them.

Galaxy interactions (section 26-7) can account for a wide variety of the peculiar galaxies which have been observed. The evidence for ring formation and for galactic bridges which results from computer models explains many of the odd features which are observed in photographs of galaxies.

Galaxy formation (section 26-8) is one area which still is very poorly understood. The detailed Big Bang theories of cosmology do not provide for a natural explanation of how galaxies form. Galaxy evolution is a somewhat different matter. With the information we have about the types of objects of which galaxies are made, it should be possible to predict how a galaxy should change with time. Observations of spectra and photometry of the integrated light from galaxies can tell us if such observations are consistent with the theoretical models. While the changes in the light from a galaxy are not likely to change over time periods of less than a few billion years, the finite speed of light permits us to observe similar galaxy types at different stages in their lifetimes. Nearby galaxies are seen in the recent past while the most remote galaxies are seen as they were in the remote past. We can search for systematic differences. Remember to point out to students that the combined light of galaxies is dominated by about the four percent of the stars which are more massive than the sun. The stars with less mass than the sun contain about 80 percent of the known mass of the galaxies and are nearly invisible. The missing mass or dark matter, which may comprise far more mass than that which is visible, is not represented by the radiation either.

Answers to Chapter 26 Computational Questions

14. We can use the Hubble law:
$$v = H_0 \, r$$
where $v = 75{,}000$ km/s and
$$r = 1.4 \times 10^9 \text{ pc} = 1400 \text{ Mpc}$$
$$H_0 = \frac{v}{r} = \frac{75{,}000 \text{ km/s}}{1400 \text{ Mpc}}$$
$$= 54 \text{ km/s/ Mpc}$$

15. We can use the equation first stated on page 343 of the text:
$$m - M = 5 \log d - 5$$
with $m = +10$ and $M = -19$. Thus
$$10 + 19 = 5 \log d - 5$$
$$5 \log d = 10 + 19 + 5 = 34$$
$$\log d = \frac{34}{5} = 6.8$$
$$d = 10^{6.8} = 6.3 \times 10^6 \text{ pc}$$
$$= 6.3 \text{ Mpc}$$

16. The distance between the galaxies is
$$a = 600 \text{ kpc} = 6 \times 10^5 \text{ pc}$$
$$= (6 \times 10^5)(3.086 \times 10^{16}) = 1.11 \times 10^{22} \text{ m}$$

To convert the orbital period ($P = 4 \times 10^{10}$ yr) into seconds, we must know the number of seconds in a year:
$$1 \text{ yr} = (365.24)(24)(60)(60) = 3.16 \times 10^7 \text{ s}$$

Thus
$$P = (4 \times 10^{10})(3.16 \times 10^7) = 1.26 \times 10^{17} \text{ s}$$

We can use Kepler's third law given in Box 4-3 on page 69:
$$M_1 + M_2 = \frac{4\pi a^3}{GP^2}$$
$$= \frac{4(3.1416)(1.11 \times 10^{22})^3}{(6.67 \times 10^{-11})(1.26 \times 10^{17})^2}$$
$$= \frac{4(3.1416)(1.37) \times 10^{66}}{(6.67)(1.59) \times 10^{23}}$$
$$= 1.6 \times 10^{43} \text{ kg} \left(\frac{1 \text{ M}_\odot}{1.99 \times 10^{30} \text{ kg}} \right)$$
$$= 8.2 \times 10^{12} \text{ M}_\odot$$

17. We shall use the Hubble law
$$v = H_0 \, r$$
with $v = 10{,}800$ km/s.
$$r = \frac{v}{H_0}$$

If $H_0 = 50$ km/s/Mpc, then
$$r = \frac{10{,}800}{50} \text{ Mpc} \approx 220 \text{ Mpc}$$

If $H_0 = 100$ km/s/Mpc, we get a distance half as large:
$$r = \frac{10{,}800}{100} \text{ Mpc} \approx 110 \text{ Mpc}$$

Using $H_0 = 75$ km/s/Mpc, we get

$$r = \frac{10{,}800}{75} \text{ Mpc} \approx 140 \text{ Mpc}$$

$$\approx 470 \text{ million light-years}$$

18. From the graph in Figure 26-25 we see that the orbital speed is

$$v = 280 \text{ km/s} = 2.8 \times 10^5 \text{ m/s}$$

The orbital period then is

$$P = \frac{2\pi r}{v} = \frac{2\pi (20 \text{ kpc})}{(2.8 \times 10^5 \text{ m/s})} \left(\frac{3.086 \times 10^{19} \text{ m}}{1 \text{ kpc}} \right)$$

$$= 1.4 \times 10^{16} \text{ s} \left(\frac{1 \text{ yr}}{3.16 \times 10^7 \text{ s}} \right)$$

$$= 4.4 \times 10^8 \text{ yr}$$

$$= 440 \text{ million years}$$

We can use the form of Kepler's law given in Box 25-2:

$$M = \frac{rv^2}{G}$$

$$= \frac{(20)(3.086 \times 10^{19})(2.8 \times 10^5)^2}{6.67 \times 10^{-11}} \text{ kg}$$

$$= 7.25 \times 10^{41} \text{ kg} \left(\frac{1 \text{ M}_\odot}{1.99 \times 10^{30} \text{ kg}} \right) = 3.6 \times 10^{11} \text{ M}_\odot$$

CHAPTER 27

QUASARS, BLAZARS, AND ACTIVE GALAXIES

Chapter Summary

This chapter provides a brief review of the early history of radio astronomy and notes the enormous radio luminosities of radio galaxies. The observed properties of quasars are described and their high luminosities and relatively small sizes are discussed. The possible links of quasars with active galaxies are explained and the evidence which might suggest the presence of supermassive black holes in these objects is reviewed.

27-1 Quasars look like stars but have huge redshifts

The problem of identifying the first quasar is related. The significance of the name is explained and the z values, defined in Boxes 26-2 and 27-1, are given for a few quasars. The distances implied by the Hubble law and typical values of the Hubble constant are noted. Some of the emission lines in the spectra of quasars are shown. The deceleration parameter is described.

Box 27-1 The relativistic redshift

The relativistic relationship between z and velocity is derived and used to compute the velocity of recession of a quasar.

Box 27-2 Redshifts and distance

Scales showing the relationship between redshift and distance for two values of the deceleration parameter are presented.

27-2 Quasars are the ultraluminous centers of distant galaxies

The nonthermal spectrum of quasars is explained. Hypotheses that the quasar redshifts may not be cosmologic in origin are examined and some of the problems associated with these theories are related. Examples of possible physical connections between quasars and galaxies of very different redshifts are discussed. The detection of some host galaxies at the same redshifts as their quasars is mentioned.

27-3 Seyfert galaxies and radio galaxies bridge the gap between normal galaxies and quasars

The discovery of and observational characteristics of Seyfert galaxies and radio galaxies are reviewed. The possible connection between Seyfert and radio galaxies and radio-quiet and radio-loud quasars respectively is established.

27-4 Quasars, blazars, Seyferts, and radio galaxies are active galaxies

The discovery of BL Lac objects or blazars is reviewed. The point that blazars may be radio galaxies seen end-on is made. Apparent superluminal motion of jets of matter from active galaxies is explained. The size limitations derived from light-travel times and observations of variability are discussed.

27-5 Supermassive black holes may be the "central engines" that power active galaxies

The evidence of condensed supermassive objects at the centers of active galaxies is reviewed and explanations involving theoretical models of supermassive black holes are examined. Galactic velocity dispersions, accretion disks and jets are described. Properties of Type I and Type II Seyfert galaxies are discussed.

Box 27-3 The plausibility of extremely massive black holes

Simple calculations of the densities of stellar mass and supermassive black holes are computed and compared. These values are related to comparable densities in nature. The density of a black hole with the mass of the entire universe is computed and discussed.

Teaching Hints and Strategies

While most of the early *quasar discoveries* (section 27-1) were a result of identifications of optical counterparts of strong radio sources, more recent surveys have relied on the very strong emission lines which are easily identified in objective-prism surveys. Although some quasars are strong radio sources, many are not. It might be helpful here to review Kirchoff's rules of spectral analysis and try to impress on students that the strong emission lines provide information about the source of radiation. You might compare quasar spectra with spectra of other objects such as stars, planetary nebulae, H II regions and normal galaxies.

Encourage students to compare the discussion of *relativistic redshifts* in Box 27-1 with the discussion in Box 26-2. Have them compute velocities using both relativistic and nonrelativistic formulae for z values of .2, .4, .6, .8, and 1.0. Point out that z values greater than or equal to 1 result in velocities which are equal to or greater than the speed of light. Such velocities are not permitted. Observed values of z much greater than 1 are not uncommon. Be sure to point out that the relativistic redshifts discussed here are not those predicted by Einstein as tests of the general theory of relativity, but are merely the result of applying special relativity to the Doppler effect.

The enormous *redshifts of quasars* has a number of interesting implications which can be noted. 1) The large redshifts frequently cause emission lines in the ultraviolet to appear in the visual spectral region. These lines are normally not seen from the surface of the Earth due to atmospheric absorption of UV radiation. 2) The large redshifts are generally accepted by most astronomers as being cosmological in origin. However, a small group of astronomers (section 27-2) is not convinced of this. Are there other explanations for the redshift which do not involve the expansion of the universe? If quasars were not very far away, their luminosities would not be unusual at all. Attempts to find physical links between quasars and more nearby objects have been the most common means used by some astronomers to demonstrate that at least some quasars are not remote objects. 3) The translation of the quasar redshift into a distance is based on the adoption of a value for the Hubble constant as is noted in the text. It also involves two assumptions which are very important. The first deals with the use of the Hubble law for quasars. Is it proper for astronomers to apply a law developed for galaxies to these objects which are not like galaxies? A less important assumption has to do with those quasars having the greatest redshifts. Is it safe to assume that the simple linear relationship between redshift and distance is valid beyond the realm of direct distance determinations? This obviously involves extrapolation which should be discussed and the dangers of such actions described. You might want to return to this problem in the last chapter when discussing the differences between open and closed universes.

A table comparing the properties of *active galaxies* (sections 27-4 and 27-5) can help to emphasize the general trends in visual appearance. You might note also that several studies published in recent years have shown that quasars which have the lowest redshifts, and should be the closest quasars, frequently show evidence of faint surrounding light suggesting galaxies when examined using image processing techniques and long exposures. A series of photographs of a normal galaxy using different exposure times will help to show how the fainter regions become visible and change the image with increasing exposure time. In order for an object to change its brightness by a large amount, all parts must change together in phase, otherwise random phases of the changes in different parts of the object will tend to cancel. Any signal that synchronizes the change must travel at or below the speed of light. This limits the size of the

216   CHAPTER 27

object so that the light-travel time is equal to the characteristic time for the observed change.

The proof of the *existence of supermassive black holes* (section 27-5) is much more difficult than for stellar mass objects in binary systems. The restrictions imposed by the observations of galactic nuclei are not nearly as limiting as in the case of the objects which represent the last stages of stellar evolution. Most research papers on the subject state that the observations are consistent with the existence of a black hole, but cannot rule out other possible explanations. The first papers reporting the observations of M 87 which indicated the presence of a supermassive object in its core were very careful to explain that a black hole model was only one possible explanation. The headlines in the newspaper reports of the study read "Black Hole Found in Center of Galaxy." You might discuss how science journalism is often less than accurate in its presentation of the subject matter. It might be interesting to use the formula for the Schwarzschild radius to compute the size of a 5 billion solar mass black hole and compare it to the size of the solar system, which is representative of the sizes of quasars and active galactic nuclei.

When discussing *the plausibility of extremely massive black holes* it might be enlightening to compute the average densities of familiar mass distributions to compare with the examples provided in Box 27-3. Use the mass of a globular cluster and a typical radius to compute its mean density. Use the data in Figure 25-17 with the equation in Box 25-2 to compute the mass in the disk of the Galaxy at representative points. Apply information about the thickness of the disk to estimate the mean densities in these regions.

Answers to Chapter 27 Computational Questions

12. We can make use of the relativistic equation for the redshift stated in Box 27-1:

$$z = \left(\frac{c+v}{c-v}\right)^{1/2} - 1$$

Since $v = 0.99c$ we get

$$z = \left(\frac{c + 0.99c}{c - 0.99c}\right)^{1/2} - 1$$

$$= \left(\frac{1 + 0.99}{1 - 0.99}\right)^{1/2} - 1$$

$$= \left(\frac{1.99}{0.01}\right)^{1/2} - 1 = \sqrt{199} - 1$$

= 14.1−1

= 13.1

13. We can make use of the equation given in Box 27-1:

$$\frac{v}{c} = \frac{(z+1)^2 - 1}{(z+1)^2 + 1}$$

Since $z = 4.73$ we get

$$\frac{v}{c} = \frac{(4.73+1)^2 - 1}{(4.73+1)^2 + 1}$$

$$= \frac{(5.73)^2 - 1}{(5.73)^2 + 1}$$

$$= \frac{32.83 - 1}{32.83 + 1}$$

$$= \frac{31.83}{33.83}$$

$$= 0.942$$

The quasar seems to be receding from us at a speed of 94% the speed of light.

14. We make use of the equation given in Box 27-1:

$$\frac{v}{c} = \frac{(z+1)^2 - 1}{(z+1)^2 + 1}$$

with $z = 6.0$ we get

$$\frac{v}{c} = \frac{(6.0+1)^2 - 1}{(6.0+1)^2 + 1}$$

$$= \frac{(7.0)^2 - 1}{(7.0)^2 + 1}$$

$$= \frac{49 - 1}{49 + 1}$$

$$= \frac{48}{50}$$

$$= 0.96$$

The quasar would seem to be receding from us at a speed of 96% the speed of light.

218 CHAPTER 27

15. We can use the equation given in Box 27-3:

$$\rho = \frac{3c^6}{32 \pi G^3 M^2}$$

with $\rho = 1 \text{ g/cm}^3 = 1000 \text{ kg/m}^3$.

Solving for M^2 we get

$$M^2 = \frac{3c^6}{32 \pi G^3 \rho}$$

$$= \frac{3(3 \times 10^8)^6}{32 \pi (6.67 \times 10^{-11})^3 (10^3)} \text{ kg}^2$$

$$= \frac{3 \times 729 \times 10^{48}}{32 \pi \times 296.7 \times 10^{-33} \times 10^3} \text{ kg}^2$$

$$= \frac{2187 \times 10^{48}}{29830 \times 10^{-30}} \text{ kg}^2$$

$$= \frac{21.87 \times 10^{50}}{2.983 \times 10^{-26}} \text{ kg}^2$$

$$= 7.33 \times 10^{76} \text{ kg}^2$$

Taking the square root of both sides of the preceding equation:

$$M = (7.33 \times 10^{76})^{1/2} \text{ kg}$$

$$= 2.7 \times 10^{38} \text{ kg} \left(\frac{1 \text{ M}_\odot}{1.99 \times 10^{30} \text{ kg}}\right)$$

$$= 1.36 \times 10^8 \text{ M}_\odot$$

$$\approx 1.4 \times 10^8 \text{ M}_\odot$$

It would take 140 million solar masses of water at a density of 1 g/cm³ to make a black hole.

16. We use the equation stated in Box 27-3:

$$\rho = \frac{3c^6}{32 \pi G^3 M^2}$$

In Box 8-1 on page 146 we saw that the mass of the Earth is:

$$M = 5.976 \times 10^{24} \text{ kg}$$

and so:

$$\rho = \frac{3 \, (3 \times 10^8)^6}{32 \, \pi \, (6.67 \times 10^{-11})^3 \, (5.976 \times 10^{24})^2} \, \text{kg/m}^3$$

$$= \frac{2187 \times 10^{48}}{1{,}065{,}000 \times 10^{-33+48}} \, \text{kg/m}^3$$

$$= \frac{2187}{1065 \times 10^{-30}} \, \text{kg/m}^3$$

$$= 2.05 \times 10^{30} \, \text{kg/m}^3$$

The required density is 2×10^{30} kg/m^3.

17. The equation for the Schwarschild radius was first given on page 443 and repeated in Box 27-3:

$$R_{Sch} = \frac{2GM}{c^2}$$

For a billion-solar-mass black hole

$$M = 10^9 \, M_\odot \left(\frac{1.99 \times 10^{30} \, \text{kg}}{1 \, M_\odot} \right) = 1.99 \times 10^{39} \, \text{kg}$$

and so:

$$R_{Sch} = \frac{2 \, (6.67 \times 10^{-11}) \, (1.99 \times 10^{39})}{(3 \times 10^8)^2} \, \text{m}$$

$$= \frac{26.55 \times 10^{28}}{9 \times 10^{16}} \, \text{m}$$

$$= 2.95 \times 10^{12} \, \text{m} \left(\frac{1 \, \text{AU}}{1.496 \times 10^{11} \, \text{m}} \right)$$

$$= 19.7 \, \text{AU}$$

$$\approx 20 \, \text{AU}$$

18. As an order of magnitude estimate, assume that the most luminous galaxies shine with the brilliance of 100 billion Suns. Thus

$$L_{Gal} = 10^{11} \, L_\odot$$

The Sun's absolute magnitude is about +5. And so we can calculate the galaxy's absolute magnitude (M_{Gal}) as follows:

$$M_{Gal} - M_\odot = 2.5 \log \left(\frac{L_\odot}{L_{Gal}} \right)$$

$$M_{Gal} - 5 = 2.5 \log \left(\frac{L_\odot}{10^{11} L_\odot} \right)$$

$$M_{Gal} = 5 + 2.5 \log \left(10^{-11} \right)$$

$$= 5 + 2.5 \, (-11)$$

$$= -22.5$$

Now we must calculate the apparent magnitude of the galaxy viewed from a distance of

$$d = 10^{10} \text{ ly} \left(\frac{1 \text{ pc}}{3.26 \text{ ly}} \right) = 3.07 \times 10^9 \text{ pc}$$

To do this, we use the equation first stated in Box 19-2:

$$m - M = 5 \log d - 5$$

$$m + 22.5 = 5 \log (3.07 \times 10^9) - 5$$

$$= 5 \log (3.07) + 5 \log (10^9) - 5$$

$$= 5 \log (3.07) + 5 \times 9 - 5$$

$$= 5 \times 0.487 + 45 - 5$$

$$= 2.44 + 40 = 42.44 \approx 42.5$$

$$m = 42.5 - 22.5 = 20$$

As we read on page 49, this is the visual limit of a large Earth-based telescope. With careful photography using a CCD camera, a giant elliptical galaxy with a magnitude of −22.5 should be detected at ten billion light-years.

20. The distance to the Andromeda galaxy is

$$d = 2.25 \times 10^6 \text{ ly} = 0.69 \times 10^6 \text{ pc}$$

We can use the smaller-angle formula to calculate the linear distance from the center of M31 that corresponds to 1.1 arc sec:

$$D = \frac{\alpha d}{206{,}265} = \frac{(1.1)(0.69 \times 10^6)}{206{,}265} \text{ pc}$$

$$= 3.68 \text{ pc} \left(\frac{3.086 \times 10^{16} \text{ m}}{1 \text{ pc}} \right)$$

$$= 1.14 \times 10^{17} \text{ m} = 1.14 \times 10^{14} \text{ km}$$

The orbital period (P) for an object moving at a velocity of 110 km/s at this distance from the galaxy's center is:

$$P = \frac{2\pi D}{v}$$

$$= \frac{2\pi (1.14 \times 10^{14})}{110} \text{ s}$$

$$= 6.51 \times 10^{12} \text{ s} \left(\frac{1 \text{ yr}}{3.156 \times 10^7 \text{ s}}\right) = 2.06 \times 10^5 \text{ yr}$$

To make use of Kepler's third law, we first convert the orbital radius (D) into AU:

$$D = 3.68 \text{ pc} \left(\frac{206{,}265 \text{ AU}}{1 \text{ pc}}\right) = 7.59 \times 10^5 \text{ AU}$$

And now we can write

$$M = \frac{a^3}{P^2} = \frac{D^3}{P^2}$$

$$= \frac{(7.59 \times 10^5)^3}{(2.06 \times 10^5)^2} M_\odot$$

$$= \frac{(7.59)^3}{(2.06)^2} \times 10^5 M_\odot$$

$$= 103 \times 10^5 M_\odot$$

$$\approx 10 \text{ million solar masses}$$

21. The Eddington limit is

$$L_{\text{Edd}} = 30{,}000 \left(\frac{M}{M_\odot}\right) L_\odot$$

If $M = 10^7 M_\odot$ then the maximum luminosity is:

$$L_{\text{Edd}} = 30{,}000(10^7) L_\odot = 3 \times 10^{11} L_\odot$$

22. From Figure 5-3 on page 81 we see that red light has a wavelength of about 700 nm and green light has a wavelength of about 500 nm. Thus $\lambda_0 = 700$ nm and $\lambda = 500$ nm, and so

$$\Delta\lambda = \lambda - \lambda_0 = 500 - 700 = -200 \text{ nm}$$

with the minus sign indicating approach. The required redshift is therefore

$$z = \frac{\Delta\lambda}{\lambda_0} = -\frac{200}{700} = -\frac{2}{7}$$

In Box 27-1 we find:

$$\frac{v}{c} = \frac{(z+1)^2 - 1}{(z+1)^2 + 1} = \frac{(-2/7 + 1)^2 - 1}{(-2/7 + 1)^2 + 1} = \frac{(5/7)^2 - 1}{(5/7)^2 + 1}$$

$$= \frac{25 - 49}{25 + 49} = -\frac{24}{74} = -0.32$$

You must approach the traffic light at nearly 1/3 the speed of light.

CHAPTER 28

COSMOLOGY: THE CREATION AND FATE OF THE UNIVERSE

Chapter Summary

This chapter covers the observational basis for the expanding universe theories and Hubble's law. The cosmic microwave background radiation is discussed in the context of the Big Bang theory. The role of the mean density in the determination of the future of the universe is examined in terms of the rate of expansion and the possible curvature of space.

28-1 We live in an expanding universe

Olber's paradox is reviewed. Basic assumptions of Newtonian mechanics are contrasted with special and general relativity. Einstein's cosmology is noted and his introduction of the cosmological constant is related. Hubble's law is discussed and the explanation of the universal expansion is described.

28-2 The expanding universe probably emerged from an explosive event called the Big Bang

The Big Bang theory is introduced and the upper limit to the age of the universe is computed from the Hubble constant. Resolution of Olber's paradox due to the finite age of the universe and to the cosmological redshift, which is generated by the expansion, is examined. The concept of a cosmic singularity is introduced and the Planck time is defined.

Box 28-1 The Planck time: A limit of knowledge

The quantum mechanical limits imposed on knowledge of the earliest stages of the Big Bang are derived and described. The Compton wavelength, Planck mass, Planck length, and Planck time are defined.

28-3 The microwave radiation that fills all space is evidence for a hot Big Bang

The prediction of the microwave background radiation and its detection are reviewed. Motion of the Earth through this radiation is related to anisotropies which have been observed.

28-4 Cosmic background radiation dominated the universe during the first million years

The mass density is contrasted with the radiation density of the early universe. The domination of the universe by matter at present is explained as being due to the low energies of the microwave photons in spite of their overwhelming numbers. The era of recombination is defined.

28-5 The future of the universe will be determined by the average density of matter in it

The role of the matter density in binding the universe is reviewed. The current evidence that observable matter cannot result in a bound universe is noted and the impact of dark matter on this problem is discussed.

Box 28-2 Cosmology with a non-zero cosmological constant

The results of adding a non-zero cosmological constant to the equations of general relativity are reviewed. Upper and lower limits on the possible values of the cosmological constant are noted.

28-6 The rate of deceleration of the universe can be determined by observing extremely distant galaxies

The use of the Hubble diagram to determine whether the universe is bounded or unbounded is reviewed. The deceleration parameter and the density parameter are defined and the size of the universe with time is displayed for various values of the deceleration parameter.

28-7 The shape of the universe is another indicator of its ultimate fate

The curvature of space is described and related to the values of density described above. The resulting fate of each of these types of universe is noted. The number density of galaxies is defined and discussed as evidence in support of a nearly flat universe.

28-8 If the universe is bounded, we are living inside a universe that will ultimately collapse in a Big Crunch

The future of a bound universe is explored. The ultimate ionization of all matter and the subsequent limitation of quantum mechanics are noted in a reversal of the events following the Big Bang.

28-9 If the universe expands forever, black-hole evaporations will occur in the extremely distant future

The potential impact of black-hole evaporation in an ever-expanding universe is examined. The concept of white holes as information sources is introduced.

Box 28-3 The evaporation of black holes

Primordial low-mass black holes are described. The quantum mechanical evaporation of black holes is explained. Relations between black-hole temperature and black-hole evaporation time and mass are presented. Temperatures and evaporation times are displayed graphically for a broad range of mass.

Essay by Stephen W. Hawking
The Edge of Spacetime

Essay Summary

A brief history of cosmology from early times to the present is provided. Contributions of the ancient Greeks, Galileo, Newton, Einstein, and Friedmann are noted. Special attention is devoted to the concept of limitations to our knowledge about the universe and whether there are clearly defined boundaries to space and time.

Chapters 28 and 29 in the text provide an excellent discussion of modern cosmology and related topics. Students generally find this material very interesting, but are not well prepared to understand it. This essay deals with some of the more philosophical aspects of the limits to our ability to understand and describe space and time. Stephen Hawking is a remarkable person with a physical handicap who has made enormous contributions to mankind.

Teaching Hints and Strategies

To permit students to appreciate the nature of the *expansion of the universe* (section 28-1) and the redshifts of galaxies which result, it is useful to review the early concepts of cosmology. Early cosmologies were geocentric in nature. The Earth was assumed to occupy a preferred place in the universe. The shift to the heliocentric cosmologies in the seventeenth century removed the Earth from its special place and replaced it with the Sun. Investigations of our stellar system became possible in the late eighteenth century and early nineteenth century due to improvements in the quality of telescopes and modern astrophysics grew rapidly during the latter nineteenth and early twentieth centuries. The investigations of the Milky Way up until the 1920s generally maintained that our Sun was centrally located in the Milky Way. The discovery that the Sun was far from the center of the Galaxy at last showed that our position in the universe is no better or worse than that of any other object. It is rather interesting that one consequence of curved space is the possibility of eliminating the concepts of edges and a center and still having a finite universe. You might want to describe a circle as a one dimensional curved space that has no beginning or end and, therefore, no center. The surface of a sphere is an example of a two dimensional curved space with no edges and no center (of the surface). A three dimensional analog is impossible to describe because we have run out of dimensions in which to illustrate the curvature.

Point out that the *Big Bang theory* (section 28-2) leads to some interesting interpretations and applications of Hubble's law. Remind students that the Hubble constant is determined from data gathered from the nearest galaxies and, therefore, represents the recent past to the present. The relationship may be linear into the distant past or may turn up or down. From classical analysis you can describe how the expansion should have slowed down due to the gravitational attraction which was much stronger in the past when objects were closer together. You can review orbital motion noting how the speed of one object attracted by another decreases as they separate due to conservation of energy. The estimates for the age of the universe obviously are dependent upon the accepted value of the Hubble constant and can be no more accurate than the value of the Hubble constant.

The discussion of the *Planck time* (Box 28-1) presents an interesting facet of modern science. The limitations to our knowledge as identified by modern quantum theory are worth noting.

It should be stressed that the cosmic *microwave radiation* (section 28-3 and 28-4) produces a limit to the horizon of the universe which can be observed in visible light. The conditions which existed at more remote times and distances were not capable of producing large amounts of visible light. The anisotropy of the background radiation is interpreted as due to the motion of the Earth with respect to the average motion in the observable universe. You might review the concept of the center of the universe and the assumption of its fixed position.

The role of the *density* (section 28-5) in determining the *fate of the universe* can be discussed by using an analogy with Newtonian mechanics and orbital motion. If you described the relative values of kinetic and potential energies for the various conic sections in terms of gravitational force as the binding force, you can now expand upon that idea in the context of the binding of the universe on the cosmic scale. The very low densities involved result from the enormous intergalactic distances and should be compared to the densities of other celestial objects. List the densities of neutron stars, the Sun, Jovian planets, the Earth, the Earth's atmosphere, an H II region, a globular cluster, a galaxy and the universe as a whole. Be sure to impress upon students that most of the universe is virtually empty space. The "billions and billions of stars" occupy an insignificant part of it.

The concept of *deceleration of the universe* (section 28-6) is a little difficult for the students to grasp. A demonstration of the deceleration produced by objects thrown up near the surface of the Earth can provide some insight. A discussion of the effects of varying the launch speed and the gravitational force exerted by the Earth in the context of satellites and projectiles will help to reinforce and extend these ideas.

For those students who have a tendency to revere noted scientists, it is interesting to note that even Albert Einstein made mistakes. His adoption of a *non-zero cosmological constant* (Box 28-2) was probably one of his most famous. The fact that he and his first wife divorced indicates that he made at least one other.

In conjunction with the introduction of concepts of *curved space* (section 28-7), you might point out that the advance of mathematics has been closely linked with the advance of astronomy and the other physical sciences. The development of calculus by Newton was instrumental in his success in generating a viable theory of mechanics where many others had failed. The development of nonEuclidean geometries led to the explosive impact of relativity and relativistic cosmologies. The curvature of space permits the possibility of a universe which has no edges or center and yet is finite. You might discuss how the models in modern science have become increasingly mathematical and less physical. This trend toward the abstract formalism of mathematics leads to more precise descriptions of what is observed while it tends to become less comprehensible in terms of our everyday experiences. This is to be expected when we deal with the largest and smallest objects in the universe. It should not be surprising that our very limited realm of existence is inadequate to describe all that we experience. You might review the scope of our immediate experience using powers-of-ten notation in conjunction with the scale of the microscopic realm and the universe as a whole to illustrate this point.

The *evaporation of black holes* (Box 28-3) is considered by many students to be in the realm of science fiction. Point out that the same principles permit us to explain radioactivity and are commonly applied in modern physics.

Answers to Chapter 28 Computational Questions

11. The maximum age (T) of the universe is related to the Hubble constant (H_o) as follows:

$$T = \frac{1}{H_o}$$

For $H_o = 50$ km/s/Mpc we get

$$T = \frac{1}{50 \text{ km/s/Mpc}} = \frac{1 \text{ Mpc}}{50 \text{ km/s}} \left(\frac{3.09 \times 10^{19} \text{ km}}{1 \text{ Mpc}} \right)$$

$$= 6.18 \times 10^{17} \text{ s} \left(\frac{1 \text{ yr}}{3.156 \times 10^7 \text{ s}} \right) = 1.96 \times 10^{10} \text{ yr}$$

$$\approx 20 \text{ billion years}$$

For $H_o = 75$ km/s/Mpc we can write

$$T = \frac{50}{75} (20 \text{ billion years}) \approx 13 \text{ billion years}$$

Similarly, for $H_o = 100$ km/s/Mpc we get

$$T = \frac{50}{100} (20 \text{ billion years}) = 10 \text{ billion years}$$

Obviously the universe must be older than the oldest stars in globular clusters. Therefore values of H_o which give T less than the ages of these stars are suspect.

12. The age of the universe (T) is related to the Hubble constant (H_o) by:

$$H_o = \frac{1}{T}$$

$$= \frac{1}{6000 \text{ yr}} \left(\frac{1 \text{ yr}}{3.156 \times 10^7 \text{ s}} \right)$$

$$= 5.28 \times 10^{-12} \text{ s} \left(\frac{3.09 \times 10^{19} \text{ km}}{1 \text{ Mpc}} \right)$$

$$= 1.63 \times 10^8 \text{ km/s/Mpc}$$

Turning to the Hubble law and setting $v = c$ we see

$$r = \frac{v}{H_o} = \frac{c}{H_o}$$

$$= \frac{3 \times 10^5 \text{ km/s}}{1.63 \times 10^8 \text{ km/s/Mpc}} = 1.8 \times 10^{-3} \text{ Mpc}$$

$$= 1.8 \text{ kpc} \approx 6000 \text{ light years}$$

Therefore all of the stars in our Galaxy more than 6000 light years from Earth would have to be receding from us at speeds greater than the speed of light!

13. We can use the equation given on page 534:

$$\rho_{rad} = \frac{4 \sigma T^4}{c^3}$$

and so

$$T^4 = \frac{c^3 \rho_{rad}}{4 \sigma}$$

where

$$\sigma = 5.67 \times 10^{-8} \, Wm^{-2}K^{-4}$$

If we take $\rho_{rad} = 300 \times 7 \times 10^{-31}$ kg/m^3 = 2.1×10^{-28} kg/m^3 we get

$$T^4 = \frac{(3 \times 10^8)^3 (2.1 \times 10^{-28})}{4 \times 5.67 \times 10^{-8}} \, K^4$$

$$= 2.5 \times 10^4 \, K^4$$

And so

$$T = (2.5 \times 10^4)^{1/4} \, K = 12.6 \, K$$

14. We can use the equation given on page 536:

$$\rho_c = \frac{3 H_0^2}{8 \pi G}$$

First of all, we must convert the units of H_0 to seconds as follows:

$$H_0 = 100 \, km/s/Mpc = 10^5 \, \frac{m}{s \, Mpc} \left(\frac{1 \, Mpc}{3.09 \times 10^{22} \, m} \right)$$

$$= 3.24 \times 10^{-18} \, seconds$$

And now we can write

$$\rho_c = \frac{3 (3.24 \times 10^{-18})^2}{8 \pi (6.67 \times 10^{-11})} \, kg/m^3$$

$$= 1.88 \times 10^{-26} \, kg/m^3$$

$$\approx 1.9 \times 10^{-26} \, kg/m^3$$

230   CHAPTER 28

15. We can make use of the equation given in Box 28-1 on page 531:

$$\lambda_c = \frac{h}{mc}$$

where $h = 6.625 \times 10^{-34}$ Js

$$\lambda_c = \frac{6.625 \times 10^{-34}}{1 \times 3 \times 10^8} \text{ m}$$

$$= 2.2 \times 10^{-42} \text{ m}$$

This length is far too small to be detected.

18. We can use the equation given in Box 28-3 on page 544:

$$t = 10{,}240 \; \pi^2 \left(\frac{G^2 M^3}{hc^4} \right)$$

Since $M = 10 \; M_\odot = 1.99 \times 10^{31}$ kg, we get:

$$t = 10.240 \; \pi^2 \left[\frac{(6.67 \times 10^{-11})^2 (1.99 \times 10^{31})^3}{(6.625 \times 10^{-34})(3 \times 10^8)^4} \right]$$

$$= 1.01 \times 10^5 \left[\frac{350.6 \times 10^{71}}{536.6 \times 10^{-2}} \right]$$

$$= 6.6 \times 10^{77} \text{ s} \left(\frac{1 \text{ yr}}{3.156 \times 10^7 \text{ s}} \right)$$

$$= 2.1 \times 10^{70} \text{ years}$$

19. We can use the equation given in Box 28-3 on page 544:

$$T = \frac{hc^3}{16 \; \pi^2 k G M}$$

and so:

$$M = \frac{hc^3}{16 \; \pi^2 k G T}$$

With $T = 3$ K we get:

$$M = \frac{(6.625 \times 10^{-34})(3 \times 10^8 \text{ m/s})^3}{16 \; \pi^2 (1.38 \times 10^{-23})(6.67 \times 10^{-11})(3)} \text{ kg}$$

$$= \frac{178.9 \times 10^{-10}}{4361 \times 10^{-34}} = \frac{1.789}{0.4361} \times 10^{+22} \text{ kg}$$

$$= 4.1 \times 10^{22} \text{ kg}$$

20. According to the formula in Box 28-1 on page 531, the Planck time is given by:

$$t_P = \left(\frac{Gh}{c^5}\right)^{1/2}$$

$$= \left(\frac{(6.67 \times 10^{-11})(6.625 \times 10^{-34})}{(3 \times 10^8)^5}\right)^{1/2} \text{ s}$$

$$= \left(\frac{44.18 \times 10^{-45}}{243 \times 10^{40}}\right)^{1/2} \text{ s}$$

$$= \left[\left(\frac{4.418}{2.43}\right) \times 10^{-86}\right]^{1/2} \text{ s}$$

$$= 1.35 \times 10^{-43} \text{ seconds}$$

21. The age of the universe is about 15 billion years. Converting to seconds, we get

$$t = 15 \times 10^9 \text{ yr} \left(\frac{3.156 \times 10^7 \text{ s}}{1 \text{ yr}}\right)$$

$$= 4.73 \times 10^{17} \text{ s}$$

We can now use the equation given in Box 28-3:

$$t = 10{,}240 \, \pi^2 \left(\frac{G^2 M^3}{hc^4}\right)$$

and so:

$$M^3 = \frac{hc^4 t}{10{,}240 \, \pi^2 \, G^2}$$

$$= \frac{(6.625 \times 10^{-34})(3 \times 10^8)^4 (4.73 \times 10^{17})}{(1.01 \times 10^5)(6.67 \times 10^{-11})^2} \text{ kg}^3$$

$$= \frac{2538 \times 10^{15}}{44.93 \times 10^{-17}} = 56.49 \times 10^{32} \text{ kg}^3$$

$$M = (5.65 \times 10^{33})^{1/3} \text{ kg} = 1.78 \times 10^{11} \text{ kg}$$

$$\approx 2 \times 10^{11} \text{ kg}$$

CHAPTER

29

EXPLORING THE EARLY UNIVERSE

Chapter Summary

This chapter discusses some of the recent developments in elementary particle physics and in the unification of the forces of nature in the context of cosmology. The problems of the isotropy and flatness of the universe are noted and inflationary theories which have been introduced to explain them are examined. The four basic forces of nature are compared and contrasted with a discussion of the progress of grand unification theories.

29-1 The flatness of the universe and the isotropy of the microwave background suggest that the newborn universe had a brief period of vigorous expansion

The evidence suggesting that the universe is very nearly flat and that the density is almost exactly the critical density is reviewed. The flatness problem and the isotropy problems are noted and the concept of an inflationary epoch is presented. A distinction is drawn between the expansion of space and the motion of objects through space.

29-2 During inflation, all the mass and energy in the universe burst forth from the vacuum of space

A brief review of the Heisenberg uncertainty principle is presented in the context of pair production and annihilation and virtual particles. The transition from virtual particles to real particles is explained.

29-3 As the primordial fireball cooled off, most of the matter and antimatter in the early universe annihilated each other

Equilibrium conditions and threshold temperatures for particles and gamma-ray photons are discussed. The problem of a matter dominated universe with virtually no antimatter is noted and symmetry breaking and thermal equilibrium are defined.

Box 29-1   Elementary particles and threshold temperatures

The energy required for rest mass in pair production is interpreted in terms of a threshold temperature in the radiation field.

29-4   A background of neutrinos and most of the helium in the universe are relics of the primordial fireball

Nuclear reactions involving neutrons, protons and neutrinos are presented. The instability of deuterium in high-energy radiation fields is identified as the factor limiting formation of helium in the early universe. The background of neutrinos and antineutrinos which remains from the early universe is discussed. The possible detection of neutrino masses is reviewed and proposed as a possible solution to the dark-matter problem.

29-5   Galaxies were formed from density fluctuations in the early universe

The isotropy of the early universe is examined and the role of density fluctuations in generating galaxies, clusters of galaxies and superclusters of galaxies is described. The Jeans length is defined. The discovery of voids in space and the overall bubbly distribution of galaxies are discussed.

29-6   Grand unified theories explain that all the forces had the same strength immediately after the Big Bang

The four basic forces of nature are listed and described. Exchange particles are presented for each force and examples of each force are given. The theories of quantum electrodynamics and grand unification are discussed. Attempts to construct supergrand unified theories are reviewed and additional phases in the early evolution of the universe are identified.

29-7   Cosmic strings and other oddities may be relics of the early universe

The false vacuum is defined and symmetry breaking is discussed in the context of the possible generation of cosmic strings during the evolution of the universe. Voids and filaments in the observed distributions of galaxies are examined as possible manifestations of cosmic strings.

29-8   Kaluza–Klein theories and supergravity predict that the universe may have 11 dimensions

The multidimensional theories developed by Kaluza to integrate the forces of nature with four-dimensional space–time are described. Modifications by Klein to introduce quantum mechanics into these theories are noted. The eleven-dimensional Kaluza–Klein theories are examined and compared with the theory of supergravity. The speculative nature of these exotic theories is discussed.

Teaching Hints and Strategies

The *flatness and isotropy problems* (section 29-1) with evolutionary cosmologies may be difficult for students to recognize as they derive from the statistical properties of systems and the tendency of such systems to display disorganization rather than uniformity. The same type of problem can be described for the solar system. Why do the orbits of the planets and asteroids move in a thin plane rather than orbits oriented at random? The answer is that physical conditions controlled by the laws of nature almost ensure the flattening of a spherical cloud into a thin disk. The same type of problem exists for the universe but no answer was available until the inflationary universe theories were developed. Impress upon students that disorder is expected and that ordered systems require explanations.

The concept of *virtual particles and virtual pairs* (section 29-2) seems rather abstract and unreal to most students. This is true for many of the theories in this chapter. Try to concentrate on observational tests of these abstract theories whenever possible. The Heisenberg uncertainty principle is often believed by students to be due to some ineptness of scientists or deficiency in our theories. Stress the fact that this principle expresses the limitation that nature imposes on itself about the precision of physical quantities. We know about the principle because we measure its consequences.

When discussing symmetry breaking and the resulting discrepancy between the amounts of *matter and antimatter* (section 29-3), stress the concept of threshold temperature, as it appears several times in related contexts in this chapter. You might discuss the antimatter applications and problems frequently mentioned in the Star Trek series and movies. Be sure to emphasize that antimatter is real and not a creation of science fiction. Note also that matter–antimatter annihilation is total mass-to-energy conversion, unlike nuclear fusion reactions which involve only a small percentage of conversion. Refer to section 18-6 and Box 18-3. Point out that positrons produced in the interiors of stars are destroyed by pair annihilation with electrons to produce pure energy.

The natural explanation of the *cosmic helium abundance* (section 29-4) found in stars is one the achievements of the Big Bang theory. Remind students that the Big Bang was not responsible for the generation of chemical elements heavier than helium. Such elements are generated in the processes of stellar evolution noted in sections 21-2, 22-5 and 22-6.

The problem of *galaxy formation* (section 29-5) is clearly one of the most interesting in modern astrophysics. Note the destinction between dark matter and cold dark matter. While neutrinos may provide enough mass to solve the dark-matter problem in some areas such as formations of clusters of galaxies, they do not provide solutions to the problem of individual galaxy formation. The supersymmetric partners predicted by supergravity theories, described in the previous section and shown in Table 29-2, are candidates for cold dark matter. The supercomputer simulations shown in Figure 29-9 are just one more example of the dramatic power of supercomputers in modern astronomy.

Table 29-1 summarizing the properties of the *four basic forces of nature* (section 29-6) should be discussed in some detail to reinforce the similarities and differences which exist among these forces and particles. It is very interesting that the understanding of the origin of the entire universe is dependent upon our knowledge of the smallest particles of which it is made. Note also that the energies required for experiments to test the more advanced GUTs and superGUTs require very expensive high-energy accelerators or are just not possible at all. You might review the recent cancellation of funding for the Superconducting Super Collider (SSC) and its impact on basic physics research in general. The predictions of the possible observable consequences of such theories in cosmology may represent our only means of testing these exotic theories.

The discussion of the *structural elements of the universe* (section 29-7) and their possible relationship to cosmic strings is somewhat reminiscent of our attempts to understand the structure of the Milky Way Galaxy and connections to spiral density wave theories. Although the formation of stars from massive clouds of gas and dust is reasonably well understood on theoretical grounds, the theory of the formation of galaxies is much more elusive. Up until recently it has been difficult to understand why galaxies ever formed at all. It might be interesting to speculate about the voids and filaments in Figure 29-13 and their similarities to maps of stars and nebulae along the Milky Way. How certain are we that the voids are real and not due to absorption? What types of materials might be responsible for such absorption and how could the existence of such materials be tested?

The *eleven-dimensional space-time theories* (section 29-8) really stretch the imagination to the limits. Remind students that theory may often seem abstract and unreal at first, but in many cases it provides testable insights into nature which prove to be quite enlightening. Ask students how they think that Isaac Newton might react if he were presented with what we regard as safe, ordinary science of today. He probable would recoil in stark terror from relativity, quantum mechanics, nuclear physics, etc. It might be interesting here to note the high esteem which ancient Greek philosophers had for pure reason in matters of "natural philosophy." We have passed through a couple of millenia in which observational and experimental methods have improved dramatically. The development of relativity and quantum mechanics at the beginning of this century have clearly demonstrated that "common sense" physical models often are not adequate. The application of pure reason to modern problems is central to the advance of modern science.

Answers to Chapter 29 Computational Questions

12. We can make use of the relation given on page 556:

$$\Delta E \times \Delta t \geq h/2\pi$$

With $\Delta t = 1.0 \times 10^{-8}$ s, we get:

$$\Delta E = \frac{h}{2\pi \Delta t}$$

$$= \frac{6.625 \times 10^{-34}}{2\pi \times 10^{-8}} \text{ J}$$

$$= 1.05 \times 10^{-26} \text{ J}$$

$$\approx 1.0 \times 10^{-26} \text{ J}$$

13. We can make use of the equation given on page 556:

$$\Delta m \times \Delta t \geq h/2\pi c^2$$

The mass of a proton is 1.67×10^{-27} kg, and so

$$\Delta m = 2 \times 1.67 \times 10^{-27} \text{ kg} = 3.34 \times 10^{-27} \text{ kg}$$

$$\Delta t = \frac{h}{2\pi c^2 \Delta m} \text{ s}$$

$$= \frac{6.625 \times 10^{-34}}{2\pi (3 \times 10^8)^2 (3.34 \times 10^{-27})} \text{ s}$$

$$= \frac{662.5 \times 10^{-36}}{188.9 \times 10^{-11}} \text{ s} = 3.51 \times 10^{-25} \text{ s}$$

$$\approx 3.5 \times 10^{-25} \text{ seconds}$$

14. According to the discussion in Box 29-1 on page 559, threshold temperature (T) is related to mass (m) of a particle by:

$$mc^2 = kT$$

Thus

$$T = \frac{mc^2}{k}$$

Each particle of an electron-antielectron pair has a mass of 9.11×10^{-31} kg, and so:

$$T = \frac{mc^2}{k}$$

$$= \frac{(9.11 \times 10^{-31})(3 \times 10^8)^2}{1.38 \times 10^{-23}}$$

$$= 5.9 \times 10^9 \text{ K}$$

15. According to the formula on page 521, the Jean's length is given by:

$$L_J = \left(\frac{\pi k T}{m G \rho}\right)^{1/2}$$

where m = mass of hydrogen atom = 1.67×10^{-27} kg. We therefore get:

$$L_J = \left(\frac{\pi (1.38 \times 10^{-23})(3000)}{(1.67 \times 10^{-27})(6.67 \times 10^{-11})(10^{-18})}\right)^{1/2} \text{ m}$$

$$= \left(\frac{13010 \times 10^{-23}}{11.14 \times 10^{-56}}\right)^{1/2} = (1.167 \times 10^{36})^{1/2} \text{ m}$$

$$= 1.08 \times 10^{18} \text{ m} \left(\frac{1 \text{ ly}}{9.46 \times 10^{15} \text{ m}}\right)$$

$$= 114 \text{ ly} \approx 100 \text{ ly}$$

To calculate the mass inside a sphere whose diameter is L_J, we realize:

$$M = \rho V = \frac{4}{3} \pi \rho \left(\frac{L_J}{2}\right)^3$$

$$= \frac{\pi}{6} \rho (L_J)^3$$

$$= \frac{\pi}{6} (10^{-18})(1.08 \times 10^{18})^3 \text{ kg}$$

$$= 0.66 \times 10^{-18+54} \text{ kg}$$

$$= 6.6 \times 10^{35} \text{ kg} \left(\frac{1 \text{ M}_\odot}{1.99 \times 10^{30} \text{ kg}}\right)$$

$$= 3.3 \times 10^5 \, M_\odot$$

16. The relationship between the Hubble constant (H_o) and the critical density (ρ_c) was given in Chapter 28 on page 536:

$$\rho_c = \frac{3H_o^2}{8\pi G}$$

We first convert the units of H_o entirely to seconds:

$$H_o = 50 \text{ km/s/Mpc} = 50 \, \frac{\text{km}}{\text{s Mpc}} \left(\frac{1 \text{ Mpc}}{3.086 \times 10^{19} \text{ km}} \right)$$

$$= 16.2 \times 10^{-19} \text{ s}$$

$$= 1.62 \times 10^{-18} \text{ seconds}$$

The critical density for $H_o = 50$ km/s/Mpc is therefore:

$$\rho_c = \frac{3(1.62 \times 10^{-18})^2}{8\pi(6.67 \times 10^{-11})}$$

$$= 4.7 \times 10^{-27} \text{ kg/m}^3$$

Since the critical density is proportional to the square of the Hubble constant, for $H_o = 100$ km/s/Mpc, we get

$$\rho_c = 4 \times 4.7 \times 10^{-27} \text{ kg/m}^3$$

$$= 1.9 \times 10^{-26} \text{ kg/m}^3$$

On page 534 we saw that the density of luminous matter in the universe is 5×10^{-28} kg/m^3.

For $H_o = 50$ km/s/Mpc, the density of dark matter is thus

$$4.7 \times 10^{-27} \text{ kg/m}^3 - 5 \times 10^{-28} \text{ kg/m}^3$$

$$= 4.2 \times 10^{-27} \text{ kg/m}^3$$

The number of neutrinos in a cubic meter of space is

$$100 \, \frac{1}{\text{cm}^3} \left(\frac{10^6 \text{ cm}^3}{\text{m}^3} \right) = 10^8$$

So the mass of each neutrino must be

$$\frac{4.2 \times 10^{-27}}{10^8} \text{ kg} = 4.2 \times 10^{-35} \text{ kg}$$

For $H_0 = 100$ km/s/Mpc, the density of dark matter is

$$1.9 \times 10^{-26} \text{ kg/m}^3 - 5 \times 10^{-28} \text{ kg/m}^3$$
$$\approx 1.8 \times 10^{-26} \text{ kg/m}^3$$

So the mass of each neutrino must be

$$\frac{1.8 \times 10^{-26}}{10^8} \text{ kg} = 1.8 \times 10^{-34} \text{ kg}$$

The mass of the neutrino is often expressed in electron volts (eV). We can convert kilograms to electron volts as follows:

For 1 kg

$$E = mc^2 = 1 \times (3 \times 10^8)^2 = 9 \times 10^{16} \text{ J}$$

But in Box 5-4 on page 89 we saw that

$$1 \text{ eV} = 1.602 \times 10^{-19} \text{ J}$$

So 1 kg $= \dfrac{9 \times 10^{16}}{1.6 \times 10^{-19}}$ eV $= 5.6 \times 10^{35}$ eV

And so the lower limit of the mass of a neutrino is:

$$4.2 \times 10^{-35} \text{ kg} \left(\frac{5.6 \times 10^{35} \text{ eV}}{1 \text{ kg}} \right) = 23.5 \text{ eV}$$

and the upper limit is

$$1.8 \times 10^{-34} \left(\frac{5.6 \times 10^{35} \text{ eV}}{1 \text{ kg}} \right) = 101 \text{ eV}$$

In round figures, we conclude that the mass of a neutrino would have to be in the range of 20 to 100 eV.

AFTERWORD

THE SEARCH FOR EXTRATERRESTRIAL LIFE

Afterword Summary

This brief supplement reviews information relating to the existence of life beyond the Earth. Organic molecules are identified, the Miller–Urey experiment is discussed and radio astronomy searches for evidence of other civilizations beyond the solar system are described. The Drake equation is used to estimate the number of technologically advanced civilizations in the Galaxy. Future plans to search for extraterrestrial life are noted.

Teaching Hints and Strategies

When discussing the role of carbon in forming the vast number of *organic molecules* it can be of interest to refer to the periodic table of the elements (Figure 7-6). Explain how the table is arranged with each column being composed of elements having similar chemical and physical properties. Many science fiction works have suggested the existence of silicon-based life forms. Note that silicon is the element immediately below carbon in the table. You might also review the information about the abundance of silicon in the universe. As shown in Table 7-4 silicon is much less abundant in the universe. This is due to the fact that carbon is more readily generated in nuclear reactions during stellar evolution as noted in sections 21-2 and 22-4 on stellar evolution. Be sure to note also that vast amounts of the minerals in the terrestrial planets contain silicon. Sand is primarily silicon!

Students should be cautioned repeatedly when discussing the *Drake equation* that we have very little firm information about the value of *any* of the terms needed to compute N. This material can provide an opportunity to review our current state of knowledge about a wide variety of topics covered in earlier chapters. It leads naturally to discussions of the nature of the solar system and the differences between terrestrial and Jovian planets, to applications of information about stellar evolution and to analysis of our current state of information about the Galaxy and how it might vary with time due to the evolution of its stars and nebulae.

The *star formation rate* in the Galaxy is determined for our immediate location and can only be assumed to vary with time in some fashion. Stress that we have what amounts to a snapshot of the current stellar population in a small portion of the Galaxy centered on the Sun. We must try to account for the numbers of various stars based on the information we have about the evolution of stars. If we assume that stars of given masses form at some specified rate and evolve as expected, we can predict the relative numbers of stars and compare with our current stellar census. There is no guarantee that there is not more than one way to account for the current stellar population. The very great differences that exist within spiral galaxies should cause us to be very skeptical about an assumption that local star formation rates can be assumed to be the same throughout the Galaxy.

Our direct information about the *number of planets which might be suitable for life* is worthy of some discussion as well. Remind students that planets in this category are terrestrial planets and not Jovian planets. The terrestrial planets are low mass objects which make them much more difficult to detect than Jovian planets. We have not yet detected even Jovian planets around any other star with certainty. It might be wise at this point to talk briefly about the definitions of planets and stars as suggested in Chapters 7 and 19. Although we have not yet detected objects having masses like those of planets around other stars, our ability to measure proper motions and radial velocities has improved dramatically over the past twenty years and we may now be capable of such measurements. Studies designed to detect accelerations of stars produced by planets require several years of observations due to the orbital periods of such systems.

The *remaining terms in the Drake equation* are even less certain than the first two. While one might speculate that the probabilities are nearly one, is may well be that any or all of them has a near zero probability. Note that just because one can produce a logical formula for the calculation of N is no assurance that we have sufficient knowledge about the individual terms to make reliable calculations. When discussing the funding of the projects which are designed to search for extraterrestrial life such as *Project Cyclops and SETI*, be sure to indicate that the instruments used for such activities often are used for other scientific research projects as well.

RESOURCE GUIDE

Chapter 1
Astronomy and the Universe

Reading Material for the Student

Asimov, I. *The Measure of the Universe*. 1983, Harper & Row. An extensive journey through the cosmos in half powers of ten.
Chaisson, E. *Cosmic Dawn*. 1981, Little Brown. Eloquent introduction to the universe and our place in it.
Gore, R. "The Once and Future Universe" in *National Geographic*, June 1983, p. 704. A colorful introduction to modern astronomy.
Hartmann, E. & Miller, R. *Cycles of Fire*. 1987, Workman. Illustrated tour of the stellar and extragalactic realms.
Jastrow, R. *Red Giants and White Dwarfs*, 3rd ed. 1990, Norton. A basic introduction to cosmic evolution.
Loudon, J. "Learning the Sky by Degrees" in *Astronomy*, Dec. 1986, p. 54. An introduction to angular measurement.
Morrison, P & P. *Powers of Ten*. 1982, Freeman. A tour of the universe, where each step corresponds to a power of ten increase or decrease in distance.
Preiss, B. & Fraknoi, A. *The Universe*. 1987, Bantam. A collection of introductory essays on astronomy and science fiction stories inspired by them.
Preston, R. *First Light*. 1987, Atlantic Monthly Press. Beautifully written vignettes about the practice of astronomy, using Palomar scientists as examples.
Robinson, L. "Conversations with Astronomer-Astronaut Karl Henize" in *Sky & Telescope* Nov. 1986, p. 446.
Snow, T. *The Cosmic Cycle*. 1984, Darwin.

Recent Audio-Visual Aids

Cosmos (14 one-hour episodes on 7 video tapes, Astronomical Society of the Pacific) Popular introduction to astronomy by Carl Sagan.
Of Stars and Men (53 min. video program from Harlow Shapley's book, made in the early 1960's in cartoon format, narrated by Shapley; Walt Disney Home Video)
Powers of Ten (famous 8 minute film about the cosmic and microscopic realms, both versions now available on home video; Astronomical Society of the Pacific)
Splendors of the Universe (Astronomical Society of the Pacific.) Three sets of 15 beautiful color images taken with large telescopes in Australia by David

Malin, whose color photographs can be found throughout the text. Many of the images are specially processed to bring out subtle color differences and detail.

Universe. (Videotape, 1976, Astronomical Society of the Pacific.) General introduction to the cosmos, narrated by William Shatner.

Topics for Discussion and Papers

1. Powers of Ten

After discussing the powers of ten notation, many instructors show the film *Powers of Ten*, which was made by the brilliant film-making team of Charles and Ray Eames from a children's book by Kees Boeke called *The Universe in 40 Jumps*. The film is available in two versions, a preliminary one (which has better narration and sports relativistic clocks) and a final color one, which is simpler but not as clearly narrated. At long last, both have been made available on an inexpensive VHS video tape produced by Pyramid Films, and available from the Astronomical Society of the Pacific. Students seem to be very intrigued by the film, and might be good to have them discuss it or write a brief essay about their reactions. The books *Powers of Ten* and *Measure of the Universe*, cited in the references, above, expand the themes of the film.

2. Astronomy as a Profession

For more information on careers in astronomy, see the booklet *Understanding the Universe: A Career in Astronomy*, available free if you send a long stamped self-addressed envelope to: Education Office, American Astronomical Society, c/o Dr. Charles Tolbert, McCormick Observatory, University of Virginia, P. O. Box 3818, University Station, Charlottesville, VA 22903.

The following recent books and articles by or about astronomers could help students appreciate what it's like to do astronomy in the closing decade of the 20th century:

Boslaugh, S. *Stephen Hawking's Universe*. 1985, Morrow.
Cohen, M. *In Quest of Telescopes*. 1980, Cambridge U. Press and Sky Publishing.
Graham-Smith, F. & Lovell, B. "How We Became Astronomers" in *Pathways to the Universe*. 1988, Cambridge U. Press.
Greenstein, G. *Frozen Star*. 1984, Freundlich Books.
Hill, E. *My Daughter Beatrice [Tinsley]*. 1986, American Physical Society.
Lightman, A. & Brawer, R. *Origins: The Lives and Worlds of Modern Cosmologists*. 1990, Harvard U. Press.
Preston, R. *First Light*. 1987, Atlantic Monthly Press.

Smith, D. "From Cambridge to Cosmology: An Interview with Martin Rees" in *Sky & Telescope*, Aug. 1983, p. 107.

Smith, H. "Bright Stars and Big Money" (reminiscences by Harlan Smith) in *Astronomy*, July 1990, p. 13.

3. Scaling Astronomical Distances

After reading about the definition of the astronomical unit and the light year, students can get a good appreciation of cosmic distances by constructing a scale model of the solar system and the galaxy. For example, as suggested in question 17, they might imagine the Sun to be the size of a basketball or beach ball and then calculate the size and distance of Earth, Jupiter, Pluto, the nearest star, and the Galactic Center.

4. Pseudoscience

In Section 1-8, Kaufmann emphasizes the idea that the "universe is rational". Many students, brought up on sensational media stories about ghosts, psychic powers, and the spirit world will struggle against the idea of a rational universe. This can lead to spirited and useful discussions about the nature of scientific evidence and proof, the standards of evidence in the media, and the reason that irrational and paranormal stories have such appeal.

For those interested in examining paranormal claims and the evidence against them further, some good readings include:

Abell, G. & Singer, B., eds. *Science and the Paranormal*. 1981, Scribner's. An excellent anthology of skeptical articles.

Frazier, K., ed. *Paranormal Borderlands of Science*. 1981, Prometheus Books; *Science Confronts the Paranormal*. 1986, Prometheus Books. Two collections of articles from *Skeptical Inquirer* magazine.

Gardner, M. *Science: Good, Bad, and Bogus*. 1981, Prometheus Books.

A longer bibliography of books and articles debunking astronomical topics in pseudoscience (astrology, UFO's, ancient astronauts, the face on Mars, etc.) is available for a donation of $3 from the Astronomical Society of the Pacific.

5. A Beginning Survey

I like to take a little survey the first day of class to get some sense of the students' background and knowledge. Among the questions that have sparked good discussion have been:

a. Can you name a living astronomer? (Other than the instructor, the author of the text, and Carl Sagan.) How many astronomers can you name? (Com-

pare that to the number of football players, rock stars, movie actors, chemists, classical violinists, or philosophers the students can name.)
b. Can you explain why the sky is blue?
c. Can you give the mechanism which causes the Sun to shine?
d. Do you believe in astrology? Why or why not?
e. How old is the solar system?
f. How do you think the universe began?

Chapter 2
Knowing the Heavens

Reading Material for the Student

For readings about observing the constellations and stars, see Appendix 4.

Bartky, I. "The Bygone Era of Time Balls" in *Sky & Telescope*, Jan. 1987, p. 32.

Bartky, I. & Harrison, E. "Standard and Daylight Savings Time" in *Scientific American*, May 1979.

Brown, H. *Man and the Stars*. 1978, Oxford U. Press. Includes good sections on the history of astronomy and timekeeping and its effects on human culture.

DiCicco, D. "Breaking the New Moon Record" in *Sky & Telescope*, Sep. 1989, p. 322.

Gingerich, O. "Notes on the Gregorian Calendar Reform" in *Sky & Telescope*, Dec. 1982, p. 530.

Krupp, E. *Beyond the Blue Horizon: Myths and Legends of the Sun, Moon, Stars, and Planets*. 1991. Harper Collins. Superb, humerous book on the stories ancient cultures constructed.

Kunitzsch, P. "How We Got our Arabic Star Names" in *Sky & Telescope*. Jan. 1983, p. 20.

Landes, D. *Revolution in Time: Clocks and the Making of the Modern World*. 1983, Harvard U. Press.

Lovi, G. "Sidelights on Constellation Art" in *Sky & Telescope*, Oct. 1989, p. 391.

MacRobert, A. "Names of the Stars" in *Sky & Telescope*. Sep. 1992, p. 278

MacRobert, A. "Time and the Amateur Astronomer" in *Sky & Telescope*, Apr. 1989, p. 378.

Marschall, E. "A Matter of Time: Adding a Leap Second" in *Science*, Dec. 18, 1987, p. 164.

Mayer, B. "Observational Astrology" in *Mercury*, Jul /Aug. 1987, p. 111. On observing "your" constellation in the real sky.
Motz, L. & Nathanson, C. *The Constellations: An Enthusiast's Guide to the Night Sky*. 1988, Doubleday. Mythology, astronomy, and observing information organized by constellations.
Moyer, G. "The Gregorian Calendar" in *Scientific American*, May 1982.
Moyer, G. "Luigi Lilio and the Gregorian Reform of the Calendar" in *Sky & Telescope*, Nov. 1982, p. 418.
O'Fee, B. "Sundials" in *Astronomy*, Jan. 1986, p. 47.
Siletti, W. "The Changing Immutability of the Heavens [on precession] in *Astronomy*, June 1981, p. 18.

Reading Material for the Instructor

Chapman, A. *Dividing the Circle: The Development of Critical Angular Measurement in Astronomy (1500-1850)*. 1990. Horwood/Prentice Hall.
Coyne, G., et al. *Gregorian Reform of the Calendar*. 1985, Specola Vaticana.
Debarbat, S., et al., eds. *Mapping the Sky: Past Heritage and Future Directions*. 1988, Kluwer. Proceedings of IAU Symposium 133, with many useful historical papers.
DeVorkin, D. *Practical Astronomy*. 1986, Smithsonian Institution Press. A slightly mathematical introduction to time, coordinates, motions, eclipses.
Howse, D. *Greenwich Time and the Discovery of Longitude*. 1980, Oxford U. Press.
Kluepfel, C. "How Accurate is the Gregorian Calendar?" in *Sky & Telescope*, Nov. 1982, p. 417.
Pederson, O. *Early Physics and Astronomy: A Historical Introduction*. rev. ed. 1993. Cambridge U. Press.

Recent Audio-Visual Aids

Star Maps (40 slides, Astronomical Society of the Pacific) Photos and diagrams of the Northern Hemisphere Constellations.
Stars & Telescopes (21 slides, American Association of Physics Teachers) Naked-eye views of the sky and telescopes.
MMI Corporation carries a variety of slide sets illustrating the concepts in this chapter, including *The Earth's Motions*, *Concepts of Time*, and *The Celestial Sphere*.

Topics for Discussion and Papers

1. Naming Celestial Objects

Several companies now advertise in the popular media offering to name stars for anyone who sends in a fee between $30 and $40. While this is not illegal, many customers who purchase such a "star name" are very disappointed after they realize that no-one—not professional astronomers, not amateurs, not reference books—absolutely no-one will ever know or use the name they purchased.

But bringing up this issue can then lead to an interesting discussion of how newly discovered objects in the sky are named, and how the International Astronomical Union (IAU) was given the responsibility for assigning names to those objects that can be named. For more on this subject, see:

Masursky, H. & Strobell, M. "Memorials on the Moon" in *Sky & Telescope*, Mar. 1989, p. 265.
Millman, P. "Names on Other Worlds" in *Sky & Telescope*, Jan. 1984, p. 23.
Sagan, C. "A Planet Named George" in *Broca's Brain*. 1979, Random House.

The origins of star and constellation names from ancient times are discussed in:

Allen, R. *Star Names: Their Lore and Meaning*. 1899, reprinted by Dover Books.
Gallant, R. *The Constellations: How They Came to Be*. 1979, Four Winds Press.
Gingerich, O. "The Origin of the Zodiac" in *Sky & Telescope*, Mar. 1984, p. 218.
Kunitzsch, P. "How We Got our Arabic Star Names" in *Sky & Telescope*, Jan. 1983, p. 20.
Mozel, P. "The Real Berenice's Hair" in *Sky & Telescope*, May 1990, p. 485.
Ridpath, I. *Star Tales*. 1988, Universe Books.

An interesting paper topic for students interested in the folktales and legends of other cultures is to research what names the constellations have been given in non-western countries.

2. The Sun at High Latitudes

Students tend to be "mid-latitude chauvinists", since few have had any experience with the cycles of the Sun at very low or high latitudes (and its effect on human behavior). An interesting project is to have students find examples of the effects of the long days and nights at high latitudes in literature, music, or art.

One interesting example is the song "The Sun Won't Set" in Stephen Sondheim's musical *A Little Night Music*, which is itself adapted from an Ingmar Bergman film entitled *Smiles of a Summer Night*. Recordings exist on both Columbia and RCA records and compact disks.

3. Constructing Star Catalogs

The laborious construction of detailed catalogs makes a fascinating topic for student papers, particularly if students can compare what it took in earlier centuries to what it takes now. Some useful references include the book edited by Debarbat, et al., cited in the readings for instructors and:

Ashbrook, J. "How the BD [Star Catalog] Was Made" in *Sky & Telescope*, Apr. 1980, p. 300.
Gingerich, O. "J.L.E. Dryer and his NGC" in *Sky & Telescope*, Dec. 1988, p. 621.
Ridpath, I. *Norton's 2000.0 Star Atlas and Reference Handbook*. 1989, Longman's/John Wiley.

Students who are amateur astronomers often enjoy researching the story of Charles Messier's catalog. An avid comet hunter, Messier kept mistaking nebulae and galaxies for the new comets he was seeking. He thus made his catalog to figure out once and for all, the objects most easily mistaken for comets.

4. Precession and Astrology

Students who take astrology seriously (and there are many more of them in our classes than we would like to believe) are often shocked to learn that precession has moved the vernal equinox — since this means that the *sun-signs they follow in newspaper astrological columns are no longer aligned with the constellations after which they were named about two thousand years ago.*

This can lead into useful discussions about astrology itself and why scientists are skeptical about a system born in an age of superstition, when the objects in the sky were seen as manifestations of the gods.
Some useful references:

Astronomical Society of the Pacific: *Astronomy versus Astrology*. This 20-page information packet is available for a donation of $4.
Culver, R. & Ianna, P. *Astrology: True or False*. 1988, Prometheus Books.
Dean, G. "Does Astrology Need to be True?" in *The Skeptical Inquirer*, Winter 1986-87, p. 166; Spring 1987, p. 257.

Fraknoi, A. "Your Astrology Defense Kit" in *Sky & Telescope*, Aug. 1989, p. 146.
Kelly, I. "The Scientific Case Against Astrology" in *Mercury*, Nov/Dec. 1980, p. 135.
Kurtz, P., et al. "Astrology and the Presidency" in *The Skeptical Inquirer*, Fall 1988, p. 3.
Lovi, G. "Zodiacal Signs Versus Constellations" in *Sky & Telescope*, Nov. 1987, p. 507.

The reference by Kurtz, above, gives full information about the sad episode in which it was revealed that an astrologer in San Francisco named Joan Quigley had so great an influence over Nancy Reagan (who was distraught about the assassination attempt on her husband's life and thus easy prey), that for several years, the White House schedule was set up according to Quigley's astrological charts. Students might want to discuss the state of science education in a country where so many people (including White House staff members) saw nothing wrong with this.

CHAPTER 3
Eclipses and Ancient Astronomy

Reading Material for the Student

On the Astronomy of Ancient Civilizations

Aveni, A. *Archaeoastronomy in the New World*. 1982, Cambridge U. Press.
Aveni, A. "Archaeoastronomy: Past, Present and Future" in *Sky & Telescope*, Nov. 1986, p. 456.
Canby, T. "The Anasazi: Riddles in the Ruins" in *National Geographic*, Nov. 1982, p. 554.
Hadingham, E. *Early Man and the Cosmos*. 1984, Walker & Co.
Krupp, E. *Echoes of the Ancient Skies*. 1983, Harper & Row.
Maranto, G. "Stonehenge: Can It Be Saved?" in *Discover*, Dec. 1985, p. 60.

On Eclipses and the Moon

Allen, D. & C. *Eclipse*. 1987, Allen & Unwin. A nice introduction to eclipse history, chasing, and science.
Codona, J. "The Enigma of Shadow Bands [During Eclipses]" in *Sky & Telescope*. May 1991, p. 483.
Krisciunas, K. "Eclipses Science from Hawaii" in *Sky & Telescope*. Dec. 1991,

p. 597. (The whole issue is devoted to reports on the July 1991 eclipse of the Sun).
Littman, M. & Willcox, K. *Totality*. 1991, U. of Hawaii Press. Best lay introduction to science and lore of eclipses.
Marschall, L. "Shadow Bands: Solar Eclipse Phantoms" in *Sky & Telescope*, Feb. 1984, p. 116.
Meeus, J. "Solar Eclipse Diary: 1985 - 1995" in *Sky & Telescope*, Oct. 1984, p. 296.
Seltzer, A. "How to Observe and Photograph the Annular Solar Eclipse" in *Astronomy*, May 1984, p. 50.
Stephenson, S. "Historical Eclipses" in *Scientific American*, Oct. 1982.
Whiteman, M. "Eclipse Prediction on Your Computer" in *Astronomy*, Nov. 1986, p. 67.

On Eratosthenes, Aristarchus, Hipparchus, etc.

Bronowski, J. *The Ascent of Man*. 1973, Little Brown. See the section entitled "The Music of the Spheres"
Gingerich, O. "From Aristarchus to Copernicus" in *Sky & Telescope*, Nov. 1983, p. 410.
Gingerich, O. "Aristarchus of Samos" in *Sky & Telescope*, Nov. 1980, p. 376.
Heath, T. *Aristarchus of Samos*. 1913; reprinted by Dover Books.
Sagan, C. *Cosmos*. 1980, Random House. The chapter entitled "The Shores of the Cosmic Ocean" recounts Eratosthenes' experiment; "The Backbone of Night" chapter focuses on the Greek thinkers.

Reading Material for the Instructor

Aveni, A. "Native American Astronomy" in *Physics Today*, June 1984, p. 24.
Burl, A. *The Stone Circles of the British Isles*. 1977, Yale Univ. Press.
Chamberlain, V. *When Stars Came Down to Earth: Cosmology of the Skidi Pawnee Indians*. 1982, Ballena Press.
Cook, A. *The Motion of the Moon*. 1988, Adam Hilger.
Dicks, D. *Early Greek Astronomy to Aristotle*. 1985, Cornell U. Press.
Dreyer, J. *History of Astronomy from Thales to Kepler*. 1953, reprinted by Dover Books.
Gingerich, O. "Astronomy in the Age of Columbus" in *Scientific American*. Nov. 1992, P. 100.
Krupp, E., ed. *Archaeoastronomy and the Roots of Science*. 1984, Westview Press.
Stephenson, S. & Houlden, M. *Atlas of Historical Eclipse Maps 1500 BC to AD 1900*. 1986, Cambridge U. Press.
Williamson, R. *Living the Sky: The Cosmos of the American Indian*. 1984, Houghton Mifflin.

Recent Audio-Visual Aids

The "Backbone of Night" and "Shores of the Cosmic Ocean" episodes of *Cosmos* are now available on home video from Turner Home Entertainment Corp. They can be ordered from the Astronomical Society of the Pacific.

Topics for Papers and Discussion

1. Stonehenge

Stonehenge is perhaps the best-known of the many astronomical monuments left by ancient civilizations. Because it is so well studied, it has engendered much controversy: astronomers and archeologists have often disagreed about each others' interpretations. Many of the books on archaeoastronomy cited above have sections on Stonehenge; here are a few references to get students started in investigating further:

Hawkins, G. *Stonehenge Decoded*. 1965, Delta. Hawkins was the astronomer who first popularized the astronomical interpretations of Stonehenge.
Hoyle, F. *From Stonehenge to Modern Cosmology*. 1972, W. H. Freeman.
Maranto, G. "Stonehenge: Can It Be Saved?" in *Discover*, Dec. 1985, p. 60.

2. Eclipses Elsewhere

It is an interesting coincidence that the apparent size of the Moon and Sun as seen from Earth are such that *exactly* total eclipses can occur. Michael Mendillo and Richard Hart did some calculations to see whether such a match occurs for Sun-satellite combinations on any other planet in the Solar System, and found that it did not. From the fact that the only planet whose satellite can produce this impressive phenomenon is also the only planet bearing intelligent life, they infer (with tongue planted firmly in cheek) that God must exist. (See *Physics Today*, Feb. 1974, p. 73.) In response, the August 1974 issue of *Physics Today* has a letter from E. Wicher, pointing out that tidal interactions are changing the distance between the Earth and the Moon; thus God did not exist until about 280 million B.C.!

3. Eclipse Chasing

There are large numbers of people who are passionate about seeing total eclipses of the Sun and try to travel to eclipse paths whenever they can. Eclipse travelers often tell interesting stories, especially when the eclipse path covers less populated areas. Some references:

Anderson, J. "Eclipse Prospects for the 1990's" in *Astronomy*, Feb. 1989, p. 71.
Ashbrook, J. "A Solar Eclipse Expedition of 1883" in *Sky & Telescope*, Mar. 1978, p. 211.
Bracher, K. "The Famous Eclipse of June 8, 1918" in *Sky & Telescope*, Nov. 1979, p. 411.
Harris, J. "Confessions of an Eclipse Addict" in *Astronomy*, Jan. 1988, p. 62.
Kundu, M. "Observing the Sun During Eclipses" in *Mercury*, Jul/Aug. 1981, p. 108.
Menzel, D. & Pasachoff, J. "Solar Eclipse: Nature's Superspectacular" in *National Geographic*, Aug. 1970.
Silverman, S. & Mullen, G. "Eclipses: A Literature of Misadventures" in *Natural History*, Jun/Jul. 1972.

4. Repeating Eratosthenes' Experiment

Several college groups have tried to repeat Eratosthenes' measurement of the size of the Earth. Students may want to see what was involved and what results were obtained. See:

Norton, O. "Repeating Eratosthenes' Observations" in *Mercury*, May/June 1974, p. 14.

5. Finding the Earliest Visible Phase of the Moon

In recent years, a number of amateur astronomy groups have tried to break the record of the newest moon that can be spotted with the naked eye. Students may want to try this game themselves or report on what others have done. See:

DiCicco, D. "Breaking the New Moon Record" in *Sky & Telescope*, Sep. 1989, p. 322.
Doggett, L., et al. "Moonwatch—July 14, 1988" in *Sky & Telescope*, July 1988, p. 34; Apr. 1989, p. 373.

6. Phases of the Moon in Art and Literature

An interesting project for the ambitious student is finding some examples of correct and incorrect depictions of the Moon in stories, films, cartoons, and serious art. See, for example:

North, J. *Chaucer's Universe*. 1988, Oxford U. Press.
Olson, D. & Doescher, R. "Lincoln and the Almanac Trial" in *Sky & Telescope*, Aug. 1990, p. 184.

Olson, D. & Doescher, R. "Van Gogh, Two Planets, and the Moon" in *Sky & Telescope*, Oct. 1988, p. 408.
Olson, D. & Jasinksi, L. "Chaucer and the Moon's Speed" in *Sky & Telescope*, Apr. 1989, p. 376.

CHAPTER 4
Gravitation and the Motions of Planets

Reading Material for the Student

General References on Renaissance Astronomy
Christianson, G. *This Wild Abyss*. 1978, Free Press.
Durham, F. & Purrington, R. *Frame of the Universe*. 1983, Columbia U. Press.
Ferris, T. *Coming of Age in the Milky Way*. 1988, Morrow. Eloquent, award-winning history of the development of astronomical thought.
Gillies, G. & Sanders, A. "Getting the Measure of Gravity" in *Sky & Telescope*. Apr. 1993, p. 28.
Sagan, C. *Cosmos*. 1980, Random House. See the chapter entitled "The Harmony of the Worlds"

On Copernicus
Banville, J. *Doctor Copernicus*. 1976, Godine. A novelization of Copernicus' life.
Gingerich, O. "Copernicus and Tycho" in *Scientific American*, Dec. 1973.
Kuhn T. *The Copernican Revolution*. 1957, Harvard U. Press.
Ravetz, J. "The Origin of the Copernican Revolution" in *Scientific American*, Oct. 1966.
Rosen, E. "Copernicus' Place in the History of Astronomy" in *Sky & Telescope*, Feb. 1973, p. 72.
Rosen, E. *Copernicus and the Scientific Revolution*. 1984, Krieger.

On Tycho Brahe
Christianson, G. "The Celestial Palace of Tycho Brahe" in *Scientific American*, Feb. 1961.
Gingerich, O. "Copernicus and Tycho" in *Scientific American*, Dec. 1973.

On Kepler
Banville, J. *Kepler: A Novel*. 1981, Godine.

Casper, M. *Kepler*. 1959, Collier.
Gee, B. "400 Years: Johannes Kepler" in *Physics Teacher*, vol. 9, p. 510 (1971).
Wilson, C. "How Did Kepler Discover His First Two Laws" in *Scientific American*, Mar. 1972.

On Galileo

Dickson, D. "Was Galileo Saved by a Plea Bargain" in *Science*, Aug. 8, 1986, p. 612.
Drake, S. *Galileo Studies*. 1970, U. of Michigan Press.
Gingerich O. "How Galileo Changed the Rules of Science" in *Sky & Telescope*. Mar. 1993, p. 32.
Gingerich, O. "The Galileo Affair" in *Scientific American*, Aug. 1982.
Gingerich, O. "Galileo and the Phases of Venus" in *Sky & Telescope*, Dec. 1984, p. 520.
Hoskin, M. "Galileo: An Imaginary Interview" in *The Mind of the Scientist*. 1971, Taplinger.
Lerner, L. & Gosselin, E. "Galileo and the Specter of Bruno" in *Scientific American*, Nov. 1986.
Ronan, C. *Galileo*. 1974, Putnam's.

On Newton

Christianson, G. "Newton's *Principia*: A Retrospective" in *Sky & Telescope*, July 1987, p. 18.
Cohen, I. "Newton's Discovery of Gravity" in *Scientific American*, Mar. 1981.
Cohen, I. *The Newtonian Revolution*. 1981, Cambridge U. Press.
Fauvel, J., et al., eds. *Let Newton Be: A New Perspective on His Life and Works*. 1988, Oxford U. Press.
Gingerich, O. "Newton, Halley, and the Comet" in *Sky & Telescope*, Mar. 1986, p. 230.
Hoskin, M. "Newton: An Imaginary Interview" in *The Mind of the Scientist*. 1971, Taplinger.

On Einstein and General Relativity

Gardner, M. *The Relativity Explosion*. 1976, Vintage Books.
Kaufmann, W. *Relativity and Cosmology*, 2nd ed. 1977, Haper & Row.
Schwartz, J. & McGuiness, M. *Einstein for Beginners*. 1979, Pantheon. Einstein's work explained with cartoons and basic text; accurate physics, but somewhat political.
Will, C. *Was Einstein Right?* 1986, Basic Books.
Zirker, J. "Testing Einstein's General Relativity During Eclipses" in *Mercury*, Jul/Aug. 1985, p. 98.

Reading Material for the Instructor

On Renaissance Astronomy

Bork, A. "Newton and Comets" in *American Journal of Physics*, Dec. 1987, p. 108.

Christiansen, G. *In the Presence of the Creator: Isaac Newton and His Times.* 1984, Free Press/Macmillan.

Cohen, I. *Revolutions in Science.* 1985, Harvard U. Press.

Gingerich, O., ed. *The Nature of Scientific Discovery.* 1975, Smithsonian Institution Press. Proceedings of a conference on Copernicus and the Copernican Revolution.

Field, J. *Kepler's Geometrical Cosmology.* 1988, U. of Chicago Press.

Harrison, E. "Newton and the Infinite Universe" in *Physics Today*, Feb. 1986, p. 24.

Murdin, L. *Under Newton's Shadow: Astronomical Practices in the 17th Century.* 1985, Adam Hilger.

Neyman, J. "Nicholas Copernicus: An Intellectual Revolutionary" in *The Heritage of Copernicus.* 1974, MIT Press.

Parker, G. "Galileo, Planetary Atmospheres, and Prograde Revolution" in *Science*, Feb. 8, 1985, p. 597.

Patter, P. *The Annus Mirabilis of Sir Isaac Newton.* 1970, MIT Press.

Taton, R. & Wilson, C., eds. *Planetary Astronomy from the Renaissance to the Rise of Astrophysics.* 1989, Cambridge U. Press.

Toulmin, S. & Goodfield, J. *The Fabric of the Heavens: the Development of Astronomy and Dynamics.* 1961, Harper.

van Helden, A., ed. *Siderius Nuncius or The Sidereal Messenger of Galileo.* 1989, U. of Chicago Press. A new annotated translation.

For other references, see:

DeVorkin, D.: *A History of Modern Astronomy and Astrophysics: A Selected and Annotated Bibliography.* 1982, Garland.

On Einstein and General Relativity

French, A., ed. *Einstein: A Centenary Volume.* 1979, Harvard U. Press.

Hoton, G. & Elkana, Y., eds. *Albert Einstein: Historical and Cultural Perspectives.* 1979, Princeton U. Press.

Kaufmann, W. *Cosmic Frontiers of General Relativity.* 1977, Little Brown. Best nonmathematical introduction to general relativity and black holes.

Pais, A. *Subtle is the Lord.* 1982, Oxford U. Press. Detailed scientific biography of Einstein.

Rindler, W. *Essential Relativity*, 2nd ed. 1977, Springer Verlag.

Stern, F. "Einstein and Germany" in *Physics Today*, Feb. 1986, p. 40.
Woolf, H. *Some Strangeness in Proportion: A Centennial Symposium to Celebrate the Achievements of Albert Einstein*. 1980, Addison-Wesley.

Recent Audio-Visual Aids

Astronomers of the Past (50 slides, Astronomical Society of the Pacific) Portraits of noted astronomers, including Copernicus, Galileo, Brahe, and Einstein; with capsule biographies.
Coming of Age in the Milky Way (audiotape excerpts from the book, read by T. Ferris, Astronomical Society of the Pacific)
Copernicus (1973, 10 min. film, Pyramid Films) A short film from the Eames studio, examining Copernicus through documents and artifacts.
Galileo: The Challenge of Reason (1969, 23 min., Learning Corp. of America film) Superb dramatization of Galileo's life and ideas. (Available through many university film rental services)
The Harmony of the Worlds (episode 3 of *Cosmos*; available on videotape from Turner Home Entertainment; distributed by the Astronomical Society of the Pacific)
Motion: Newton's Laws (24 min. film produced by Open University, BBC-TV; distributed by University Media)
Newton: The Mind that Found the Future (1970, 21 min., Learning Corp. of America) Edmund Halley narrates a superb dramatization of Newton's life and work. (Available through many university film rental services)
PASCO's Drawings of Famous Physicists (20 slides, American Association of Physics Teachers)

Topics for Discussion and Papers

1. Kepler and Music

In working out his laws of planetary motion, Kepler continued the Greek Pythagorean tradition of searching for connections between mathematical and musical harmony. He actually devised a complex system for the "harmony of the spheres"—the "music" that the planets "played" as they orbited the Sun.

In 1957, the composer Paul Hindemith (himself a follower in the Pythagorean tradition) wrote a little-known opera about the life and ideas of Kepler called *The Harmony of the World*. Several recordings of a suite from this opera have been available at various times on small import labels.)

In 1977, Willie Ruff (a professor of music at Yale and a student of Hindemith's) and John Rodgers (a geologist and amateur musician) programmed a computer to synthesize Kepler's "harmony of the spheres" from 100 years of data on the actual motions of the planets. Students with an interest in music can find the full story (and information on how to order a recording) on p. 286 of the May/June 1979 issue of *American Scientist* magazine.

2. Copernicus, Music, and Fred Hoyle

In 1966, astronomer (and science fiction writer) Fred Hoyle published a fascinating novel called *October the First Is Too Late*, in which one element of the complex plot revolves around the question: Who would be of greater interest to the people of the distant future — a 20th century physicist or a 20th century composer? The novel took form after Hoyle and his friend composer Leo Smit were discussing this question during hikes in the Scottish highlands. After the book was published, Hoyle and Smit even gave a concert and lecture at Caltech based on the ideas and music discussed in the book.

In 1973, the 500th anniversary of Copernicus' birth, Smit and Hoyle were commissioned to write a piece of music in honor of the great astronomer. They produced a suite called *Copernicus — Narrative and Credo*, based on the music of Copernicus' time and chronicling some of the highlights of his life. The piece ends with a fascinating "credo" written and spoken by Hoyle, reflecting on the meaning of the Copernican revolution for our time and offering a modern astronomers personal reactions to the universe. The recording was once available on the Desto label, number DC 7178.

Students could analyze the novel (which requires some familiarity with Everett and Wheeler's many-worlds interpretation of quantum mechanics) and/or try their hand at writing their own "credo" based on the material they have learned in the course. Ambitious students majoring in music may even want to try their hand at writing or at least outlining a musical piece that draws its inspiration from astronomical themes. To see other examples of such music, they can refer to:

Fraknoi, A. "The Music of the Spheres" in *Mercury*, May/June 1977, p. 15;
Fraknoi, A. "More Music of the Spheres" in *Mercury*, Nov/Dec. 1979, p. 128.
  (These are reprinted in a booklet entitled *Interdisciplinary Approaches to Astronomy*, available from the Astronomical Society of the Pacific.)

3. Galileo and the Law

In the 1980's the Vatican actually reopened the "Galileo case" and rescinded some of the Church's censure of his work. Students interested in the law may want to take a look at the initial trial of Galileo and the recent developments. See:

Dickson, D. "Was Galileo Saved by a Plea Bargain?" in *Science*, Aug. 8, 1986, p. 612.
Hansen, J. "The Crime of Galileo" in *Science 81*, Mar. 1981, p. 14.

Katz, C. "The Prosecution Never Rests: The Trials of Galileo Galilei" in *California State Bar Journal*, Mar./Apr. 1977.
Redondi, P. *Galileo: Heretic*. 1987, Princeton U. Press.

4. The Copernican Revolution as an Archetype

The noted historian of science Thomas Kuhn has proposed that the Copernican revolution in our thinking about the universe can serve as an archetype for revolutionary changes (paradigm shifts) in our scientific world view. Students interested in the history of science or the history of ideas would enjoy reading Kuhn's books *The Structure of Scientific Revolutions* (2nd ed., 1970, U. of Chicago Press) and *The Copernican Revolution* (1957, Harvard U. Press).

To celebrate the 500th anniversary of Copernicus' birth, a conference was held to examine "Copernican revolutions" in science. Its proceedings can be found in the book *The Heritage of Copernicus* edited by J. Neyman (1974, MIT Press.) Students with an interest in other sciences might like to compare revolutions in other fields with those in astronomy.

CHAPTER 5
The Nature of Light and Matter

Reading Material for the Student

Augensen, H. & Woodbury, J. "The Electromagnetic Spectrum" in *Astronomy*, June 1982, p. 6.
Bova, B. *The Beauty of Light*. 1988, Wiley. A readable introduction to all aspects of the production and decoding of light by a science writer.
Darling, D. "Spectral Visions: The Long Wavelengths" in *Astronomy*, Aug. 1984, p. 16; "The Short Wavelengths" in *Astronomy*, Sep. 1984, p. 14.
Rowan-Robinson, M. *Cosmic Landscape: Voyages Back Along the Photon's Track*. 1980, Oxford U. Press. Looks at the universe in each wavelength band.
Sobel, M. *Light*. 1987, U. of Chicago Press. An excellent nontechnical introduction to all aspects of light by a physicist.

Reading Material for the Instructor

Buchwald, J. *The Rise of the Wave Theory of Light*. 1989, U. of Chicago Press.
Hansch, T., et al. "The Spectrum of Atomic Hydrogen" in *Scientific American*, Mar. 1979.
Hearnshaw, J. *The Analysis of Starlight*. 1986, Cambridge U. Press. A history of spectroscopy.

Jungnickel, C. & McCormach, R. *The Now Mighty Theoretical Physics: 1870-1925*. 1986, U. of Chicago Press.
Tolstoy, I. *James Clerk Maxwell*. 1981, U. Of Chicago Press.
Wroblewski, A. "De Mora Luminis" in *American Journal of Physics*, July 1985, p. 620. On Roemer's determination of the speed of light.

Recent Audio-Visual Aids

Electron Distributions in the Hydrogen Atom (25 slides, showing orbitals from n=1 to n=6, American Association of Physics Teachers)
Lights on Campus (20 slides showing spectra of different types of street lights found on college campuses, American Association of Physics Teachers)
Master of Light (25 min. film or videocassette on the life and work of Albert Michelson; Cinema Guild, 1997 Broadway, New York, NY 10019)
The Sky at Many Wavelengths (11 slides, Astronomical Society of the Pacific) Views of the sky in all the bands of the electromagnetic spectrum.
Spectra of the Stars (36 slides, Astronomical Society of the Pacific) Selected by James Kaler from his book of the same name.
Steel, Stars, and Spectra (24 min. film on the use of spectra in a steel mill and in astronomy, produced by Open University, BBC-TV; distributed by University Media)

Topics for Discussion and Papers

The History of our Understanding of Light and Spectra

The text gives a necessarily brief overview of the fascinating history of how we came to understand the complex nature of light and the interpretation of spectra. The following sources can help students delve more deeply into the people and ideas involved:

Baker, A. *Modern Physics and Antiphysics*. 1970, Addison-Wesley.
Boorse, H., et al: *The Atomic Scientists*. 1989, John Wiley.
Cline, B. *Men Who Made a New Physics*. 1965, Signet.
Gingerich, O. "Unlocking the Chemical Secrets of the Cosmos" in *Sky & Telescope*, July 1981, p. 13. On Kirchhoff and Bunsen.
Meadows, A. "The New Astronomy" in Gingerich, O., ed. *Astrophysics and 20th Century Astronomy to 1950*. 1984, Cambridge U. Press.
Snow, C. *The Physicists*. 1981, Macmillan.

The Universe with Other Eyes

An interesting exercise is to have students consider what the universe would look like if our eyes (or equivalent sense organs) were tuned to other bands of the electromagnetic spectrum. What objects would be visible and what objects we are used to seeing in the sky would be invisible? This topic might be discussed twice during the semester—once when the spectrum is introduced and then again toward the end of the course when the students are more familiar with celestial objects that radiate mainly in nonvisible regions of the spectrum. For a good reference in this area, see:

Henbest, N. & Marten, M. *The New Astronomy*. 1983, Cambridge U. Press.

Measuring the Speed of Light

Section 5-1 introduces the history of measuring the speed of light — a measurement that occupied humanity for more than three centuries. Students who want to look further into the way in which the speed was measured can consult:

Boyer, C. "Early Estimates of the Velocity of Light" in *Isis*, vol. 33, p. 24 (1941).
Cohen, I, "Roemer and the First Determination of the Velocity of Light" in *Isis*, vol. 31, p. 327 (1939).
DeBray, M. "The Velocity of Light: The History of Its Determination from 1849 to 1933" in *Isis*, vol. 25, p. 437 (1936).
Hansen, J. "Ole Roemer" in *A.S.P. Leaflet*, No. 187, Sep. 1944.
Keeports, D. "Looking for Ghosts to Measure the Speed of Light" in *The Physics Teacher*, Sep. 1990, p. 398.
Livingston, D. *The Master of Light: A Biography of Albert Michelson*. 1973, Scribners.

Chapter 6
Optics and Telescopes

Reading Material for the Student

On Telescopes in General

Clark, D. *The Cosmos from Space*. 1987, Crown.

Introduction to Instruments in Space, Present and Future.

Bunge, R. "Dawn of a New Era: Big Scopes" in *Astronomy*. Aug. 1993, p. 48.
Cohen, M. *In Quest of Telescopes*. 1980, Cambridge U. Press. On what it's like to observe with large astronomical telescopes.

Cornell, J. & Carr, J., eds. *Infinite Vistas: New Tools for Astronomy*. 1985, Scribners. Essays on new telescopes & projects.
Cornell, J. & Gorenstein, P., eds. *Astronomy from Space*. 1983, MIT Press. Articles about various space observatories.
Davies, J. *Satellite Astronomy*. 1988, Horwood/Wiley. Introduction to instruments doing astronomy above the Earth.
Eicher, D. "A New Era in Space: Space Astronomy for the 1990's" in *Astronomy*, Jan. 1990, p. 22.
Haynes, R. "How We Get Pictures from Space" in *Mercury*, May/June 1990, p. 77.
Krisciunas, K. *Astronomical Centers of the World*. 1988, Cambridge U. Press. History of and guide to major observatories.
Learner, R. *Astronomy Through the Telescope*. 1981, Van Nostrand Reinhold. Illustrated history of telescopes and discoveries.
Maran, S. "Little Missions, Big Returns" in *Astronomy*, Jan. 1989, p. 34. Introduction to some modest space instruments.
Marx, S. & Pfau, W. *Observatories of the World*. 1982, Van Nostrand Reinhold. A tour guide.
Robinson, L. "Monster Telescopes for the 1990s" in *Sky & Telescope*, May 1987, p. 495.
Ronan, C. "The Invention of the Reflecting Telescope" in *1993 Yearbook of Astronomy*, ed. by P. Moore, 1992, Norton, p. 129.
Thomsen, D. "Taking the Measure of the Stars" in *Science News*, Jan. 3, 1987, p. 10. On optical interferometry.
Tucker, W. & K. *Cosmic Inquirers*. 1986, Harvard U. Press. An excellent introduction to the VLA, Einstein x-ray observatory, and other major instruments.
Waldrop, M. "The New Art of Telescope Making" in *Science*, Dec. 19, 1986, p. 149.
Waldrop, M. "Troubled Times Ahead for Telescope Makers" in *Science*, Apr. 1, 1988, p. 28.
Waldrop, M. "The Mirror Maker: Roger Angel" in *Discover*, Dec. 1987, p. 12.
Wolff, S. "The Search for Aperture: A Selective History of Telescopes" in *Mercury*, Sep/Oct 1985, p. 139.

On the Hubble Space Telescope

Beatty, J. "Hubble Space Telescope: Astronomy's Greatest Gambit" in *Sky & Telescope*, May 1985, p. 409; "Will Space Telescope Be Ready?" in *Sky & Telescope*, Feb. 1987, p. 146.
Chaisson, E. & Villard, R. "HST: The Mission" in *Sky & Telescope*, Apr. 1990, p. 378.
Chien, P. "The Launch of HST" in *Astronomy*, Jul. 1990, p. 30.

Cole, S. "Space Telescope Comes to Life" in *Astronomy*, Oct. 1988, p. 42; "Raising Hubble" in *Astronomy*, Aug. 1990, p. 38.

Cowen, R. "Space Telescope: A Saga of Setbacks" in *Science News*, vol. 137, p. 8. (Jan 6, 1990).

Davidson, G. "How we'll Fix Hubble" in *Astronomy*. Feb. 1993, p. 42.

Field, G. & Goldsmith, D. *Space Telescope: Eyes Above the Atmosphere*. 1990, Contemporary Books. The best intro to HST.

Field, G. & Goldsmith, D. "Space Telescope" in *Mercury*, Mar/Apr. 1990, p. 34.

Fienberg, R. "Hubble's Road to Recovery" in *Sky & Telescope*. Nov. 1993, p. 16.

Fienberg, R. "Hubble Space Telescope Sees First Light" in *Sky & Telescope*, Aug. 1990, p. 140; "HST Takes Wing" in *Sky & Telescope*, July 1990, p. 31; "HST: Astronomy's Discovery Machine" in *Sky & Telescope*, Apr. 1990, p. 366.

Greenstein, G. "First Light for the Hubble" in *Air & Space*, Mar/Apr. 1990, p. 52.

Hanle, P. "Astronomers, Congress, and the Large Space Telescope" in *Sky & Telescope*, Apr. 1985, p. 300. (A good history.)

Hawley, S. "Delivering HST to Orbit" in *Sky & Telescope*, Apr. 1990, p. 373.

Hoffman, J. "How we'll Fix the Hubble Space Telescope" in *Sky & Telescope*. Nov. 1993, p. 23.

Kanipe, J. "Hubble's Brave New Universe" in *Astronomy*, Sep. 1990, p. 44.

Longair, M. *Alice and the Space Telescope*. 1989, Johns Hopkins U. Press. Overly cute, popular-level intro to HST, using imagery from *Alice in Wonderland*.

Longair, M. "The Scientific Challenge of Space Telescope" in *Sky & Telescope*, Apr. 1985, p. 306.

Maran, S. "Hubble Illuminates the Universe" in *Sky & Telescope*. June 1992, p. 619.

Maran, S. "The Promise of Space Telescope" in *Astronomy*, Jan. 1990, p. 38.

O'Dell, R. "Building the Space Telescope" in *Sky & Telescope*, July 1989, p. 31.

Sherill, T. "The Space Telescope in Eclipse" in *Astronomy*, Feb. 1990, p. 36.

Tresch-Fienberg, R. "First Light for Space Telescope" in *Sky & Telescope*, Dec. 1986, p. 562.

Villard, R. "The Space Telescope at Work" in *Astronomy*, June 1989, p. 38. How the HST will operate.

Villard, R. "The World's Biggest Star Catalog" in *Sky & Telescope*, Dec. 1989, p. 583. (On the HST Guide Star Catalog)

Waldrop, M. "Astronomers Survey Hubble Damage" in *Science*. 13 July 1990, p. 112.

Waldrop, M. "Space Telescope" in *Science*, Jul. 15, 1983, p. 249; Aug. 5, 1983,

p. 534; "Great Telescope, Bad Service Plan" in *Science*, Dec. 22, 1989, p. 1551.

Watson, G. "Building the Space Telescope's Optical System" in *Astronomy*, Jan. 1986, p. 15.

On Specific Telescopes on the Ground

Berry, R. "The Telescope That Defies Gravity: The Keck Telescope" in *Astronomy*, July 1988, p. 42.

Evans, D. & Mulholland, J. *Big and Bright*. 1986, U. of Texas Press. A history of the McDonald Observatory.

Fischer, D. "A Telescope for Tomorrow" in *Sky & Telescope*, Sep. 1989, p. 249. (About ESO's 3.6-meter New Technology Telescope.)

Gustafson, J. & Sargent, W. "The Keck Observatory: 36 Mirrors Are Better Than One" in *Mercury*, Mar/Apr. 1988, p. 43.

Gustafson, J. "Pioneering Astronomy at Lick Observatory" in *Astronomy*, May. 1988, p. 6.

Henbest, N. "British Astronomy Reaches New Heights on La Palma" in *Astronomy*, Apr. 1985, p. 6.

Horodyski, J. "Reach for the Stars: The Story of Mt. Wilson Observatory" in *Astronomy*, Dec. 1986, p. 6.

Jastrow, R. & Baliunas, S. "Mount Wilson: America's Observatory" in *Sky & Telescope*. Mar. 1993, p. 18.

Keel, W. "Galaxies Through a Red Giant" in *Sky & Telescope*. June 1992, p. 626. [The Soviet 6-meter reflector.]

Krisciunas, K. "Mauna Kea and La Palma: Astronomical Centers of the World" in *Mercury*, Mar/Apr. 1989, p. 34.

Learner, R. "The Legacy of the 200-inch" in *Sky & Telescope*, Apr. 1986, p. 349.

Norris, R. "The Australia [Radio] Telescope" in *Sky & Telescope* Dec. 1988, p. 615.

Osterbrock, D., et al. *Eye On the Sky: Lick Observatory's First Century*. 1988, U. of California Press. Interesting history of the first "big-science" astronomy institution.

Osterbrock, D., et al. "Lick Observatory: The First Century" in *Mercury*, Mar/Apr. 1988, p. 34.

Reeves, R. "Science at McDonald Observatory" in *Astronomy*, July 1987, p. 6.

Ridpath, I. "The William Herschel Telescope" in *Sky & Telescope*, Aug. 1990, p. 136.

Shore, L. "VLA: The Telescope That Never Sleeps" in *Astronomy*, Aug. 1987, p. 15.

Sinnott, R. "The Keck Telescope's Giant Eye" in *Sky & Telescope*, July 1990, p. 15.

Smith, D. "The Submillimeter Giants" in *Sky & Telescope*, Aug. 1985, p. 119.

Teske, R. "Starry, Starry Night: Observing on Kitt Peak" in *Mercury*. Jul/Aug. 1991, p. 115.
Tucker, W. & K. "The Mushrooms of San Augustin: The VLA" in *Mercury*, Sep/Oct 1986, p. 130; Nov/Dec 1986, p. 162.
Wampler, J. "The 153-inch Anglo-Australian Telescope" in *Sky & Telescope*, Oct. 1975, p. 225.
West, R. "Europe's Astronomy Machine" in *Sky & Telescope*, May 1988, p. 471. On the European Southern Observatory's instruments and future plans.

On Specific Telescopes in Orbit

Blair, W. & Gull, T.: "Astro: Observatory in a Shuttle" in *Sky & Telescope*, June 1990, p. 591.
Bohm-Vitense, E. "Observing with IUE" in *Mercury*, Mar/Apr. 79, p. 29.
Cordova, F. & Mason, K. "Exosat: Europe's New X-Ray Satellite" in *Sky & Telescope*, May 1984, p. 397.
Darius, J. "Scanning the Ultraviolet Sky" in *Science 82*, Jul/Aug. 1982, p. 80. Results from IUE.
Engle, M. "Astronomy from the Shuttle" in *Astronomy*, Apr. 1984, p. 66.
Giacconi, R. "The Einstein X-ray Observatory" in *Scientific American*, Feb. 1980.
Kondo, Y., et al. "IUE: 15 Years and Counting" in *Sky & Telescope*. Sep. 1993, p. 30.
Maran, S. "ASTRO: Science in the Fast Lane" in *Sky & Telescope*. June 1991, p. 591. [On the ASTRO mission of the Space Shuttle.]
Overbye, D. "The X-ray Eyes of Einstein" in *Sky & Telescope*, June 1979, p. 527.
Robinson, L. "The Frigid World of IRAS" in *Sky & Telescope*, Jan. 1984, p. 4.
Shore, L. "IUE: Nine Years of Astronomy" in *Astronomy*, Apr. 1987, p. 14.
Tucker, W. & Giacconi, R. "The Birth of X-ray Astronomy" in *Mercury*, Nov/Dec. 1985, p. 178; Jan/Feb. 1986, p. 13.
Tucker, W. *The Star Splitters*. 1984, NASA Special Publication. A book on the HEAO satellites and their results.

Reading Material for the Instructor

Bahcall, J., et al. "What the Longest Exposures from HST Will Reveal" in *Science*, Apr. 13, 1990, p. 178.
Burbidge, G. & Hewitt, A., eds. *Telescopes for the 1980's*. 1981, Annual Reviews.
Carter, W. & Robertson, D. "Studying the Earth by Very-Long-Baseline Interferometry" in *Scientific American*, Nov. 1986.

Chaisson, E. "Early Results from the Hubble Space Telescope" in *Scientific American*. June 1992, p. 44.
Clark, G. "New Instruments for High-energy Astrophysics" in *Physics Today*, Nov. 1982, p. 27.
Coulman, C. "Fundamental and Applied Aspects of Astronomical 'Seeing'" in *Annual Reviews of Astronomy and Astrophysics*, vol. 23, p. 19 (1985).
Garstang, R. "Status and Prospects for Ground-based Observatory Sites" in *Annual Reviews of Astronomy & Astrophysics*, vol. 27, p. 19 (1989)
Hearnshaw, J. *The Analysis of Starlight*. 1986, Cambridge U. Press. A history of spectroscopy.
Kellermann, K. "The Very Long Baseline Array" in *Science*, July 12, 1985, p. 123.
Kondo, Y., et al., eds. *Exploring the Universe with the IUE Satellite*. 1987, Kluwer.
O'Dell, C. "The Hubble Space Telescope Observatory" in *Physics Today*, Apr. 1990, p. 32.
Rohlfs, K. *Tools of Radio Astronomy*. 1986, Springer-Verlag.
Van Allen, J. "Space Science, Space Technology, and the Space Station" in *Scientific American*, Jan 1986.

Recent Audio-Visual Aids

The Countdown to the Invisible Universe (58 min. film or videotape on IRAS from the *Nova* TV series; Coronet Films)
The Hubble Space Telescope (25 slides, Astronomical Society of the Pacific) Images of the launch, instruments, and first scientific photos, with captions and background material.
Hubble Space Telescope Guide Star Catalog (information on 18 million objects on CD-ROM, with software; available from the Astronomical Society of the Pacific)
The Infared Universe: An IRAS Gallery (25 slides, Astronomical Society of the Pacific) With detailed captions and background information by Charles Beichman, IRAS Project Scientist.
Light Pollution (20 slides, Astronomical Society of the Pacific) With captions and background information by David Crawford.
Mirrors on the Universe (1979, 29 min. film, U. of Arizona Film Library) On the history of large telescopes and the building of the MMT.
Optical Effects at Home (5 slides illustrating chromatic aberration, reflections, etc; American Association of Physics Teachers)
The Radio Universe (40 slides, Astronomical Society of the Pacific) Radiographs made with the VLA; comes with 24 page background and captions book.

The Sky at Many Wavelengths (11 slides, Astronomical Society of the Pacific)
Views of the sky in all the bands of the electromagnetic spectrum.
Stars & Telescopes (21 slides, American Association of Physics Teachers)
Naked-eye views of the sky and telescopes.

Topics for Discussion and Papers

1. Spherical Aberration and the Hubble Space Telescope Mirror

The discovery of the spherical aberration error in fabricating the primary mirror for the Hubble Space Telescope led to a tremendous outpouring of media stories about the "broken telescope in space". In actuality, the situation is more complicated and less terrible than the early stories. Students who want to research the issues involved can refer to:

Begley, S., et al. "Heaven Can Wait" in *Newsweek*, July 9, 1990, p. 49.
Fienberg, R. "Space Telescope: Picking Up the Pieces" in *Sky & Telescope*, Oct. 1990, p. 352. An excellent summary.
Villard, R. "The Hubble Space Telescope: A Progress Report" in *Mercury*, Sep/Oct. 1990.
Waldrop, M. "Astronomers Survey Hubble Damage" in *Science*, 13 July 1990, vol. 249, p. 112.

2. Schmidt Telescopes and Sky Surveys

Section 6-2 introduces the use of Schmidt telescopes in taking wide-field surveys of the sky in both hemispheres. These surveys are very valuable references for finding changes in celestial objects and the story of how the surveys were done makes for interesting reading. Students can refer to:

Schombert, J. "Surveying the Northern Sky: The New Palomar Sky Survey" in *Sky & Telescope*, Aug. 1987, p. 128.
Wilson, T. "The New Improved Palomar Sky Survey" in *Astronomy*, Oct. 1986, p. 16.

3. The CCD Revolution

Section 6-3 briefly discusses one of the great changes now taking place in professional astronomy, the movement from photographic plates to CCD's. As electronic detectors become more versatile and less expensive, more and more of the large observatories are switching to CCD's as the light detecting device of choice. Students who want to chronicle the technology and sociology of this change, can refer to:

DiCicco, D. "A Versatile CCD for Amateurs" in *Sky & Telescope*, Sep. 1990, p. 250.
Janesick, J. & Blouke, M. "Sky on a Chip: The Fabulous CCD" in *Sky & Telescope*, Sep. 1987, p. 238.
Kristian, J. & Blouke, M. "Charge-Coupled Devices in Astronomy" in *Scientific American*, Oct. 1982.
McClean, I. *Electronic and Computer-Aided Astronomy: From Eyes to Electronic Sensors*. 1989, Horwood/Wiley. The first introductory book in this field.
Svec, M. "The Birth of Electronic Astronomy" in *Sky & Telescope*. May 1992, p. 497. [On the work of Joel Stebbins with photocells.]

4. The History of Radio Astronomy

The history of radio astronomy makes a fascinating case study for students interested in the changes in modern science and technology. There are many threads students might enjoy following up: Jansky and Reber pursuing their work despite the lack of interest from the astronomical community; the progress in radio (and radar) technology during World War II; the race between astronomers in different countries to detect the "spin-flip" radiation of hydrogen and to build better and better radio telescopes; etc. See:

Hey, J. *The Evolution of Radio Astronomy*. 1973, Neale Watson.
Kraus, J. "Grote Reber, Founder of Radio Astronomy" in *The Journal of the Royal Astronomical Society of Canada*, June 1988, p. 107.
Pfeiffer, J. *The Changing Universe*. 1956, Random House. Good basic introduction.
Spradley, J. "The First True Radio Telescope" in *Sky & Telescope*, July 1988, p. 28. (On Reber's first instrument.)
Sullivan, W., ed. *The Early Years of Radio Astronomy*. 1984, Cambridge U. Press.
Sullivan, W., ed. *Classics in Radio Astronomy*. 1982, Reidel. Reprints of pioneering papers; technical.
Sullivan, W. "Radio Astronomy's Golden Anniversary" in *Sky & Telescope*, Dec. 1982, p. 544.
Sullivan, W. "A New Look at Karl Jansky's Original Data" in *Sky & Telescope*, Aug. 1978, p. 101.

5. Very Long Baseline Interferometry

As discussed in section 6-5, Very Long Baseline Interferometry has made it possible for radio telescopes to attain far better resolution than telescopes with much shorter wavelengths. This has allowed astronomers to probe the details of distant galaxies and quasars with unprecedented detail. Students who want to delve further into this work, should see:

Verschuur, G. *The Invisible Universe Revealed: The Story of Radio Astronomy.* 1987, Springer-Verlag.
Bartusiak, M. "Very Large Astronomy" in *Science 84*, Jul/Aug. 1984, p. 64.
Gordon, M. "VLBA — A Continent Size Radio Telescope" in *Sky & Telescope*, June 1985, p. 487.
Kellermann, K. & Thompson, R. "The Very Long Baseline Array" in *Scientific American*, Jan. 1988.
Smith, D. "Merlin: A Wizard of a Telescope" in *Sky & Telescope*, Jan. 1984, p. 31.
Weiler, K., et al. "Radio Astronomy Looks to Space" in *Astronomy*, May 1988, p. 18.

6. The Kuiper Airborne Observatory

The telescope aboard the Kuiper Airborne Observatory briefly mentioned in section 6-6 has made a number of interesting discoveries, including the ring system around Uranus. Based in Mountain View, California, the KAO has made many trips to other parts of the world, and was one of the instruments brought to bear on Supernova 1987A. See:

Cameron, R. "NASA's 91-cm Airborne Telescope" in *Sky & Telescope*, Nov. 1976, p. 327.
Campins, H. & Lynch, D. "KAO: The Airborne Assault on Comet Halley" in *Astronomy*, Mar. 1986, p. 90.
Elliot, J., et al. "Discovering the Rings of Uranus" in *Sky & Telescope*, June 1977, p. 412.
DiCicco, D. "Fishing on Saturn with the KAO" in *Sky & Telescope*, Nov. 1980, p. 367.

7. The Gamma Ray Observatory and the Universe of Gamma Rays

Gamma-rays have so far been the least explored region of the electromagnetic spectrum, but astronomers hope this is about to change with the launch of the Gamma-Ray Observatory. Students interested in the exotic may enjoy doing a report on the past and future of gamma-ray astronomy. See:

Bignami, G. "Gamma-Ray Astronomy Comes of Age" in *Sky & Telescope*, Oct. 1985, p. 301.
Hurley, K. "What Are Gamma-Ray Bursters?" in *Sky & Telescope*, Aug. 1990, p. 143.
Kniffen, D. "The Gamma Ray Observatory" in *Sky & Telescope May 1991*, p. 488.

Leventhal, M. & MacCallum, C. "Gamma-Ray-Line Astronomy" in *Scientific American*, July 1980.
Neal, V., et al. "Gamma Ray Observatory: The Next Great Observatory in Space" in *Mercury*, Jul/Aug. 1990, p. 98.
Taubes, G. "The Great Annihilator" in *Discover*, June 1990, p. 69.
Weekes, T. "Gamma Rays: The Last Frontier" in *Mercury*, May/June 1981, p. 78.

8. Light Pollution

A major problem facing observatories located near major urban centers (and everyone who enjoys the dark night sky) is the unchecked growth of city and suburban lighting — which floods the sky with unwanted light (which would be far better if it were directed toward the ground.) The brighter the sky, the harder it is to make observations of the most distant (and thus faintest) objects.

Astronomers and environmentalists have joined to urge cities to change lighting ordinances to make top reflectors and less invasive lighting sources a high priority. This is an excellent paper topic for students interested in environmental issues and the connections between science and society. See:

Crawford, D. & Hunter, T. "The Battle Against Light Pollution" in *Sky & Telescope*, July 1990, p. 23.
Hunter, T. & Goff, B. "Shielding the Night Sky" in *Astronomy*, Sep. 1988, p. 47.
Taubes, G. "Twinkle, Twinkle, Great Big Bauble" in *Discover*, Nov. 1987, p. 60. About a ring of light the French wanted to put in orbit.
Sperling, N. "Light Pollution: A Challenge for Astronomers" in *Mercury*, Sep/Oct. 1986, p. 144.
Mood, J. & S. "Palomar and the Politics of Light Pollution" in *Astronomy*, Nov. 1985, p. 6.
Hendry, A. "Light Pollution: A Status Report" in *Sky & Telescope*, June 1984, p. 504.
Sullivan, W. "Our Endangered Night Skies" in *Sky & Telescope*, May 1984, p. 412.
Brunk, B. "Bright Lights Ahead" in *Astronomy*, Apr. 1982, p. 42.

A new slide and information set about light pollution has just been published by the Astronomical Society of the Pacific (in cooperation with the International-al Dark-sky Association.)

9. Mauna Kea and the Keck Telescope

In the last decade, the top of the extinct volcano called Mauna Kea (on the big island of Hawaii) has become a world center for astronomical observations. On this 14,000-foot peak, the seeing is superb, the darkness is guaranteed by

270   Resource Guide

the surrounding national park, and much of the island's weather takes place at atmospheric levels below the telescopes.

Section 6-2 mentions the soon to be completed Keck Telescope (with its unique 10-meter mirror); several other instruments, in the visible and invisible regions of the spectrum, are also in operation or being planned for the summit. Students should refer to:

Berry, R. "The Telescope That Defies Gravity: The Keck Telescope" in *Astronomy*, July 1988, p. 42.
Gustafson, J. & Sargent, W. "The Keck Observatory: 36 Mirrors Are Better Than One" in *Mercury*, Mar/Apr. 1988, p. 43.
Krisciunas, K. "Mauna Kea and La Palma: Astronomical Centers of the World" in *Mercury*, Mar/Apr. 1989, p. 34.
Sinnott, R. "The Keck Telescope's Giant Eye" in *Sky & Telescope*, July 1990, p. 15.
Smith, D. "The Submillimeter Giants" in *Sky & Telescope*, Aug. 1985, p. 119.
Telesco, C., et al. "NASA's New Infrared Eye" in *Sky & Telescope*, Dec. 1980, p. 462.
Waldrop, M. "Mauna Kea: Coming of Age" in *Science*, Dec. 4, 1981, p. 111.

10. Touring the Observatories of the World

Supposing you were a travel agent, trying to put together a world tour of the most interesting observatories in the world and money were no object. What sites would you include on your "must-see" list; which observatories would you include if you had more time? What role would accessibility play in your considerations? Students with an interest in geography or travel really enjoy reflecting on this (pardon the expression) as a paper or panel discussion topic. To help them get started, see:

"*Astronomy* Magazine's 1989 Directory of Observatories, Planetariums, and Museums" in *Astronomy*, May 1989, p. 50.
"A Visitor's Guide to NASA" in *Sky & Telescope*, Feb. 1985, p. 102.
Kirby-Smith, H. *U.S. Observatories: A Directory and Travel Guide*. 1976, Van Nostrand Reinhold.
Krisciunas, K. *Astronomical Centers of the World*. 1988, Cambridge U. Press. History of and guide to major observatories.
Marx, S. & Pfau, W. *Observatories of the World*. 1982, Van Nostrand Reinhold. A tour guide.
Mayer, B. "Ghostly Drummers and Zenith Tubes" in *Astronomy*, Sept. 1981, p. 6. (On touring observatories in England.)
Stott, C. "Greenwich: Where East Meets West" in *Sky & Telescope*, Oct. 1984, p. 300.

11. Selecting Your First Telescope or Binoculars

Some students, inspired by their astronomy class, may find themselves inclined to buy a telescope or good pair of binoculars. But a visit to a large telescope store, or a quick perusal of the ads in *Sky & Telescope* and *Astronomy* may leave them very confused by the wide variety of models, claims, and prices they are likely to encounter.

Thus it can be interesting to have students research the different type of telescopes and binoculars for beginners and to do a "Consumer Report" on what type of telescopes or binoculars might be best for each purpose. Recent issues of *Sky & Telescope* and *Astronomy* have carried thorough reviews of more sophisticated instruments, and several articles have considered telescopes and binoculars for the beginner:

"Astronomy's 1991 Guide to Telescopes" in *Astronomy*, Oct. 1990, special center section.
Berry, R. "A Beginner's Guide to Telescope Types" in *Astronomy*, Aug. 1981, p. 43.
Harrington, S. "Selecting Your First Telescope" in *Mercury*, Jul/Aug. 1982, p. 106.
Levy, D. "Discovering Binocular Astronomy" in *Astronomy*, Dec. 1985, p. 44.
MacRobert, A. "How to Choose a Telescope" in *Sky & Telescope*, Dec. 1983, p. 492.
Morris, K. "Binoculars for Astronomy" in *Astronomy*, Dec. 1986, p. 71.

CHAPTER 7
Our Solar System

Reading Material for the Student

Barnes-Svarney, P. "The Chronology of Planetary Bombardments" in *Astronomy*, July 1988, p. 21.
Chapman, C. & Morrison, D. *Cosmic Catastrophes*. 1989, Plenum. Includes nice sections on the early violent history of the solar system.
Ciaccio, E. "Atmospheres" in *Astronomy*, May 1984, p. 6.
Cohen, M. *In Darkness Born: The Story of Star Formation*. 1988, Cambridge U. Press. Connects star formation to planet formation.
Couper, H. & Henbest, N. *New Worlds: In Search of the Planets*. 1986, Addison-Wesley.
Esposito, L. "The Changing Shape of Planetary Rings" in *Astronomy*, Sep. 1987, p. 6.

Fisher, D. *The Birth of the Earth*. 1987, Columbia U. Press.
Frazier, K. *The Solar System*. 1985, Time-Life. Colorful introduction by a science journalist.
Gore, R. "Between Fire and Ice: The Planets" in *National Geographic*, Jan. 1985, p. 4. A nice overview.
Hartmann, W. "Piecing Together Earth's Early History" in *Astronomy*, June 1989, p. 24.
Janos, L. "Timekeepers of the Solar System" in *Science '80*, May/June 1980, p. 44.
Kelch, J. *Small Worlds: Exploring the 60 Moons of our Solar System*. 1990, Julian Messner.
Miller, R. & Hartmann, W. *The Grand Tour: A Traveler's Guide to the Solar System*. 1981, Workman.
Morrison, D. "Planetary Astronomy in the 1990's" in *Sky & Telescope*, Feb. 1992, p. 151. [Bahcall Committee report.]
Morrison, D. "Return to the Planets: A Blueprint for the Future" in *Astronomy*, Sep. 1983, p. 6.
Morrison, D. & Owen, T. *The Planetary System*. 1988, Addison-Wesley. A superb introductory text on the solar system.
Preiss, B., ed. *The Planets*. 1985, Bantam. An anthology of articles about each planet (by noted planetary astronomers) and science fiction stories inspired by modern planetary exploration.
Reeves, H. "The Origin of the Solar System" in *Mercury*, Mar/Apr. 1977, p.
Sheehan, W. *Worlds in the Sky*. 1992. U. of Arizona Press. A history of planetary astronomy and exploration.
Stewart, J. *Moons of the solar System*. 1991. McFarland. An Encyclopedic almanac.

Reading Material for the Instructor

Baugher, J. *The Space-Age Solar System*. 1988, J. Wiley. An introductory text.
Beatty, J., et al. *The New Solar System*, 3rd ed. 1990, Sky Publishing. A compendium of authoritative articles on planetary astronomy by leading practitioners.
Briggs, G. & Taylor, F. *The Cambridge Photographic Atlas of the Planets*. 1982, Cambridge U. Press.
Carr, M., ed. *The Geology of the Terrestrial Planets*. NASA Special Publication 469.
Hunt, G., ed. *Recent Advances in Planetary Meteorology*. 1985, Cambridge U. Press.
Kasting, J., et al. "How Climate Evolved on the Terrestrial Planets" in *Scientific American*, Feb. 1988.

Klinger, J., et al., eds. *Ices in the Solar System*. 1985, Reidel.
Lada, C. "Energetic Outflows from Young Stars" in *Scientific American*, July 1982.
Lunine, J. "Origin and Evolution of Outer Solar System Atmospheres" in *Science*, July 14, 1979, p. 141.
Morrison, D. & Owen, T. *The Planetary System*. 1988, Addison-Wesley. An introductory text by two noted planetary astronomers.
Ross, J. & Aller, L. "The Chemical Composition of the Sun" in *Science*, Mar. 26, 1976, p. 122.
Smolouchowski, R. *The Solar System*. 1983, Scientific American Books.
Tatarewicz, J. *Space Technology and Planetary Astronomy*. 1990. Indiana U. Press. A careful history of planetary missions.
Taylor, S. *Solar System Evolution*. 1992. Cambridge U. Press.
Veverka, J., et al. *Planetary Geology in the 1980's*. 1985, NASA Special Publication 467.
Wetherill, G. "The Formation of the Earth from Planetesimals" in *Scientific American*, June 1981.
Wetherill, G. "The Formation of Terrestrial Planets" in *Annual Reviews of Astronomy and Astrophysics*, vol. 18, p. 77 (1979).
Young, A. "What Color is the Solar System?" in *Sky & Telescope*, May 1985, p. 399.

Recent Audio-Visual Aids

The Planetary System (100 slides, Astronomical Society of the Pacific) A revised and updated collection of the best planetary images as of 1989, assembled by Dr. David Morrison and Sherwood Harrington.
The Solar System at Your Fingertips (20 slides, Hansen Planetarium) A collection of telescope and spacecraft images of planets and satellites.
"Tales from Other Worlds" from *Planet Earth* series (1985, video from Intellimation, P. O. Box 1922, Santa Barbara, CA 93116) Part of a series on the Earth in a cosmic context.
Worlds in Comparison, 3rd ed. (20 slides, Astronomical Society of the Pacific) 3rd edition has several slides comparing giant planets and their satellites.

Topics for Discussion and Papers

1. The Age of the Earth and the Solar System

Only a century or so ago, many well-educated people would have laughed at the notion—lying at the heart of this chapter—that the Earth and other planets formed some five billion years ago. It wasn't that long ago (in the 17th century) that Archbishop James Ussher suggested that the date of all creation was Oct. 23, 4004 BC—a proposal which seemed quite reasonable to the thinkers of his time. (Alas, even today, there are a few people, especially those trying to argue desperately against evolution, who find that 6,000-year age for the Earth very appealing.)

The progress we have made in dating the Earth and the solar system is a fascinating topic for further exploration (in class or in a paper). See:

Albritton, C. *The Abyss of Time*. 1980, Freeman, Cooper. A scholarly study tracing our ideas about the age of the Earth from 1666 to 1960.

Reese, L., et al. "The Chronology of Archbishop Ussher" in *Sky & Telescope*, Nov. 1981, p. 404.

2. Science Fiction and the Solar System

Traveling to the other planets and satellites of our solar system is one of the main themes of science fiction stories and films. It is fascinating to see how our increasing understanding of the worlds with which we share our solar system is reflected in science fiction. Students who like science fiction might enjoy comparing a modern science fiction view of Mars or Venus to one from the 1920's or 1930's.

Some references to get them started:

Asimov, I., et al. *The Science Fictional Solar System*. 1982, Granada. Scientifically reasonable stories about each planet.

Fraknoi, A. "Science Fiction with Reasonable Astronomy" in *Mercury*, Jan/Feb. 1990, p. 26.

Nicholls, P., ed. *The Science Fiction Encyclopedia*. 1979, Doubleday. A good reference book that points the reader to stories dealing with many topics in science.

Preiss, B., ed. *The Planets*. 1985, Bantam. Science essays and scientifically reasonable stories by leaders in each field.

CHAPTER 8
Our Living Earth

Reading Material for the Student

Note: In finding readings on individual solar system objects you can also consult the appropriate sections in the books cited for the solar system in general in Chapter 7.

Briggs, "Aurora Patrol" in *Air & Space*, Mar. 1989, p. 60. On airborne observations of the aurorae.
Calder, N. *The Restless Earth*. 1972, Viking. Somewhat dated, but a superb introduction to our planet.
Cattermole, P. & Moore, P. *The Story of the Earth*. 1985, Cambridge U. Press.
Chyba, C. "The Cosmic origins of Life on Earth" in *Astronomy*, Nov. 1992, p. 28.
Comins, N. "Life on an Older Earth" in *Astronomy*, Mar. 1993, p. 40.
Cordell, B. "Mars, Earth, and Ice" in *Sky & Telescope*, July 1986, p. 17.
Gillett, S. "The Rise and Fall of the Early Reducing Atmosphere" in *Astronomy*, July 1985, p. 66.
Hartmann, W. "Piecing Together Earth's Early History" in *Astronomy*, June 1989, p. 24.
Heppenheimer, T. "Journey to the Center of the Earth" in *Discover*, Nov. 1987, p. 86.
Lanzerotti, L. & Uberoi, C. "Earth's Magnetic Environment" in *Sky & Telescope*, Oct. 1988, p. 360.
Overbye, D. "The Shape of Tomorrow: Predicting Continental Drift" in *Discover*, Nov. 1982, p. 20.
Reddy, F. "Celestial Winds, Polar Lights" in *Astronomy*, Aug. 1983, p. 6. On the Dynamics Explorer satellite and our magnetosphere.
Strain, P. & Engle, F. *Looking at earth*. 1992, Turner Publishing.
Toon, O. & Olson, S. "The Warm Earth" in *Science '85*, Oct. 1985, p. 50. Why the planet Earth turned out habitable.
Wahr, J. "The Earth's Inconstant Rotation" in *Sky & Telescope*, June 1986, p. 545.
Weisburd, S. "Seismic Journey to the Center of the Earth" in *Science News*, July 5, 1986, p. 10.

Reading Material for the Instructor

Note: In finding readings on individual solar system objects you can also consult the appropriate sections in the books cited for the solar system in general in Chapter 7.

The September 1983 and September 1989 issues of *Scientific American* were devoted to the Earth.

Akasofu, S. "The Dynamic Aurorae" in *Scientific American*, May 1989.
Bloxham, J. & Gubbins, D. "The Evolution of the Earth's Magnetic Field" in *Scientific American*, Dec. 1989.
Brush, S. "The Discovery of the Earth's Core" in *American Journal of Physics*, Sep. 1980, p. 705.

Burgess, E. & Torr, D. *Into the Thermosphere: The Atmosphere Explorers*. 1987, NASA SP-490. History of the exploration of the Earth's upper atmosphere.

Carr, M. & Greeley, R. *Volcanic Features of Hawaii: A Basis for Comparison with Mars*. 1980, NASA SP-403.

Carter, W., et al. "Variations in the Rotation of the Earth" in *Science*, June 1, 1984, p. 957.

Cook, A. *Interiors of the Planets*. 1980, Cambridge U. Press.

Courtillot, V. & Besse, J. "Magnetic Field Reversals, Polar Wanderings, and Core-Mantle Coupling" in *Science*, Sep. 4, 1987, p. 114.

Hoffman, K. "Ancient Magnetic Reversals: Clues to the Geodynamo" in *Scientific American*, May 1988.

Hones, E. "The Earth's Magnetotail" in *Scientific American*, Mar. 1986.

Jacobs, J. *Reversals of the Earth's Magnetic Field*. 1984, Heyden & Sons/Adam Hilger.

Mutter, J. "Seismic Images of Plate Boundaries" in *Scientific American*, Feb. 1986.

Nance, R., et al. "The Supercontinent Cycle" in *Scientific American*, July 1988.

Repetto, R. "Deforestation in the Tropics" in *Scientific American*, Apr. 1990.

van Allen, J. *Origins of Magnetospheric Physics*. 1983, Smithsonian Institution Press.

Wahr, J. "The Earth's Rotation" in *Annual Reviews of Earth and Planetary Sciences*, vol. 16 (1988).

Wetherill, G. "Formation of the Earth" in *Annual Reviews of Earth and Planetary Sciences*, vol. 18 (1990).

Yates, H., et al. "Terrestrial Observations from NOAA Operational Satellites" in *Science*, Jan. 31, 1986, p. 463.

Recent Audio-Visual Aids

Atmospheric Phenomena (Two slide sets, 14 and 5 slides, American Association of Physics Teachers)

The Earth from Space (25 slides, from the hansen Planetarium or the Astronomical Society of the Pacific)

Planet Earth (1985 public television video series, available from Intellimation, P.O. Box 1922, Santa Barbara, CA 93116)

Shuttle Views the Earth: Geology from Space (40 slides, Lunar & Planetary Institute)

Topics for Discussion and Papers

1. Telling from Space Whether the Earth Is Inhabited

An interesting exercise or paper topic is to ask how and at what distance an alien visitor could learn that the Earth bears life. What experiments could be done from the distance of the nearest star, and what can only be understood from orbit around our planet? See:

Shklovskii, I. & Sagan, C. *Intelligent Life in the Universe*. 1966, Holden Day. Chapter 18.
Marvin, U. "The Rediscovery of Earth" in Cornell, J. & Gorenstein, P., eds. *Astronomy from Space*. 1983, MIT Press.
Goldsmith, D. & Owen, T. *The Search for Life in the Universe*. 1980, Benjamin Cummings. See the beginning of Chapter 14.

2. The Acceptance of Continental Drift

Both critics and supporters of current methods of dealing with controversial new ideas in science like to point the story of Alfred Wegener and the notion of continental drift. Supporters of the status quo point to the eventual acceptance of continental drift (once strong, clear evidence became available) as evidence that the system works.

Critics charge that the long period that passed between the time Wegener first proposed his ideas and the time they were generally accepted shows that science—like many human institutions—tends to resist change and is needlessly hostile to unconventional new ideas. Students may want to look into what happened and then compare the situation to the response to controversial new ideas in astronomy.

For more on Wegener, see:

Hallam, A. "Alfred Wegener and the Hypothesis of Continental Drift" in *Scientific American*, Feb. 1975.
Hurley, P. "The Confirmation of Continental Drift" in *Scientific American*, Apr. 1968.
Mauskopf, S., ed. *The Reception of Unconventional Science*. 1979, Westview Press. Has long chapter on continental drift.

3. Continental Drift and Quasars

Today, astronomers and geologists can use radio signals from orbiting satellites and remote cosmic radio sources to make extremely precise measurements of the rate of continental drift. This is a great paper or report topic for students interested in the interaction between sciences. See:

Silverberg, E. & Byrd, D. "A Mobile Telescope for Measuring Continental Drift" in *Sky & Telescope*, May 1981, p. 405.
Wong, C. "Watching Earth Move from Space" in *Sky & Telescope*, Mar. 1978, p. 198.
Carter, W. & Robertson, D. "Studying the Earth by Very-Long-Baseline Interferometry" in *Scientific American*, Nov. 1986.
Schwarzschild, B. "Studying Tectonics with Quasars" in *Physics Today*, Apr. 1981, p. 9. (advanced treatment)

4. The Music of the Earth's Magnetic Field

In 1970, composer Charles Dodge and three physicists produced an unusual recording called *The Earth's Magnetic Field* (Nonesuch records H-71250), in which the musical notes are based on the "Bartels' Diagram" for the Earth's magnetic field activity for the year 1961. Because the changes in the magnetic index of the Earth are caused by solar activity, the composers can claim (on the record jacket) that "...in a real sense, the music...represents the Sun playing on the magnetic field of the Earth." It is strange music, but a clever idea — and probably the only record a student will ever purchase with a brief physics lecture on the cover!

CHAPTER 9
Our Barren Moon

Reading Material for the Student

Note: In finding readings on individual solar system objects you can also consult the appropriate section in the books cited for the solar system in general in Chapter 7.

Beatty, J. "The Making of a Better Moon" in *Sky & Telescope*, Dec. 1986, p. 558.
Brownlee, S. "A Whacky Theory of the Moon's Birth" in *Discover*, Mar. 1985, p. 65.
Cadogan, P. "The Moon's Origin" in *Mercury*, Mar/Apr. 1983, p.34.
Compton, W. *Where No Man Has Gone Before*. 1989, NASA SP-4214. A good history of the Apollo program.
Cooper, H. *Apollo on the Moon* and *Moon Rocks*. 1970, Dial. By a respected science journalist, chronicling the Apollo missions and the analysis of the returned samples.

Cordell, B. "Searching for the Lost Lunar Lakes" in *Astronomy*, Mar. 1993, p. 26.
Cortright, E., ed. *Apollo Expeditions to the Moon*. 1975, NASA SP-350. Story of Apollo as told by astronauts and other NASA personnel involved with the missions.
French, B. *The Moon Book*. 1977, Penquin.
Hartmann, W. "The Moon's Early History" in *Astronomy*, Sep. 1976, p. 6.
Hockey, T. *The Book of the Moon*. 1986, Prentice Hall.
Hurt, H. "I'm at the Foot of the Ladder" in *Astronomy*, July 1989, p. 22. (Review of the Apollo missions).
Kitt, M. "Sculpting the Moon" in *Astronomy*, Feb. 1987, p. 82.
McConnell, D. "Basic Lunar Astrophotography" in *Astronomy*, Dec. 1985, p. 69.
Moore, P. *The Moon*. 1980, Rand McNally. A reference atlas.
Morrison, D. & Owen, T. "Our Ancient Neighbor the Moon" in *Mercury*, May/June 1988, p. 66; Jul/Aug. 1988, p. 98.
Murray, C. & Cox C. *Apollo: The Race to the Moon*. 1989, Simon & Schuster. Popular-level account, from many interviews.
Nichols, R. "From Footprint to Foothold" in *Astronomy*, July 1989, p. 48 (on future Moon missions.)
Register, B. "The Fate of the Moon Rocks" in *Astronomy*, Dec. 1985, p. 15.
Rubin, A. "Whence Came the Moon" in *Sky & Telescope*, Nov. 1984, p. 389.
Schmitt, H. "Exploring Taurus-Littrow: Apollo 17" in *National Geographic*, Sep. 1973. Schmitt was a geologist astronaut.
Stern, A. "Where the Lunar Winds Blow Free" in *Astronomy*, Nov. 1993, p. 36.
Weaver, K. "First Explorers on the Moon: The Incredible Story of Apollo 11" in *National Geographic*, Dec. 1969.
Washburn, M. "The Moon — A Second Time Around" in *Sky & Telescope*, Mar. 1985, p. 209. On lunar bases for the 21st century.
Wood, J. "Exploration of the Moon" in Cornell, J. & Gorenstein, P., eds. *Astronomy from Space*. 1983, MIT Press.

Reading Material for the Instructor

Note: In finding readings on individual solar system objects you can also consult the appropriate section in the books cited for the solar system in general in Chapter 7.

Anderson, D. "The Interior of the Moon" in *Physics Today*, Mar. 1974.
Boss, A. "The Origin of the Moon" in *Science*, Jan. 24, 1986, p. 341.
Cadogan, P. *The Moon: Our Sister Planet*. 1981, Cambridge U. Press.

Cook, A. *The Motion of the Moon*. 1988, Adam Hilger.
El-Baz, F. "Surface Geology of the Moon" in *Annual Reviews of Astronomy & Astrophys.*, vol. 11 (1974).
Goldreich, P. "Tides and the Earth-Moon System" in *Scientific American*, Apr. 1972.
Hartmann, W. *Moons and Planets*, 2nd ed. 1983, Wadsworth.
Hartmann, W., et al., eds. *Origin of the Moon*. 1986, Lunar & Planetary Institute.
Runcorn, S. "The Moon's Ancient Magnetism" in *Scientific American*, Dec. 1987.
The initial scientific reports on the Apollo missions appeared in *Science* magazine, as follows:

Apollo 11: Sep. 19, 1969 and Jan. 30, 1970
Apollo 12: Mar. 6, 1970
Apollo 14: Aug. 20, 1971
Apollo 15: Jan. 28, 1972
Apollo 16: Jan. 5, 1973
Apollo 17: Nov. 16, 1973

Wilhelms, D. *To a Rocky Moon: A Geologist's History of Lunar Exploration*. 1993, U. of Arizona Press.

Recent Audio-Visual Aids

The Origin of the Moon (10 min. film or videotape, Charles Barnett, Motion Picture/Video Production, Los Alamos National Laboratory, Los Alamos, NM 87545) A Cray supercomputer simulation of the recent protoplanetary collision hypothesis for the origin of the Moon.

The Moon Kit (18 slides and extensive booklet; Astronomical Society of the Pacific) Slides emphasize the geology and surface of the Moon, but the booklet has information on phases, folklore, etc.

NASA: The 25th Year (1984, Astronomical Society of the Pacific) A video summarizing the history of NASA with a concise summary of the steps leading to the moon landings and the landings themselves.

One Small Step (1978 episode of NOVA TV series, 60 min, Astronomical Society of the Pacific) On the space program leading to the landing on the Moon.

Topics for Discussion and Papers

1. Observing the Moon Through Telescopes and Binoculars

Since it is the easiest object in the sky to observe, no student should go through an astronomy class without having seen the Moon close-up for him and herself. For those who want to do more extensive Moon observing, here are some useful guides:

Chaikin, A. "A Guided Tour of the Moon" in *Sky & Telescope*, Sep. 1984, p. 211.
DiCicco, D. "Breaking the New Moon Record" in *Sky & Telescope*, Sep. 1989, p. 322.
Kitt, M. "Observe the Apollo Landing Sites" in *Astronomy*, July 1989, p. 66.
Kitt, M. "Eight Lunar Wonders" in *Astronomy*, Mar. 1989, p. 66.
MacRobert, A. "Close-up of an Alien World" in *Sky & Telescope*, July 1984, p. 29.
McConnell, D. "Basic Lunar Astrophotography" in *Astronomy*. Dec. 1985, p. 69.
Price, F. *The Moon Observer's Handbook*. 1989, Cambridge U. Press.
Spain, D. "A Lunar Sampler" in *Astronomy*, Feb. 1984, p. 35. A primer for observers.

2. Were the Landings Worth It?

In this chapter, Kaufmann echoes the sentiments of many scientists when he says "almost everything astronauts did on the moon could have been accomplished years earlier at greatly reduced cost using robots." This makes for a great class discussion topic, especially after students do some research in the books and articles about the Apollo program (see reading list). It can be exciting to have a debate about the value of the manned Moon landings in front of the class, with two teams of well-prepared debaters and then a vote in the audience.

3. Lunar Poetry and Music

The Moon and the Moon landings have inspired their share of poetry and music and students with an interest in these fields can have a good time finding references or recordings. Here are a few suggestions to get them started:

Poetry:

Philips, R., ed. *Moonstruck: An Anthology of Lunar Poetry*. 1974, Vanguard.
Vas Dias, R., ed. *Inside Outer Space*. 1970, Doubleday/Anchor.

Music:

Norgard, Per: *Luna for Orchestra* (Dansk Musik Antologi 018)
Vangelis: "Mare Tranquilitatis" on *Albedo 0.39* (RCA LPL 1-5136)
Journey to the Moon (recordings from the Apollo 11 mission, mixed with music by a rock group) (Buddah BDS 5045)
Cosmology: "Phases of the Moon" on *Cosmology* (Vanguard VSD 79394)

4. Lunacy and the Moon

We know that the Moon, given its proximity and brightness, has some effects on terrestrial life. Biologists have long been familiar with "lunar rhythms" in a variety of organisms, including fireworms, grunions, and salmanders. The human menstrual cycle is another example of a biological rhythm that is timed with the Moon's orbital period.

A much more controversial question is whether the full Moon is connected with increased incidence of strange or aberrant behavior. Anecdotal evidence abounds for the hypothesis and "folk wisdom" is very attached to the idea (note the origin of the word lunacy, for example.) However, careful statistical studies of hospital admissions, homicide records, and other data-bases of human behavior have failed to show any connection between crazy or violent activity and the phases of the Moon. This is a good area for investigation for students majoring in psychology or interested in delusion. Some references:

Abell, G. "Moon Madness" in G. Abell & B. Singer, eds. *Science and the Paranormal*. 1981, Scribners.
Campbell, D. & Beets, J. "Lunacy and the Moon" in *Psychological Bulletin*, vol. 86, p. 1123 (1978).
Culver, R., et al. "Moon Mechanisms and Myths: A Critical Appraisal of Explanations of Purported Lunar Effects on Human Behavior" in *Psychological Reports*, vol. 62, p. 683 (1988).
Martens, R., et al. "Lunar Phase and Birthrate: A 50-year Critical Review" in *Psychological Reports*, vol. 63, p. 923 (1988).
Rotton, J., et al. "Detecting Lunar Periodicities" in *Psychological Reports*, vol 52, p. 111 (1983).
Rotton, J. & Kelly, I. "The Lunacy of It All: Lunar Phases and Human Behavior" in *Mercury*, May/June 1986, p. 73.

Chapter 10
Sun-Scorched Mercury

Reading Material for the Student

Note: In finding readings on individual solar system objects you can also consult the appropriate section in the books cited for the solar system in general in Chapter 7.

Beatty, J. "Mercury's Cool Surprise" in *Sky & Telescope*, Jan. 1992, p. 35.
Chapman, C. "Mercury's Heart of Iron" in *Astronomy*, Nov. 1988, p. 22.
Cordell, B. "Mercury: The World Closest to the Sun" in *Mercury*, Sep/Oct 1984, p. 136.
Davies, M. *Atlas of Mercury*. 1978, NASA SP-423.
Dunne, J. & Burgess, E. *The Voyage of Mariner 10: Mission to Venus and Mercury*. 1978, NASA SP-424.
Hartmann, W. "The Significance of the Planet Mercury" in *Sky & Telescope*, May 1976, p. 307.
Kunzig, R. "Iron Planet" in *Discover*, Feb. 1989, p. 66.
Murray, B. & Burgess, E. *Flight To Mercury*. 1977, Columbia U. Press.
Strom, R. *Mercury: The Elusive Planet*. 1987, Smithsonian Institution Press.
Strom, R. "Mercury: The Forgotten Planet" in *Sky & Telescope*, Sep. 1990, p. 256.

Reading Material for the Instructor

Note: In finding readings on individual solar system objects you can also consult the appropriate section in the books cited for the solar system in general in Chapter 7.

Gault, D., et al. "Mercury" in *Annual Reviews of Astron. & Astrophys.*, vol. 15, p. 97 (1977).
Gingerich, O. "How Astronomers Finally Captured Mercury" in *Sky & Telescope*, Sep. 1983, p. 203.
Hoff, D. & Schmidt, G. "The Rotation of Mercury" in *Sky & Telescope*, Sep. 1979, p. 219. A lab exercise.
Murray, B. "Mercury" in *Scientific American*, May 1976.
The Mariner 10 first mission reports were in *Science* magazine, vol. 185, p. 141. (1974)

Recent Audio-Visual Aids

See the slide sets and videos recommended in Chapter 7.

Topics for Discussion and Papers

1. Double Sunsets on Mercury and "Earth Chauvinism"

Kaufmann's description of a "double sunset" on Mercury (in section 10-2) could serve as good springboard for a discussion of "Earth chauvinism" and not taking for granted what happens on Earth. As students read more about the planets in the next chapters, it might be instructive to make a list of what "normal" Earth phenomena would be different if they lived on one of the other worlds with which we share the solar system.

2. Mercury's Perihelion and the General Theory of Relativity

As discussed in section 4-7, explaining the advance of Mercury's perihelion was one of the great triumphs of Einstein's General Theory. A variety of other ideas had been put forward to explain Mercury's anomalous motion, including the presence of another planet closer to the Sun. The hypothesized planet was even given a name — Vulcan, after the Roman god of fire (a name that "Star Trek" fans in your classes will recognize.) Students who want to explore this interesting history can consult:

Fontenrose, R. "In Search of Vulcan" in *Journal for the History of Astronomy*, vol. 4, p. 145 (1973).

CHAPTER 11
Cloud-Covered Venus

Reading Material for the Student

Note: In finding readings on individual solar system objects you can also consult the appropriate section in the books cited for the solar system in general in Chapter 7.

Allen, D. "Laying Bare Venus' Dark Secrets" in *Sky & Telescope*, Oct. 1987, p. 350.
Basilevsky, A. "The Planet Next Door" in *Sky & Telescope*, Apr. 1989, p. 360.
Beatty, J. "A Radar Tour of Venus" in *Sky & Telescope*, June 1985, p. 507.

Beatty, J. "A Soviet Space Odyssey" in *Sky & Telescope*, Oct. 1985, p. 310. About the Vega missions.
Beatty, J. "Radar Views of Venus" in *Sky & Telescope*, Feb. 1984, p. 110; "Report from a Torrid Planet," May 1982, p. 452.
Beatty, J. "Working Magellan's Magic" in *Sky & Telescope*, Aug. 1993, p. 16.
Burgess, E. *Venus: An Errant Twin*. 1985, Columbia U. Press.
Burnham, R. "What Makes Venus Go?" in *Astronomy*, Jan 1993, p. 40.
Chapman, C. "The Vapors of Venus and Other Gassy Envelopes" in *Mercury*, Sep/Oct. 1983, p. 130.
Cordell, B. "Venus" in *Astronomy*, Sep. 1982, p. 6.
Esposito, L. "Does Venus Have Active Volcanoes" in *Astronomy*, July 1990, p. 42.
Fimmel, R., et al. *Pioneer Venus*. 1982, NASA SP- 461.
Kunzig, R. "Voyage to Venus" in *Discover*, Apr. 1989, p. 54. On the Magellan mission.
Mims, S. "Revealing the Venusian Secrets" in *Astronomy*, July 1982, p. 66.
Montoya, E. & Fimmel, R. "Pioneers in Space" in *Mercury*, Mar/Apr. 1988, p. 56.
Sagan, C. "Heaven and Hell" in *Cosmos*. 1980, Random House. Eloquent chapter comparing Venus and Earth.
Saunders, S. "The Exploration of Venus: A Magellan Progress Report" in *Mercury*, Sep/Oct. 1991, p. 130.
Saunders, S. "Venus: A Hellish Place Next Door" in *Astronomy*, Mar. 1990, p. 18.
Stofan, E. "The New Face of Venus" in *Sky & Telescope*, Aug. 1993, p. 22.
Stooke, P. "The Earliest Maps of Venus" in *Sky & Telescope*, Aug. 1992, p. 157.
Wall, S. "Venus Unveiled" in *Astronomy*, Apr. 1989, p. 26.

Reading Material for the Instructor

Note: In finding readings on individual solar system objects you can also consult the appropriate section in the books cited for the solar system in general in Chapter 7.

Barbato, J. & Ayer, E. *Atmospheres*. 1981, Pergamon Press.
Baskilevsky, A. & Head, J. "The Geology of Venus" in *Annual Reviews of Earth and Planetary Sciences*, vol. 16 (1988).
Hubbard, W. *Planetary Interiors*. 1984, Van Nostrand Reinhold.
Hunten, D., et al. *Venus*. 1983, U. of Arizona Press.
Kasting, J., et al. "How Climate Evolved on the Terrestrial Planets" in *Scientific American*, Feb. 1988.

Pettengill, G., et al. "The Surface of Venus" in *Scientific American*, Aug. 1980.
Pieters, C., et al. "The Color of the Surface of Venus" in *Science*, Dec. 12, 1986.
Prinn, R. "The Volcanoes and Clouds of Venus" in *Scientific American*, Mar. 1985.
Schubert, G. & Covey, C. "The Atmosphere of Venus" in *Scientific American*, July 1981.

Recent Audio-Visual Aids

"Heaven and Hell" (one part of the *Cosmos* television series, available in home video format from the Astronomical Society of the Pacific)
Venus Kit (6 slides and booklet, Astronomical Society of the Pacific) Newly processed images and 28-page booklet of background, captions, activities, and resources.
Venus Pioneer (1979, NASA film NAV-042, 28 min) On the Pioneer Venus mission and early results.
Venus Unveiled: The Magellan Images (20 slides, Astronomical Society of the Pacific) Images and extensive informatio from the first year of Magellan radar studies of Venus.

Topics for Discussion and Papers

1. Venus in Science Fiction

For decades, science fiction writers (and scientists) thought of Venus as a slightly hotter twin of Earth. Edgar Rice Burroughs (the creator of Tarzan), for example, portrayed Venus as a lush jungle planet, where lightly clad humans and unusual beasts moved about with ease. As students who are science fiction fans will attest, it is much harder to do stories about Venus as we understand it today. You might ask students to find and evaluate such stories, or to try their own hand at writing one. Some examples to start with:

Niven, L. "Becalmed in Hell" in *All the Myriad Ways*. 1971, Ballantine.
Sheffield, C. "Dinsdale Dissents" in *Vectors*. 1979, Ace.
Varley, J. "In the Bowl" in *The Persistence of Vision*. 1978, Dell.

2. Magellan Results

As the current editions of *Universe* and this Instructor's Guide were going to press, the Magellan spacecraft, despite occasional communications problems, was sending back superb high-resolution radar images of the Venusian surface. Students may want to look in recent news and science magazines, and put together a special report or panel on what new discoveries have been made since the textbook came out. Such an exercise can also serve as good reminder that our understanding of the cosmos is continuously changing as new instruments provide new data against which to test our hypotheses.

3. The Transits of Venus

Section 11-1 gives a brief introduction to the epochal measurements that have been made during the transits of Venus across the face of the Sun. For students interested in the history of science, the transit expeditions and measurements can make fascinating topics for research. Here are a few references to get them started:

Badger, G., ed. *Captain Cook, Navigator and Scientist.* 1970, Humanities Press. Includes a chapter by R. Woolley on transits of Venus.
Fernie, D. *The Whisper and the Vision.* 1976, Clarke Irwin. See the chapter entitled "Transits and Tribulations".
Ferris, T. "A Plumb Line to the Sun" in *Mercury*, May/June 1989, p. 66.
Woolf, H. *The Transits of Venus.* 1959, Princeton U. Press.

CHAPTER 12
The Martian Invasions

Reading Material for the Student

Note: In finding readings on individual solar system objects you can also consult the appropriate section in the books cited for the solar system in general in Chapter 7.

Ashbrook, J. "Asaph Hall Finds the Moons of Mars" in *Sky & Telescope*, July 1977, p. 20.
Beatty, J. "The Amazing Olympus Mons" in *Sky & Telescope*, Nov. 1982, p. 420.
Burnham, R. "New Views of Mars and Phobos" in *Astronomy*, Sep. 1989, p. 28.
Carr, M. "The Surface of Mars: A Post-Viking View" in *Mercury*, Jan/Feb. 1983, p. 2.
Carroll, M. "Digging Deeper for Life on Mars" in *Astronomy*, Apr. 1988, p. 6.
Carroll, M. "The Changing Face of Mars" in *Astronomy*, Mar. 1987, p. 6.
Carroll, M. "The First Colony on Mars" in *Astronomy*, June 1985, p. 6.
Cattermole, P. *Mars.* 1992, Chapman & Hall.

Chaikin, A., et al. "Mars or Bust" in *Discover*, Sep. 1984, p. 12. (About ideas for a manned mission.)
Cordell, B. "Mars, Earth, and Ice" in *Sky & Telescope*, Jul. 1986, p. 17.
Disk, S. "Discovering the Moons of Mars" in *Sky & Telescope*, Sep. 1988, p. 242.
Edgett, K., et al, "The Sands of Mars" in *Astronomy*, June 1993, p. 26.
Goldman, S. "The Legacy of Phobos 2" in *Sky & Telescope*, Feb. 1990, p. 156.
Hartmann, W. "What's New on Mars" in *Sky & Telescope*, May 1989, p. 471.
Horowitz, N. *To Utopia and Back*. 1986, Freeman. On the search for life on Mars with the Viking lander instruments.
Kargel, J. & Strom, R. "Ice Ages of Mars" in *Astronomy*, Dec. 1992, p. 40.
Kieffer, H., et al, eds. *Mars*. 1992, U. of Arizona Press.
McKay, C. "Did Mars Once Have Martians?" in *Astronomy*, Sep. 1993, p. 26.
McKay, C. "Return to the Red Planet: Mars Observer" in *Mercury*, Sep/Oct. 1992, p. 146.
Moore, P. "Mars—Then and Now" in *Mercury*, Mar/Apr. 1980, p. 23. Good history of how we learned about Mars.
Preiss, B., ed. *The Planets*. 1985, Bantam. Has a nice chapter on Mars by M. Carr.
Robinson, M. "Surveying the Scars of Ancient Martian Floods" in *Astronomy*, Oct. 1989, p. 38.
Sagan, C. "Blues for a Red Planet" in *Cosmos*. 1980, Random House.
Sinnott, R. "Mars Mania of Oppositions Past" in *Sky & Telescope*, Sep. 1988, p. 244.
Squyres, S. "Searching for the Waters of Mars" in *Astronomy*, Aug. 1989, p. 20.
Stooke, P. "Sizing up Phobos" in *Sky & Telescope*, May 1989, p. 477.

Reading Material for the Instructor

Note: In finding readings on individual solar system objects you can also consult the appropriate section in the books cited for the solar system in general in Chapter 7.

Arvidson, R., et al. "The Surface of Mars" in *Scientific American*, Mar. 1978.
Baker, V. *The Channels of Mars*. 1982, U. of Texas Press.
Batson, R., et al. *The Atlas of Mars*. 1979, NASA SP-438.
Carr, M. *The Surface of Mars*. 1981, Yale U. Press.
Carr, M. *The Geology of the Terrestrial Planets*. 1984, NASA Special Publication 469.
Ezell, E. & L. *On Mars: Exploration of the Red Planet 1958-1978*. 1984, NASA Special Publication 4212. A history of NASA's Mars missions.

Gingerich, O. "The Orbit of Mars: A Lab Exercise" in *Sky & Telescope*, Oct. 1983, p. 300.

Gingerich, O. "The Satellites of Mars: Prediction and Discovery" in *Journal for the History of Astronomy*, vol. 1, p. 109 (1970).

Haeberle, R. "The Climate of Mars" in *Scientific American*, May 1986.

Kasting, J., et al. "How Climate Evolved on the Terrestrial Planets" in *Scientific American*, Feb. 1988.

Leovy, C. "Martian Meteorology" in *Annual Reviews of Astron. & Astrophys.*, vol. 17, p. 387 (1979).

McKay, C., ed. *The Case for Mars II*. 1985, Univelt/American Astronautical Society. Proceedings of another conference on future Mars exploration.

Schuta, P. "Polar Wandering on Mars" in *Scientific American*, Dec. 1985.

Wells, R. *Geophysics of Mars*. 1979, Elsevier.

The early results from the Viking mission were reported in two special issues of *Science* magazine: Aug. 27, 1976 and Oct. 1, 1976.

Recent Audio-Visual Aids

Astronomers of the Past (50 slides, Astronomical Society of the Pacific) Includes portraits of Barnard, Hall, Lowell, et al.

"Blues for a Red Planet" (episode of the *Cosmos* TV series, now available in inexpensive home video format from the Astronomical Society of the Pacific)

Mars (20 Viking slides, Planetary Society)

Mars Kit (6 slides with background material and classroom activities, Astronomical Society of the Pacific)

The Planetary System (100 slides plus captions book by David Morrison, Astronomical Society of the Pacific) Includes many of the best Mars images.

Star Wars or Mission to Mars? (60 min. audiotape of a speech by Carl Sagan, National Public Radio Cassette Publishing, 2025 M St., NW, Washington, DC 20036) Order number NP-860925.

Volcanoes on Mars (20 slides, Lunar & Planetary Institute)

Topics for Discussion and Papers

1. Schiaparelli, Lowell, and the Martian Canals

The history of how the Martian canals were "discovered" and how Percival Lowell brought them so effectively into the public consciousness makes for instructive and thought-provoking reading for anyone interested in the way scientific (and not-so-scientific) ideas are debated and decided. Some references:

Hoyt, W. *Lowell and Mars*. 1976, U. of Arizona Press. The definitive study.
Sheehan, W. *Planets and Perception*. 1988, U. of Arizona Press. An interesting study, with much material on Mars, on how our eyes and minds can play tricks in the way we see the planets.
Sheehan, W. "Mars 1909: Lessons Learned" in *Sky & Telescope*, Sep. 1988, p. 247.
Shklovksii, I. & Sagan, C. *Intelligent Life in the Universe*. 1966, Holden-Day. See Chapter 20 for a good discussion of the search for life on Mars.
Tenn, J. "Giovanni Schiaparelli" in *Mercury*, Jul/Aug. 1990, p. 116.

Lowell's works had a direct influence on H.G. Wells whose popular novel *War of the Worlds* further enhanced the public suspicion that Mars must be inhabited. And if Lowell led to Wells, then Wells led to Welles—Orson Welles, whose infamous 1938 Halloween Eve radio dramatization of *War of the Worlds* managed to frighten many East Coast listeners into believing the Martians had landed. A nice study of the broadcast by a psychologist is:

Cantril, H. *The Invasion from Mars: A Study in the Psychology of Panic*. 1940, 1966, Harper Torchbook.

The Welles radio drama can be found on records and tapes (e.g. Murray Hill Records S44217); a recent re-enactment by the staff at National Public Radio was not as effective, but may be easier to find.

2. The Viking Project

Students may want to do further research on the Viking project for several reasons. It is an excellent example of a successful program that was able to accomplish a great deal of scientific research with robot devices. Its planning and execution have been documented in great detail. But also, it is a case study in the way an attractive (even romantic) hypothesis—the presence of some life in the Martian soil—is confronted by new evidence and forces its adherents to re-evaluate their views—and, if possible, regroup.

Cooper, H. *The Search for Life on Mars*. 1980, Holt Rinehart and Winston.
Ezell, E. & L. *On Mars: Exploration of the Red Planet 1958-1978*. 1984, NASA Special Publication 4212. A history of NASA's Mars missions.
Gore, R. "Sifting for Life in the Sands of Mars" in *National Geographic*, Jan. 1977.
Horowitz, N. *To Utopia and Back*. 1986, W. H. Freeman.
Washburn, M. *Mars at Last!* 1977, Putnams.

3. Manned Missions to Mars

Space advocacy groups are currently mounting a strong campaign that the U.S. set a policy goal of sending a manned mission to Mars early in the next century. Several scientists and groups have recommended doing such a mission not as a national project but as a world-wide cooperative venture, taking advantage of the new world order that is emerging from the changes in Europe. On the other hand, many people feel that at a time of constrained federal budgets and serious world problems, the expense of a manned Mars mission is best left to future generations.

Students who want to do a paper or panel discussion on this topic can consult the references below for information. (Note that they mainly present the perspective of the advocates of a manned mission; it will be important to search out articles in news magazines—using periodical indexes—that present the views of the side that urges robotic instead of human missions if we want to explore the red planet further.)

Boston, P., et al, eds. *The Case for Mars*. 1984, Univelt/American Astronautical Soc. Proceedings of a conference on manned missions to Mars. (Technical in places.)
Carroll, M. "First Colony on Mars" in *Astronomy*, June 1985, p. 6.
Chaikin, A., et al. "Mars or Bust" in *Discover*, Sep. 1984, p. 12.
Cole, S., et al. "Return to Mars" in *Astronomy*, Nov. 1987, p. 26.
McKay, C., ed. *The Case for Mars II*. 1985, Univelt/American Astronautical Society. Proceedings of another conference on future Mars exploration. (Somewhat technical.)
Oberg, J. *Mission to Mars*. 1982, Stackpole Books.
Powers, R. *Mars: Our Future on the Red Planet*. 1986, Houghton-Mifflin. A polemic for the manned exploration of Mars.

4. Seeing Mars in 3-D

The Viking cameras were able to produce a series of stereoscopic images of the planet (in black and white), which can be examined with special glasses to give the first-ever 3-D views of another planet. Two special NASA publications include 3-D glasses for just this purpose:

The Martian Landscape (1978, NASA SP-425)
Viking Orbiter Views of Mars (1980, NASA SP-441)

Both are out of print, but can sometimes still be found at Government Printing Office (federal) bookstores and through used book-dealers. Also, many university libraries have Government Document departments which carry the full line of (unsung) NASA Special Publications.

Also see:

Wall, S., et al. "Stereoscopic Views of Mars" in *Mercury*, Sep/Oct. 1977, center insert.

5. Mars in Science Fiction

Mars has been a favorite setting for science fiction stories even before H. G. Wells. But the influence of Schiaparelli and Lowell dominated for much of the latter part of the 19th century and the first half of the 20th century. It is interesting to see how the science fiction views of Mars evolved in the second half of the 20th century as the scientific exploration of Mars showed a far less hospitable fourth planet than science fiction fans had hoped. A good reference and some stories to consult:

Hipolito, J. & McNelly, W., eds. *Mars We Love You*. 1971, Pyramid Books. An excellent collection of science and science fiction about Mars from earlier decades.
Pesek, L. *The Earth is Near*. 1970, Dell. Novel about a realistic expedition to Mars and the problems they would face.
Pohl, F. *Man Plus*. 1976, Random House. A novel in which humans are bioengineered to survive on Mars.
Varley, J. "In the Hall of Martian King" in *The Persistence of Vision*. 1978, Dell. A fanciful story about Mars adapting to Earth colonists.

CHAPTER 13
Jupiter: Lord of the Planets

Reading Material for the Student

Note: In finding readings on individual solar system objects you can also consult the appropriate section in the books cited for the solar system in general in Chapter 7.

Bennett, G. "Return to Jupiter" in *Astronomy*, Jan. 1987, p. 6. Burgess, E. *By Jupiter*. 1982, Columbia Press.
Chapman, C. "The Discovery of Jupiter's Red Spot" in *Sky & Telescope*, May 1968, p. 276.
Davis, J *Flyby*. 1987, Atheneum. About the Voyager mission, with emphasis on Uranus.
Elliott, J. & Kerr, R. "How Jupiter's Ring Was Discovered" in *Mercury*, Nov/Dec 1985, p. 162.

Elliot, J. & Kerr, R. *Rings: Discoveries from Galileo to Voyager.* 1985, MIT Press.
Esposito, L. "The Changing Shape of Planetary Rings" in *Astronomy*, Sep. 1987, p. 6.
Fimmel, R., et al. *Pioneer: First To Jupiter, Saturn, and Beyond.* 1980, NASA SP-446.
Gore, R. "Voyager Views Jupiter" in *National Geographic*, Jan. 1980.
Montoya, E. & Fimmel, R. "Pioneers in Space: The Story of the Pioneer Missions, part 2" in *Mercury*, May/June 1988, p. 81.
Morrison, D. & Samz, J. *Voyage to Jupiter.* 1980, NASA SP-439. Excellent book on the Voyager mission by one of the scientists.
Smith, B. "Voyager to the Giant Planets" in Cornell, J & Gorenstein, P., eds. *Astronomy From Space.* 1983, MIT Press.
Veverka, J. "Jupiter's World: A Colossal Realm" in Preiss, B., ed. *The Planets.* 1985, Bantam.
Washburn, M. *Distant Encounters: The Exploration of Jupiter and Saturn.* 1983, Harcourt, Brace, Jovanovich.

Reading Material for the Instructor

Note: In finding readings on individual solar system objects you can also consult the appropriate section in the books cited for the solar system in general in Chapter 7.

Atreya, S. *Atmospheres and Ionospheres of the Outer Planets and their Satellites.* 1986, Springer-Verlag.
Dessler, A., ed. *Physics of the Jovian Magnetosphere.* 1983, Cambridge U. Press.
Hunt, G. "Atmospheres of the Giant Planets" in *Annual Reviews of Earth & Planetary Sciences*, vol. 11 (1983).
Stern, D. & Ness, N. "Planetary Magnetospheres" in *Annual Reviews of Astronomy & Astrophysics*, vol. 20, p. 139 (1982).
Stevenson, D. "Interiors of the Giant Planets" in *Annual Reviews of Earth & Planetary Sciences*, vol. 10 (1982).
Yeates, C., et al. *Galileo: Exploration of Jupiter's System.* 1985, NASA Special Publication.

Recent Audio-Visual Aids

Hubble Space Telescope (2 slide sets from the Astronomical Society of the Pacific) Feature some superb images of Jupiter and Saturn taken with the HST.
Jupiter (26 Voyager 1 slides, American Association of Physics Teachers)

The Planetary System (100 slides with detailed captions by David Morrison, Astronomical Society of the Pacific) Includes some of the best Jupiter images.

"Travelers' Tales" (an episode of the *Cosmos* TV series; now available in home video format from the Astronomical Society of the Pacific)

Voyager Gallery (Videodisc with thousands of images and videos, produced by Optical Data Corp. and distributed by the Astronomical Society of the Pacific)

Voyager Missions to Jupiter and Saturn (1983, NASA Film, on video from the Astronomical Society of the Pacific and several other vendors)

Voyager 1 & 2 at Jupiter (40 slides and cassette, produced by Holiday Films, distributed by the Planetary Society)

Topics for Discussion and Papers

1. The Technology of Voyager

Students with an interest in engineering and technology may want to do a report on the Voyager mission from that perspective. It was a remarkable technological achievement, especially after the aging spacecraft got to Uranus and Neptune. See the Voyager books in the reading list above (as well as in the Saturn and Uranus chapters), as well as the following:

Laeser, R., et al. "Engineering Voyager 2 Encounter with Uranus" in *Scientific American*, Nov. 1986.

Littmann, M. "The Triumphant Grand Tour of Voyager 2" in *Astronomy*, Dec. 1988, p. 34.

Miner, E. "Voyager 2 and Uranus" in *Sky & Telescope*, Nov. 1985, p. 420. On preparations for the encounter.

Murill, M. "The Grandest Tour: Voyager" in *Mercury*, May/June 1993, p. 66.

Smith, B. "Voyage of the Century" in *National Geographic*, Aug. 1990, p. 48.

In addition, many of the articles on the Neptune encounter (see Chapter 16) emphasized the engineering feats of the mission.

2. Was Voyager Worth It?

Astronomer David Morrison has calculated that the Voyager program cost each U.S. taxpayer about the price of a six-pack of beer. Students may enjoy comparing its costs to other space missions and other types of federal programs in general. A class panel debating the value of such missions of pure exploration to a country or a species might make for an interesting and thought provoking period.

3. Jupiter in Science Fiction

It is not easy to construct a viable story set in an environment as alien and hostile as Jupiter itself, but a number of science fiction writers have tried. See, for example:

Pohl, C. & F. *Jupiter*. 1973, Ballantine. A collection of stories about the giant planet.
Moffitt, D. *The Jupiter Theft*. 1977, Ballantine. Aliens come to steal the hydrogen in Jupiter's atmosphere.
Varley, J. *The Ophiuchi Hotline*. 1977, Dell. Terrestrial and jovian planet life-forms find themselves at odds.

4. The Galileo Mission

As of this writing, the Galileo mission is successfully on its way to Jupiter. As Kaufmann suggests, students may want to consult astronomy and science news magazines about its progress. Here are some articles to get them started:

Bennett, G. "Return to Jupiter" in *Astronomy*, Jan. 1987, p. 6.
Carroll, M. "Project Galileo: The Phoenix Rises" in *Sky & Telescope*, Apr. 1987, p. 359.
Harris, J. "Return to a New World" in *Astronomy*, Apr. 1990, p. 30; "Bound for Jupiter," Jan. 1990, p. 46.
Johnson, T. & Yeates, C. "Return to Jupiter: Project Galileo" in *Sky & Telescope*, Aug. 1983, p. 99. (Now a bit dated.)
Yeates, C., et al. *Galileo: Exploration of Jupiter's System*. 1985, NASA Special Publication 479.

CHAPTER 14
The Galilean Satellites of Jupiter

Reading Material for the Student

Note: In finding readings on individual solar system objects you can also consult the appropriate section in the books cited for the solar system in general in Chapter 7.

Many of the books and articles recommended for Chapter 13 also include material on the satellites of Jupiter. They are not repeated below.

Cruikshank, D. "Barnard's Satellite of Jupiter" in *Sky & Telescope*, Sep. 1982, p. 220. [On Amalthea]

Griffin, R. "Barnard and His Observations of Io" in *Sky & Telescope*, Nov. 1982, p. 428.
Hartmann, W. "View from Io" in *Astronomy*, May 1981, p. 17.
Morrison, D. "Four New Worlds" in *Mercury*, May/June 1980, p. 53.
Morrison, D. "An Enigma Called Io" in *Sky & Telescope*, Mar. 1985, p. 198.
Morrison, N. & D. "The Volcanoes of Io: Still Erupting" in *Mercury*, Jan/Feb. 1984, p. 23.
Simon, S. "The View from Europa" in *Astronomy*, Nov. 1986, p. 98.
Talcott, R. "The Violent Volcanoes of Io" in *Astronomy*, May 1993, p. 40.
Veverka, J. "Jupiter's World: A Colossal Realm" in Preiss, B., ed. *The Planets*. 1985, Bantam.

Reading Material for the Instructor

Note: In finding readings on individual solar system objects you can also consult the appropriate section in the books cited for the solar system in general in Chapter 7.

Many of the books and articles recommended for Chapter 13 also include material on the satellites of Jupiter. They are not repeated below.

Atreya, S. *Atmospheres and Ionospheres of the Outer Planets and their Satellites*. 1986, Springer-Verlag.
Johnson, T. & Soderblom, L. "Io" in *Scientific American*, Dec. 1983.
Morrison, D., ed. *Satellites of Jupiter*. 1982, U. of Arizona Press.
Morrison, D. "The Satellites of Jupiter and Saturn" in *Annual Reviews of Astron. & Astrophys.*, vol. 20, p. 469 (1982).
Soderblom, L. "The Galilean Moons of Jupiter" in *Scientific American*, Jan. 1980.
Squyres, S. "Ganymede and Callisto" in *American Scientist*, Jan/Feb. 1983.

Recent Audio-Visual Aids

Moons of Jupiter (26 Voyager 1 slides, American Association of Physics Teachers)
The Planetary System (100 slides, with detailed captions with David Morrison, Astronomical Society of the Pacific)
Voyager 1 & 2 at Jupiter (40 slides and cassette, produced by Holiday Films, distributed by the Planetary Society)

Topics for Discussion and Papers

1. Science Fiction Stories Using the Galilean Satellites

As soon as the Voyager images and data showed the Galilean satellites to be interesting, individual worlds, each with unique characteristics, science fiction writers began to find ways to use them in stories. Students who are science fiction fans might try to see how many stories they can find, written in the last decade or so, that use the satellites of Jupiter as locales or plot elements.

Some examples:

Benford, G. *Against Infinity*. 1983, Pocket Books. Ganymede is being "terrraformed" to make it more hospitable.
Benford, G. "The Future of the Jovian System" in B. Preiss, ed. *The Planets*. 1985, Bantam. About settling and exploiting the Jupiter system.
Clarke, A. *2010*. 1982, Ballantine. Life is discovered under the ice of Europa.

Many of these stories revolve around people adapting to the harsh environment of these satellites or trying to change that environment over time. Students may want to try their hand at writing their own story (or engineering proposals) along these lines.

2. Music and the Galilean Satellites

Following in the Keplerian tradition we discussed in topic 4.1, German composer Gunter Bergmann has written a piece called *The Harmony of the World of Jupiter*, based on the orbits and motions of the Galilean satellites. It was available on an imported recording on the Schwann record label (ams studio 624), played on the organ by Winfried Berger.

3. How Short a Month?

Currently, the innermost satellite of Jupiter still holds the record for the shortest "month" in the solar system (although not by much). A number of the inner satellites of Jupiter, Uranus, and Neptune have months shorter than the "day" on their planets. Students may want to discuss what the planet might look like from such a speeding moon, and what such a moon's motion might look like from a platform floating in the atmosphere of the planet it circles.

Chapter 15
The Spectacular Saturnian System

Reading Material for the Student

Note: In finding readings on individual solar system objects you can also consult the appropriate section in the books cited for the solar system in general in Chapter 7.

Note: Many of the general books and articles on the Voyager program, listed in Chapter 13, also have sections on the exploration of Saturn.

Alexander, A. *The Planet Saturn: A History of Observations, Theory, and Discovery*. 1962, Faber & Faber, reprinted by Dover.

Beatty, J. "Reports on the Voyage Encounters with Saturn" in *Sky & Telescope*, Jan., Oct., and Nov. 1981.

Beatty, J. "Rendezvous with Saturn [Pioneer 11]" in *Sky & Telescope*, Nov. 1979, p. 401.

Berry, R. & Burnham, R. "Voyager 2 at Saturn" in *Astronomy*, Nov. 1981, p. 6.

Burnham, R. "The Saturnian Satellites" in *Astronomy*, Dec. 1981, p. 6.

Croswell, K. "The Titan/Triton Connection" in *Astronomy*, Apr. 1993, p. 26.

Cuzzi, J. "Ringed Planets: Still Mysterious" in *Sky & Telescope*, Dec. 1984, p. 511; Jan. 1985, p. 19.

Davis, J. *Flyby*. 1987, Atheneum. On the Voyager mission, with emphasis on Uranus.

Elliot, J. & Kerr, R. *Rings: Discoveries from Galileo to Voyager*. 1985, MIT Press.

Esposito, L. "The Changing Shape of Planetary Rings" in *Astronomy*, Sep. 1987, p. 6.

Gore, R. "Saturn: Riddle of the Rings" in *National Geographic*, July 1981.

Greenberg, S. "Interview with David Morrison and Dale Cruikshank: Voyager at Saturn" in *Mercury*, Jan/Feb. 1981, p. 8.

Kidger, M. "The Great White Spot of Saturn" in *1993 Yearbook of Astronomy*, ed. P. Moore, 1992, Norton, p. 176.

Morrison, D. "The New Saturn System" in *Mercury*, Nov/Dec. 1981, p. 162.

Morrison, D. *Voyages to Saturn*. 1982, NASA-SP 451. Good, well-illustrated guide by one of the Voyager scientists.

Morrison, D. "Saturn: A Ringed World" in Preiss, B., ed. *The Planets*. 1985, Bantam.

O'Meara, S. "Saturn's Great White Spot Spectacular" in *Sky & Telescope*, Feb. 1991, p. 144. On HST observations of a giant storm system.

Overbye, D. "Lord of the Rings" in *Discover*, Jan. 1981, p. 24.

Owen, T. "Titan" in *Scientific American*, Feb. 1982.
Pollack, J. & Cuzzi, J. "Rings in the Solar System" in *Scientific American*, Nov. 1981.
Robertson, D. "Cassini" in *Astronomy*, Sep. 1987, p. 20. On a proposed return mission to Saturn.
Sanchez-Lavega, A. "Saturn's Great White Spots" in *Sky & Telescope*, Aug. 1989, p. 141.
Veverka, J. "Measuring the Size of Saturn's Satellites" in *Sky & Telescope*, Dec. 1975, p. 356. How it was done before Voyager.
van Helden, A. "Saturn and His Anses" and "The Solution to the Problem of Saturn" in *Journal for the History of Astronomy*, vol. 5, p. 105,155 (1974).

Reading Material for the Instructor

Note: In finding readings on individual solar system objects you can also consult the appropriate section in the books cited for the solar system in general in Chapter 7.

Note: Many of the general books and articles on the Voyager program, listed in Chapter 13, also have sections on the exploration of Saturn.

Atreya, S. *The Atmospheres and Ionospheres of the Outer Planets and Their Satellites*. 1986, Springer-Verlag.
Batson, R., et al. *Voyager 1 and 2 Atlas of Six Saturnian Satellites*. 1984, NASA Special Publication 474.
Brahic, A., ed. *Planetary Rings*. 1984, Cepaudes-Editions. Conference proceedings.
Gehrels, T. & Matthews, M., eds. *Saturn*. 1984, U. of Arizona Press. Definitive collection of papers about our post-Voyager view.
Gingerich, O. "The Rotation of Saturn and Its Rings" in *Sky & Telescope*, Nov. 1964, p. 278. (Lab exercise)
Greenberg, R. & Brahic, A., eds. *Planetary Rings*. 1984, U. of Arizona Press.
Hunt, G. "Atmospheres of the Giant Planets" in *Annual Reviews of Earth & Planetary Sciences*, vol. 11 (1983).
Ingersoll, A. "Jupiter and Saturn" in *Scientific American*, Dec. 1981.
Klinger, J., et al., eds. *Ices in the Solar System*. 1985, Reidel. Conference proceedings.
Morrison, D. "The Satellites of Jupiter and Saturn" in *Annual Reviews of Astron. & Astrophys.*, vol. 20, p. 469 (1982).
Pollack, J. "Origin and History of the Outer Planets" in *Annual Reviews of Astron. & Astrophys.*, vol. 22, p. 389 (1984).
Soderblom, L. & Johnson, T. "The Moons of Saturn" in *Scientific American*, Jan. 1982.

Stern, D. & Ness, N. "Planetary Magnetospheres" in *Annual Reviews of Astronomy & Astrophysics*, vol. 20, p. 139 (1982).
Stevenson, D. "Interiors of the Giant Planets" in *Annual Reviews of Earth & Planetary Sciences*, vol. 10 (1982).

Recent Audio-Visual Aids

Hubble Space telescope (2 slide sets from the Astronomical Society of the Pacific) Feature some superb images of Jupiter and Saturn taken with the HST.
The Planetary System (100 slides, with detailed captions by David Morrison, Astronomical Society of the Pacific)
Saturn and Its Moons (29 Voyager slides, American Association of Physics Teachers)
Voyager Gallery (Videodisc with thousands of images and videos, produced by Optical Data Corp. and distributed by the Astronomical Society of the Pacific)
Voyager 1 Saturn Encounter (40 slides and cassette, produced by Holiday Films and distributed by the Planetary Society)
Voyager 2 Saturn Encounter (40 slides and cassette, produced by Holiday Films and distributed by the Planetary Society)
Voyager: Missions to Jupiter and Saturn (1983, NASA film, available on video from the Astronomical Society of the Pacific)

Topics for Discussion and Papers

1. Observing Saturn

Saturn is perhaps the most spectacular object to observe through a small telescope and no astronomy student should miss seeing (or at the right occasion) not seeing its rings through a telescope. Here are some useful references for Saturn observing:

Dobbins, T., et al. *Introduction to Observing and Photographing the Solar System*. 1988, Willmann-Bell.
Meeus, J. "The Triple Conjunctions of Jupiter and Saturn" in *Mercury*, Mar/Apr. 1982, p. 54. (From 100BC to 3000AD)
Schmidt, R. "Disappearances of Saturn's Ring, 1600-2100" in *Sky & Telescope*, Dec. 1979, p. 500.

2. Christiaan Huygens

Huygens (1629-1695) was an interesting and influential figure in the history of astronomy and students may enjoy doing research on his many contributions to science. In addition to his work on Saturn and Titan, he contributed to our understanding of light, made pioneering observations with both telescopes and microscopes, and speculated that there must be life on other worlds:

Ashbrook, J. "The Era of Long Telescopes" in *Sky & Telescope*, Aug. 1959, p. 559.
Bell, A. *Christiaan Huygens and the Development of Science in the 17th Century.* 1947, Longmans Green.
Sagan, C. "Traveler's Tales" in *Cosmos*. 1980, Random House.

3. Encke Division or Keeler Gap

The controversy about who saw the gap in Saturn's A ring first is a good case study about the battles for priority in science. Students might use this example (or find others) to discuss why scientists take such matters of priority so seriously, which could lead to a more general examination of rewards for scientific work. See:

Osterbrock, D. & Cruikshank, D. "Keeler's Gap in Saturn's A Ring" in *Sky & Telescope*, Aug. 1982, p. 123.
Osterbrock, D. *James Keeler: Pioneering American Astrophysicist.* 1984, Cambridge U. Press.
Also the book by Alexander listed under student readings, above.

CHAPTER 16
The Outer Worlds

Reading Material on the Outer Three Planets in General

Note: In finding readings on individual solar system objects you can also consult the appropriate section in the books cited for the solar system in general in Chapter 7.

Beatty, J. & Chaikin, A., eds. *The New Solar System*, 3rd ed. 1990, Sky Publishing.
Brown, R. & Cruikshank, D. "The Moons of Uranus, Neptune, and Pluto" in *Scientific American*, July 1985. On what can be learned from the Earth.
Burgess, E. *Uranus and Neptune: The Outer Giants*. 1988, Columbia U. Press.
Dobbins, T., et al. *Introduction to Observing and Photographing the Solar System*. 1988, Willmann-Bell.
Littmann, M. *Planets Beyond: Discovering the Outer Solar System*. 1988/89, J. Wiley. The best introduction.

Waff, C. "The Struggle for the Outer Planets" in *Astronomy*, Sep. 1989, p. 44. The history of the "Grand Tour" space probe project, which eventually became the Voyager mission.

Reading Material for the Student on Uranus

Beatty, J. "A Place Called Uranus" in *Sky & Telescope*, Apr. 1986, p. 333.
Beatty, J. "Voyager 2's Triumph" in *Sky & Telescope*, Oct. 1986, p. 337 (background and anecdotes about the encounter).
Berry, R. "Uranus: The Voyage Continues" in *Astronomy*, Apr. 1986, p. 6.
Berry, R. "Voyager: Discovery at Uranus" in *Astronomy*, May 1986, p. 6.
Chaikin, A. "Voyager Among the Ice Worlds" in *Sky & Telescope*, Apr. 1986, p. 343.
Croswell, K. "Uranus on the Eve of the Encounter" in *Astronomy*, Sep. 1985. (This issue has several Uranus features.)
Davis, J. *Flyby*. 1987, Atheneum. Book about the Voyager mission with emphasis on the Uranus flyby.
Eberhart, J. "Looking Back at Uranus" in *Science News*, vol. 130, p. 4 (July 5, 1986).
Elliot, J. & Kerr, R. *Rings: Discoveries from Galileo to Voyager*. 1985, MIT Press.
Elliot, J., et al. "Discovering the Rings of Uranus" in *Sky & Telescope*, June 1977, p. 412.
Elliot, J. "Uranus: The View from Earth" in *Sky & Telescope*, Nov. 1985.
Gore, R. "Uranus: Voyager Visits a Dark Planet" in *National Geographic*, Aug. 1986, p. 179.
Harrington, S. "New Views of Miranda" in *Mercury*, Jan/Feb 1987, p. 16. (Photo feature.)
Hunt, G. & Moore, P. *Atlas of Uranus*. 1989, Cambridge U. Press.
Kerr, R. "Voyager Finds Uranian Shepherds and a Well Behaved Flock of Rings" in *Science*, vol. 231, p. 791 (Feb. 21, 1986).
Morrison, N. & Gregory, S. "Discoveries at the Rings of Uranus" in *Mercury*, Mar/Apr 1987, p. 58.
Morrison, N. "A Refined View of Miranda" in *Mercury*, Mar/Apr. 1989, p. 55.
Nagy, T. & Dailey, S. "The Literary Moons of Uranus" in *Griffith Observer*, March 1986, p. 11. (On the origin of the satellite names).
O'Meara, S. "A Visual History of Uranus" in *Sky & Telescope*, Nov. 1985, p. 411.
Overbye, D. "Voyager Was on Target Again" in *Discover*, Apr. 1986, p. 70.
Rothery, D. *Satellites of the Outer Planets: Worlds in their Own Right*. 1992, Oxford U. Press.
Waldrop, M. "Voyage to a Blue Planet" in *Science*, vol. 231, p. 916 (Feb. 28, 1986).

Reading Material for the Student on Neptune

Beatty, J. "Welcome to Neptune" in *Sky & Telescope*, Oct. 1989, p. 358.
Beatty, J. "Getting to Know Neptune" in *Sky & Telescope*, Feb. 1990, p. 136. Excellent summary.
Berry, R. "Triumph at Neptune" in *Astronomy*, Nov. 1989, p. 20.
Berry, R. "Neptune Revealed" in *Astronomy*, Dec. 1989, p. 22.
Burgess, E. *Far Encounter: The Neptune System*. 1992. Columbia U. Press.
Gore, R. "Neptune: Voyager's Last Picture Show" in *National Geographic*, Aug. 1990, p. 35.
Kaufmann, W. "Voyager at Neptune: A Preliminary Report" in *Mercury*, Nov/Dec. 1989, p. 174.
Sohus, A. & Miner, E. "The Voyager Mission to Neptune" in *Mercury*, Sep/Oct. 1988, p. 130. [Preparing for the encounter.]
[no author] "Voyager's Last Picture Show" in *Sky & Telescope*, Nov. 1989, p. 462; "Neptune and Triton" in Feb. 1990, p. 136. Photo albums.

Reading Material for the Student on Pluto

Beatty, J. "Pluto and Charon: The Dance Goes On" in *Sky & Telescope*, Sep. 1987, p. 248; "The Dance Begins," June 1985, p. 501.
Beatty, J. & Killian, A. "Discovering Pluto's Atmosphere" in *Sky & Telescope*, Dec. 1988, p. 624.
Croswell, K. "Pluto: The Enigma at the Edge of the Solar System" in *Astronomy*, July 1986, p. 6.
Eberhart, J. "Pluto: Limits on its Atmosphere, Ice on its Moon" in *Science News*, Sep. 26, 1987, p. 207.
Eicher, D. "Pluto Emerges from the Shadows" in *Astronomy*, Sep. 1988, p. 52. Brief review of recent observations.
Eicher, D. "How to Hunt Down Pluto" in *Astronomy*, Apr. 1984, p. 35.
Levy, D. *Clyde Tombaugh: Discoverer of Pluto*. 1991. U. of Arizona Press. A flawed but interesting biography.
Moore, P. "The Naming of Pluto" in *Sky & Telescope*, Nov. 1984, p. 400.
Mulholland, D. "The Ice Planet [Pluto]" in *Science 82*, Dec. 1982, p. 64.
Sobel, D. "The Last World" in *Discover*, May 1993, p. 68.

Reading Material for the Instructor on Uranus

Atraya, S. *Atmospheres and Ionospheres of the Outer Planets and Their Satellites*. 1986, Springer-Verlag.
Cuzzi, J. & Esposito, L. "The Ring of Uranus" in *Scientific American*, July 1987.

Greenberg, R. & Brahic, A., eds. *Planetary Rings*. 1984, U. of Arizona Press.
Ingersoll, A. "Uranus" in *Scientific American*, Jan. 1987.
Johnson, T., et al. "The Moons of Uranus" in *Scientific American*, Apr. 1987.
Laeser, R., et al. "Engineering Voyager 2's Encounter with Uranus" in *Scientific American*, Nov. 1986.
The July 4, 1986 issue of *Science* magazine contains a special section of technical research reports on the Voyager/Uranus results.

Reading Material for the Instructor on Neptune

Brown, R. & Cruikshank, D. "The Moons of Uranus, Neptune, and Pluto" in *Scientific American*, July 1985.
Morrison, D. & Owen, T. *The Planetary System*. 1988, Addison-Wesley. See chapter 12.
Schenk, P. & Jackson, M. "Diapirism on Triton: A Record of Crustal Layering and Instability" in *Geology*, Apr. 1993, p. 299.
The first scientific papers summarizing the Voyager discoveries appeared in *Science* magazine, vol. 246, p. 1417 (15 Dec 1989).

Reading Material for the Instructor on Pluto

Cruikshank, D., et al. "Pluto: Evidence for Methane Frost" in *Science*, Nov. 19, 1976, vol. 194, p. 835.
Stern, S. "Evidence for a Low Surface Temperature on Pluto from Millimeter-Wave Thermal Emission Measurements" in *Science*, Sep. 24, 1993, p. 1713.
Whyte, A. *The Planet Pluto*. 1980, Pergamon Press.

Recent Audio-Visual Aids

Astronomers of the Past (slide set with portraits of Adams, Kuiper, Leverrier, Lowell, Pickering; Astronomical Society of the Pacific)
The Discovery of Pluto (39 min. videotape on the life and work of Clyde Tombaugh, narrated by Clyde Tombaugh, produced by T. Hockey, distributed by the Astronomical Society of the Pacific)
Neptune Kit (12 slides and booklet by W. Kaufmann, Astronomical Society of the Pacific) — best images from Voyager encounter
The Planet That Got Knocked on Its Side (58 min. film on the Voyager Uranus misssion, part of the *Nova* TV series, distributed by Coronet Films)
The Ring Tape (Audio cassette of the discovery of Uranus' ring aboard the

Kuiper Airborne Observatory, produced by J. Elliot and distributed by MIT Press)

Uranus Discovered (20 slides with brief captions, Hansen Planetarium)

Uranus: I Will See Such Things (29 min videotape on the Voyager results, produced by JPL, distributed by the Astronomical Society of the Pacific and the Planetary Society)

Voyager at Uranus (15 slides with a 20-page booklet of detailed caption and background material, Astronomical Society of the Pacific)

Voyager Mission to Uranus (20 slides with brief captions, produced by JPL, distributed by the Planetary Society)

Voyager Neptune Encounter Highlights (1990, videotape with 21 short films from the Voyager encounter, 33 min, Astronomical Society of the Pacific)

Topics for Discussion and Papers

1. Discovering the Outer Planets

The discoveries of Uranus, Neptune, and Pluto are fascinating episodes in the history of astronomy; the Neptune story especially is full of controversy, rivalry, human laziness and enterprise, and the power of mathematical modeling. The Pluto discovery has the poignant story of Clyde Tombaugh, straight off the farm in Kansas, coming to the Lowell Observatory to undertake a taxing observational program no professional astronomer wanted to do. For students who want to explore these stories, here are some places to start:

a. The Discovery of Uranus

Bennett, J. "The Discovery of Uranus" in *Sky & Telescope*, Mar. 1981, p. 188.

O'Meara, S. "A Visual History of Uranus" in *Sky & Telescope*, Nov. 1985, p. 411

b. The Discovery of Neptune

Drake, S. & Kowal, C. "Galileo's Sighting of Neptune" in *Scientific American*, Dec. 1980.

Grosser, M. *The Discovery of Neptune*. 1962, Harvard U. Press. Moore, P. *The Planet Neptune*. 1988, Ellis Harwood/ J. Wiley.

Moore, P. "The Discovery of Neptune" in *Mercury*, Jul/Aug. 1989, p. 98.

c. The Discovery of Pluto

Hoyt, W. *Planets X and Pluto*. 1980, U. of Arizona Press

Moore, P. "The Naming of Pluto" in *Sky & Telescope*, Nov. 1984, p. 400.

Tombaugh, C. "The Search for the Ninth Planet" in *Mercury*, Jan/Feb. 1979, p. 1.
Tombaugh, C. "The Discovery of Pluto" in *Mercury*, May/June 1986, p. 66 & Jul/Aug. 1986, p. 98.
Tombaugh, C. & Moore, P. *Out of the Darkness: The Planet Pluto*. 1980, Stackpole Books.

2. Outer Planet Science Fiction

Only recently, with the information from the Voyager program, have the outer planets become rich enough in our imaginations for good science fiction stories. Your students may be able to find a number of recent stories that use the outer planets as a realistic setting. Here are a few to get them started:

Eklund, G. *A Thunder on Neptune*. 1989, Morrow. A new novel about exploring Neptune and Triton and discovering a form of life.
Niven, L. "Wait it Out" in *Tales of Known Space*. 1975, Ballantine. About being marooned on Pluto and discovering superfluid life.
Sheffield, C. "Dies Irae" in B. Preiss, ed. *The Planets*. 1985, Bantam. Adapting life to survive in the Uranus environment.
Silverberg, R. "Sunrise on Pluto" in B. Preiss, ed. *The Planets*. 1985, Bantam. Life that can survive on Pluto.

3. The Voyager Record

Having completed its reconnaissance of Neptune, the Voyager 2 spacecraft (like its twin Voyager 1) is now heading out of the solar system and into interstellar space. As a sort of "cosmic message in a bottle," Carl Sagan, Frank Drake, and co-workers assembled an audio/video record of the "sights and sounds of Earth," a copy of which has been attached to each craft. Given the vastness of the cosmic ocean, the message may say more to us today than to any possible recipients millions of years from now. Students may want to consider what sights and sounds they would have put on such a record and then compare their suggestions to what was actually done:

Druyan, A. & Ferris, T. "Earth's Greatest Hits" in *The New York Times Magazine*, Sept. 4, 1977, p. 12.
Eberhart, J. "The World on a Record" in *Science News*, vol. 112, p. 124 (1977).
Sagan, C., et al. *Murmurs of Earth: The Voyager Interstellar Record*. 1978, Random House.

CHAPTER 17
Interplanetary Vagabonds

Reading Material for the Student on Asteroids

Note: In finding readings on individual solar system objects you can also consult the appropriate section in the books cited for the solar system in general in Chapter 7.

Beatty, J. "A Picture Perfect Asteroid [Gaspra]" in *Sky & Telescope*, Feb. 1992, p. 135. [brief]
Chapman, C. & Morrison, D. "The Minor Planets: Size and Minerology" in *Sky & Telescope*, Feb. 1974, p. 92.
Chapman, C. & Morrison, D. *Cosmic Catastrophes*. 1989, Plenum Press. Excellent introduction to collision ideas for asteroids and comets.
Cunningham, C. *Introduction to Asteroids*. 1988, Willmann-Bell. Good review, written by a knowledgeable amateur.
Davies, J. "Is 3200 Phaeton a Dead Comet? in *Sky & Telescope*, Oct. 1985, p. 317.
Durda, D. "All in the Family" in *Astronomy*, Apr. 1993, p. 36.
Gunter, J. "Asteroids and Amateur Astronomers" in *Mercury*, Jan/Feb. 1985, p. 9.
Hartmann, W. "Vesta: A World of Its Own" in *Astronomy*, Feb. 1983, p. 6.
Kerr, R. "The Great Asteroid Roast: Was it Rare or Well-done" in *Science*, Feb. 2, 1990, p. 527.
Kowal, C. *Asteroids: Their Nature and Utilization*. 1988, Ellis Horwood/Wiley. Fine introduction.
Morrison, D. "Asteroids" in *Astronomy*, June 1976, p. 6.
Morrison, N. & D. "The Geography of the Asteroid Belt" in *Mercury*, Nov/Dec. 1982, p. 195.
Ostro, S. "Radar Reveals a Double Asteroid" in *Astronomy*, Apr. 1990, p. 38.
Talcott, R. "Toutatis Seen With Radar" in *Astronomy*, Apr. 1993, p. 36.

Reading Material for the Student on Meteoroids/Meteorites

Adams, F. "Observing Fallen Stars [Meteors]" in *Mercury*, Mar/Apr. 1980, p. 31.
Bagnall, P. *The Meteorite and Tektite Collector's Handbook*. 1991. Willman-Bell.

Chaikin, A. "A Stone's Throw from the Planets" in *Sky & Telescope*, Feb. 1983, p. 122. On meteorites from the Moon and Mars.
Dodd, R. *Thunderstones & Shooting Stars: The Meaning of Meteorites.* 1986, Harvard U. Press.
Drummond, J., et al. "Two Meteor Projects for Amateurs" in *Sky & Telescope*, May 1991, p. 478.
Falk, S. & Schramm, D. "Did the Solar System Start with a Bang?" in *Sky & Telescope*, July 1979, p. 18.
Fortier, E. "Etch a Meteorite" in *Sky & Telescope*, Dec. 1981, p. 527.
Graham, A. *Catalogue of Meteorites*, 4th ed. 1986. U. of Arizona Press.
Hoyt, W. *Coon Mountain Controversies.* 1987, U. of Arizona Press. A history of the controversies and discoveries surrounding meteor crater in Arizona.
Hutchinson, R. *The Search for Our Beginnings.* 1983, Oxford U. Press. On meteorites and the solar system.
Janos, L. "Timekeepers of the Solar System" in *Science '80*, May/June 1980, p. 44. On G. Wasserburg's group at Caltech.
Kronk, G. *Meteor Showers: A Descriptive Catalog.* 1988, Enslow.
Lewis, R. & Anders, E. "Interstellar Matter in Meteorites" in *Scientific American*, Aug. 1983.
MacRobert, A. "How to Report a Fireball" in *Sky & Telescope*, Apr. 1985, p. 372.
Mark, K. *Meteorite Craters.* 1987, U. of Arizona Press.
McSween, H. *Meteorites and their Parent Planets.* 1987, Cambridge U. Press.
Rao, J. "Comet Swift-Tuttle and the Perseids" in *Astronomy*, Aug. 1983, p. 16.
Smith, F. "A Collision Over Collisions" in *Mercury*, May/June 1992, p. 97. Reviewof recent political and scientific debate about large meteors that might hit Earth.
Spratt, C. & Stephens, S. "Against All Odds: Meteorites That Have Struck [the earth]" in *Mercury*, Mar/Apr. 1992, p. 50.
Stacy, D. "Meteorites by Mail" in *Air & Space*, Mar/Apr. 1990, p. 36. On finding, buying, and selling them.
Wagner, J. "The Sources of Meteorites" in *Astronomy*, Feb. 1984, p. 6.

Reading Material for the Student on Comets

Beatty, J. "An Inside Look at Halley's Comet" in *Sky & Telescope*, May 1986, p. 438.
Beatty, J. "Comet Giacobini-Zinner: The Inside Story" in *Sky & Telescope*, Nov. 1985, p. 426.
Benningfield, D. "Where Do Comets Come From?" in *Astronomy*, Sep. 1990, p. 28.
Berry, R. "Search for the Primitive" in *Astronomy*, June 1987, p. 6. On Comet Halley results.

Berry, R. "Giotto Encounters Comet Halley" in *Astronomy*, June 1986, p. 6.
Berry, R. "ICE: Mission to Comet Giacobini-Zinner" in *Astronomy*, Dec. 1985, p. 6.
Berry, R. & Talcott, R. "What We Have Learned from Comet Halley" in *Astronomy*, Sep. 1986, p. 6.
Bond, P. "Close Encounters With a Comet" in *Astronomy*, Nov. 1993, p. 42.
Bortle, J. "A Halley Chronicle" in *Astronomy*, Oct. 1985, p. 98. A chronology of each observed pass.
Brandt, J. & Chapman, R. *The Comet Book*. 1984, Jones & Bartlett.
Brandt, J. & Chapman, R. *Rendezvous in Space*. 1992. W. H. Freeman. Excellent and up-to-date introduction.
Davies, J. "Can Comets Become Asteroids?" in *Astronomy*, Jan. 1985, p. 66.
Delsemme, A. "Whence Come Comets" in *Sky & Telescope*, Mar. 1989, p. 260.
Ferrin, I. & Guzman, E. "How a Cometary Nucleus Turns On" in *Sky & Telescope*, Aug. 1981, p. 103.
Gingerich, O. "Newton, Halley, and the Comet" in *Sky & Telescope*, Mar. 1986, p. 230.
Gore, R. "Halley's Comet '86: Much More Than Met the Eye" in *National Geographic*, Dec. 1986, p. 758.
Knacke, R. "Cosmic Dust and the Comet Connection" in *Sky & Telescope*, Sep. 1984, p. 207.
Larson, S. "Observing Comet Halley's Near Nucleus Features" in *Astronomy*, Sep. 1987, p. 90.
Liller, W. "A Halley Watch from Easter Island" in *Mercury*, Sep/Oct. 1987, p. 130.
Mardsen, B. "Comet Swift-Tuttle: Does it Threaten Earth?" in *Sky & Telescope*, Jan. 1993, p. 16.
Morris, C. "The Eye's View of Halley's Comet" in *Astronomy*, Jan. 1987, p. 90.
Morrison, D. "The VEGA Flyby and Halley's Comet: A First Person Account" in *Mercury*, Jul/Aug 1986, p. 114.
Sagan, C. & Druyan, A. *Comet*. 1986, Random House.
Weaver, K. "What You Didn't See in Comet Kohoutek" in *National Geographic*, Aug. 1974.
Whipple, F. *The Mystery of Comets*. 1985, Smithsonian Institution Press.
Whipple, F. "Flying Sandbanks Versus Dirty Snowballs: The Nature of Comets" in *Mercury*, Jan/Feb. 1986, p. 2.
Yeomans, D. *Comets: A Chronological History*. 1991, Wiley.
Sky & Telescope magazine for March 1987 was a special issue devoted to articles about what we learned from Halley's Comet.

Reading Material for the Instructor on Asteroids

Note: In finding readings on individual solar system objects you can also consult the appropriate section in the books cited for the solar system in general in Chapter 7.

Chapman, C., et al. "The Asteroids" in *Annual Reviews of Astron. & Astrophys.*, vol. 16, p. 33 (1978).

Gehrels, T. & Matthews, M., eds. *Asteroids.* 1979, U. of Arizona Press. An 1,181-page compendium.

Gehrels, T. "Asteroids and Comets" in *Physics Today*, Feb. 1985, p. 32.

Nieto, M. "The Letters Between Titius and Bonnett and the Titius-Bode Law" in *American Journal of Physics*, Jan. 1985, p. 22.

Shoemaker, E. "Asteroid and Comet Bombardment of the Earth" in *Annual Reviews of Earth & Planetary Science*, vol. 11 (1983).

Smoluchowski, R., et al., eds. *The Galaxy and the Solar System.* 1986, U. of Arizona Press. On asteroid or comet showers and other ways the Galaxy at large may influence the solar system; conference proceedings.

Reading Material for the Instructor on Meteoroids/Meteorites

Anders, E. "Meteorites and the Early Solar System" in *Annual Reviews of Astron. & Astrophys.*, vol. 9 (1971).

Balsinger, H., et al. "A Close Look at Halley's Comet" in *Scientific American*, Sep. 1988.

Burke, J. *Cosmic Debris: Meteorites in History.* 1986, U. of California Press.

Cassidy, W. & Rancitelli, L. "Antarctic Meteorites" in *American Scientist*, Mar/Apr. 1982, p. 156.

Graham, A. *Catalogue of Meteorites, 4th ed.* 1986, U. of Arizona Press.

Grossman, L. "The Most Primitive Objects in the Solar System" in *Scientific American*, Feb. 1975. On carbonaceous chondrites.

McSween, H. "Achondrites and Igneous Processes on Asteroids" in *Annual Reviews of Earth & Planetary Sciences*, vol. 17 (1989).

Wasson, J. *Meteorites: Their Record of Early Solar System History.* 1985, Freeman.

Whetherill, G. "Apollo Objects" in *Scientific American*, Mar. 1979.

Wood, J. "Chondritic Meteorites and the Solar Nebula" in *Annual Reviews of Earth & Planetary Sciences*, vol. 16 (1988).

Reading Material for the Instructor on Comets

A'Hearn, M. "Observations of Cometary Nuclei" in *Annual Reviews of Earth & Planetary Sciences*, vol. 16 (1988).

Balsinger, H., et al. "A Close Look at Halley's Comet" in *Scientific American*, Sep. 1988.

Battrick, B., et al., eds. *The Exploration of Halley's Comet*. 1986, European Space Agency Special Publication 250; 3 volumes. (ESA, 8-10 rue Mario-Nikis, 75738 Paris 15, France.)

Bork, A. "Newton and Comets" in *American Journal of Physics*, Dec. 1987, p. 108.

Brandt, J. & Chapman, R. *An Introduction to Comets*. 1981, Cambridge U. Press.

Brandt, J. & Niedner, M. "The Structure of Comet Tails" in *Scientific American*, Jan. 1986.

Carusi, A., et al., eds. *Dynamics of Comets: Their Origin and Evolution*. 1985, Reidel.

Chyba, C., et al. "Cometary Delivery of Organic Molecules to the Early Earth" in *Science*, July 27, 1990, p. 366. Interesting review article.

Kerr, R. "Could an Asteroid Be A Comet in Disguise?" in *Science*, Feb. 22, 1985, p. 930.

Klinger, J., et al., eds. *Ices in the Solar System*. 1985, Reidel.

Mendis, D. "A Postencounter View of Comets" in *Annual Reviews of Astron. & Astophys.*, vol. 26, p. 11 (1988).

Spinrad, H. "Comets and Their Composition" in *Annual Reviews of Astronomy and Astrophysics*, vol. 25, 1987, p.231.

Weissman, P. "Comets at the Solar System's Edge" in *Sky & Telescope*, Jan. 1993, p. 26.

Wilkening, L., ed. *Comets*. 1982, U. of Arizona Press.

The 15-21 May 1986 issue of *Nature* (Vol. 321, # 6067) was devoted to Comet Halley results from the five space probes and ground-based observatories.

Recent Audio-Visual Aids

Comet Halley (60 min. videotape, Planetary Society)

Halley's Comet Revealed (17 slides plus a booklet of detailed captions and background material by Dr. John C. Brandt; Astronomical Society of the Pacific)

Halley the Beautiful (20 slides with brief captions, Hansen Planetarium)

"Heaven and Hell" (an episode of the *Cosmos* TV series — with material on comets, meteorites, asteroids, Tunguska, etc. — available in home video format from the Astronomical Society of the Pacific)

Remember Halley's Comet (20 slides with brief captions, Planetary Society)

"The Third Planet" (an episode in *The Miracle Planet* TV series, from Ambrose Video or the astronomical Society of the Pacific) Focus on the asteroid/comet collision hypothesis for the demise of the dinosaurs.

Topics for Discussion and Papers

1. Bode's Law

The Titius-Bode Law controversy has fascinated astronomers, historians of science, and many students. For more on this subject, see:

Jaki, S. "The Titius-Bode Law: A Strange Bicentenary" in *Sky & Telescope*, May 1972, p. 280.

Jaki, S. "The Original Formulation of the Titius-Bode Law" in *The Journal for the History of Astronomy*, vol. 3, p. 136 (1972).

Nieto, M. *The Titius-Bode Law of Planetary Distances: Its History and Theory*. 1972, Pergamon Press.

2. Naming Asteroids

At first asteroids were given classical names, but as the number of newly discovered minor planets grew, the need for names exceeded the supply of classical characters whose names had not yet been used for celestial bodies. Today, discoverers often choose to name asteroids after colleagues, home towns, sweethearts, famous people from other fields of human endeavor, etc. New names are often reported in *Sky & Telescope* and *Astronomy* magazine. A number of the books on asteroids in the readings section have tables of asteroid names, which students might want to scan to find the most unusual names. For more, see the books on asteroids in the reading list or:

Gunter, J. "Asteroids and Amateur Astronomers" in *Mercury*, Jan/Feb. 1985, p. 9.

3. Collisions with the Earth

As Kaufmann discusses, it seems more and more clear with recent research that asteroids (as well as meteoroids and comets) have collided and will collide with the Earth. Students may want to explore the odds of such collisions, and what, if anything, we could do about them. Some good references:

Chapman, C. & Morrison, D. *Cosmic Catastrophes*. 1989, Plenum Press. Excellent introduction to collision ideas for asteroids and comets. Excerpts appeared in the Nov/Dec. 1989 and Jan/Feb. 1990 issues of *Mercury* magazine.

Helin, E. "The Discovery of an Unusual Apollo Asteroid" in *Mercury*, Sep/Oct. 1981, p. 134.
Morrison, D. & Chapman, C. "Target Earth: It Will Happen" in *Sky & Telescope*, Mar. 1990, p. 261.
Sinnott, R. "An Asteroid Whizzes Past the Earth" in *Sky & Telescope*, July 1989, p. 30.
Teske, R. "Asteroid Collisions with the Earth" in *Sky & Telescope*, Jan. 1982, p. 18.

4. The Alvarez Hypothesis and Nemesis

A tremendous amount of research, debate, and controversy have been triggered by the suggestion that an asteroid could have been a major factor in the extinction of many species 65 million years ago. Kaufmann describes the genesis of this suggestion in section 17-4.

Others have taken the Alvarez hypothesis much further, with suggestions that such extinctions by cosmic collision have happened many times, that there is periodicity to the extinctions, and that a distant unseen companion of the Sun's (nicknamed Nemesis) could be responsible for such repeated extinctions by deflecting large numbers of comet nuclei from the Oort Cloud toward the inner solar system when its orbit intersects the cloud. These later suggestions have been greeted with even greater skepticism in the scientific community, although they have received a lot of publicity in the popular press.

Students who want to follow the controversy can begin with the following sources, of which the book by Goldsmith is by far the best objective guide:

Chapman, C. & Morrison, D. *Cosmic Catastrophes*. 1989, Plenum Press.
Goldsmith, D. *Nemesis: The Death Star and Other Theories of Mass Extinction*. 1985, Walker. Good, skeptical book on the controversial subject of asteroids or comets causing regular periods of devastation on Earth.
Gould, S. "An Asteroid to Die For" in *Discover*, Oct. 1989, p. 60.
Muller, R. *Nemesis: The Death Star*. 1988, Weidenfeld & Nicolson. By the physicist author of the Nemesis hypothesis.
Russell, D. "The Mass Extinctions of the Late Mesozoic" in *Scientific American*, Jan. 1982.
Weissman, P. "Are Periodic Bombardments Real?" in *Sky & Telescope*, Mar. 1990, p. 266.

5. Hunting Comets

Comet hunting is a favorite sport of amateur astronomers, pursued not only for the inherent thrill of the chase, but also because the discoverer of a new comet gets it named after him. Many comets in recent years have been discovered by patient, dedicated observers with relatively simple equipment and often no formal training in astronomy beyond an introductory course. Students who want to find out more about comet hunting can consult:

Bortle, J. "Comets and How to Hunt Them" in *Sky & Telescope*, Feb. 1981, p. 123; Mar. 1981, p. 210; "Hunting Comets," Nov. 1988, p. 579.
Levy, D. "How To Discover a Comet" in *Astronomy*, Dec. 1987, p. 74.
Machholz, D. "Comet Hunting" in *Mercury*, May/June 1979, p. 57.
Mayer, B. "Bradfield's Dozen: A Guide to Comet Seeking" in *Astronomy*, Jan. 1982, p. 43.
Rosenthal, D. "The Comet Champion: An Interview with William Bradfield" in *Sky & Telescope*, June 1988, p. 597.

6. The Central Bureau for Comet (and Other) Discoveries

To help the orderly recording and dissemination of information on short-lived astronomical phenomena like comets, the International Astronomical Union has established a Central Bureau for Astronomical Telegrams at the Harvard-Smithsonian Center for Astrophysics in Cambridge, Massachusetts. Dr. Brian Marsden coordinates this important service which receives and sends out news not only about comets, but exploding stars, burst sources, and other discoveries which astronomers need to hear about quickly. Students may enjoy doing a report on what it's like to run this sort of bureau. See:
Green, D. "What to Do if You Discover a Comet" in *Sky & Telescope*, Oct. 1987, p. 420.
Overbye, D. "Life in the Hot Seat: An Interview with Brian Marsden" in *Sky & Telescope*, Aug. 1980, p. 92.

7. Meteor Shower Observing

As Kaufmann discusses in Section 17-8, meteor showers can tell us something about the orbits and the distribution of the debris of comets along those orbits. Thus there are a number of serious amateur observers who devote themselves to the careful monitoring of meteor showers over the years. Students who want to investigate or get involved should consult:

Bagnall, P. "Dark Skies for the Perseid Meteor Shower" in *Astronomy*, Aug. 1988, p. 78. Has some good general observing hints.
Kronk, G. "Meteor Showers" in *Mercury*, Nov/Dec. 1988, p. 162.
Kronk, G. *Meteor Showers: A Descriptive Catalog*. 1988, Enslow.
MacRobert, A. "Meteor Observing" in *Sky & Telescope*, Aug. 1988, p. 131; Oct. 1988, p. 363.
Rao, J. "Comet Swift-Tuttle and the Perseids" in *Astronomy*, Aug. 1983, p. 16.

8. The Tunguska Event

Many (sometimes bizarre) hypotheses have been advanced for what caused the 1908 Tunguska explosion, but (as the book explains) a comet fits the known facts the best. Students who want to look into the story more closely can see:

Chaikin, A. "Target: Tunguska" in *Sky & Telescope*, Jan. 1984, p. 18.
Chapman, C. & Morrison, D. *Cosmic Catastrophes*. 1989, Plenum Press.
Chapman, C. & Morrison, D. "Cosmic Catastrophes" in *Mercury*, Nov/Dec. 1989, p. 185.
Greenstein, G. "Heavenly Fire: The Tunguska Event" in *Science '85*, Jul/Aug 1985, p. 70.
Ridpath, I. "The Comet that Hit the Earth" in *Mercury*, Sep/Oct. 1977, p. 2.

9. Asteroid/Comet/Meteoroid Science Fiction

Many stories in science fiction have focused on cosmic "debris" and their encounters or collisions either with the Earth or with the characters of the story. Students should be able to find lots of examples (in films as well as stories) and might enjoy critiquing a number of them for scientific accuracy.

Here are a few stories with reasonable science to get them started:

Asimov, I., et al., eds. *Comets*. 1986, Signet. A collection of stories involving comets.
Benford, G. & Brin, D. *The Heart of the Comet*. 1986, Bantam. About a mission to land on Halley's Comet in 2061.
Fodor, R. & Taylor, G. *Impact*. 1979, Leisure Books. A giant meteorite threatens the Earth.
Hoyle, F. *Comet Halley*. 1985, St. Martin's. Life is found in the famous comet.
Niven, L. & Pournelle, J. *Lucifer's Hammer*. 1977, Fawcett. Classic disaster novel where an asteroid or comet collides with the Earth.
Preuss, P. "Small Bodies" in B. Preiss, ed. *The Planets*. 1985, Bantam. A fundamentalist preacher and a scientist find fossils on an asteroid.

CHAPTER 18
Our Star

Reading Material for the Student

Bartusiak, M. "The Sunspot Syndrome" in *Discover*, Nov. 1989, p. 44.
Eddy, J. *A New Sun*. 1979, NASA SP-402.
Eddy, J. "The Case of the Missing Sunspots" in *Scientific American*, May 1977.

Emslie, A. "Explosions in the Solar Atmosphere" in *Astronomy*, Nov. 1987, p. 18. On solar flares.
Frazier, K. *Our Turbulent Sun*. 1983, Prentice-Hall.
Friedman, H. *Sun and Earth*. 1986, W. H. Freeman.
Gibson, E. "The Sun as Never Seen Before" in *National Geographic*, Oct. 1974, p. 494.
Giovanelli, R. *Secrets of the Sun*. 1984, Cambridge U. Press.
Golub, L. "Heating the Sun's Million Degree Corona" in *Astronomy*, May 1993, p. 26.
Golub, L. "Solar Magnetism: A New Look" in *Astronomy*, Mar. 1981, p. 66.
Goodwin, J., et al. *Fire Of Life*. 1981, Smithsonian. Coffee table book about solar science and legend.
Harvey, J., et al. "GONG: To See Inside Our Sun" in *Sky & Telescope*, Nov. 1987, p. 470.
Kaler, J. "Giants in the Sky: The Fate of the Sun" in *Mercury*, Mar/Apr. 1993, p. 34.
Kanipe, J. "The Rise and Fall of the Sun's Activity" in *Astronomy*, Oct. 1988, p. 22.
Krisciunas, K. "Eclipse Science from Hawaii" in *Sky & Telescope*, Dec. 1991, p. 597.
Levine, R. "The New Sun" in Cornell, J. & Gorenstein, P., eds. *Astronomy from Space*. 1983, MIT Press.
Livingston, W., et al. "Old and New Views of Solar Prominences" in *Astronomy*, July 1987, p. 18.
LoPresto, J. "The Rotation of the Sun" in *Astronomy*, Dec. 1986, p. 107.
LoPresto, J. "Looking Inside the Sun" in *Astronomy*, Mar. 1989, p. 20. On helioseismology.
MacRobert, A. "Close-up of a Star" in *Sky & Telescope*, May 1985, p. 397.
Maxwell, A. "Solar Flares and Shock Waves" in *Sky & Telescope*, Oct. 1983, p. 285.
Mitton, S. *Daytime Star*. 1981, Scribners.
Morrison, N. & D. "The Vibrating Sun" in *Mercury*, Jan/Feb 1984, p. 23.
Mullan, D. "Tuning in to the Interior of a Star" in *Astronomy*, Dec. 1984, p. 66.
Noyes, R. *The Sun, Our Star*. 1982, Harvard U. Press.
Overbye, D. "John Eddy: The Solar Detective" in *Discover*, Aug. 1982, p. 68.
Pasachoff, J. "The Sun: A Star Close Up" in *Mercury*, May/June 1991, p. 66.
Peterson, I. "More Than Just Your Average Star" in *Science News*, July 2, 1988, p. 8.
Peterson, C., et al, "Yohkoh and the Mysterious Solar Flares" in *Sky & Telescope*, Sep. 1993, p. 20.
Robinson, L. "The Disquieting Sun: How Big, How Steady?" in *Sky & Telescope*, Apr. 1982, p. 354.

Robinson, L. "The Sunspot Cycle: Tip of the Iceberg" in *Sky & Telescope*, June 1987, p. 589.
Schaefer, B. "The Astrophysics of Suntanning" in *Sky & Telescope*, June 1988, p. 595.
Verschuur, G. "The Day The Sun Cut Loose" in *Astronomy*, Aug. 1989, p. 48. About a huge flare.
Verschuur, G. "The Many Faces of the Sun" in *Astronomy*, Mar. 1989, p. 46.
Wentzel, D. *The Restless Sun*. 1989, Smithsonian Inst. Press.
Wentzel, D. "The Solar Chimes: Searching for Oscillations Inside the Sun" in *Mercury*, May/June 1991, p. 77.
Williams, G. "The Solar Cycle in Pre-Cambrian Time" in *Scientific American*, Aug. 1986.

Reading Material for the Instructor

Akasofu, S., et al. *The Solar Wind and the Earth*. 1987, Reidel.
Bahcall, J. *Neutrino Astrophysics*. 1989, Cambridge U. Press.
Babcock, H. "The Sun's Magnetic Field" in *Annual Reviews of Astron. & Astrophys.*, vol. 1, p. 41 (1963). Includes a nice historical review.
Bai, T. & Sturrock, P. "Classification of Solar Flares" in *Annual Reviews of Astron. & Astrophys.*, vol. 27, p. 421 (1989).
Christensen-Dalsgaard, J., et al. "Seismology of the Sun" in *Science*, Sep. 6, 1985, p. 923.
DeJager. C., et al., eds. *Progress in Solar Physics*. 1986, Reidel.
Deubner, F. & Gough, D. "Helioseismology" in *Annual Reviews of Astron. & Astrophys.*, vol. 22, p. 131 (1984).
Durney, B., et al., eds. *The Internal Solar Angular Velocity*. 1987, Reidel.
Foukal, P. "The Variable Sun" in *Scientific American*, Feb. 1990.
Friedlander, G. & Weneser, J. "Solar Neutrinos: Experimental Approaches" in *Science*, Feb. 13, 1987, p. 760.
Gingerich, O. & Tresch-Fienberg, R. "The Rotation of the Sun" in *Sky & Telescope*, Nov. 1982, p. 433. A lab exercise.
Howard, R. "Solar Rotation" in *Annual Reviews of Astron. & Astrophys.*, vol. 22, p. 131 (1984).
Hudson, H. "Observed Variability of the Solar Luminosity" in *Annual Reviews of Astron. & Astrophys.*, vol. 26, p. 473 (1988).
Hufbauer, K. *Exploring the Sun: Solar Science Since Galileo*. 1991, Johns Hopkins U. Press. A detailed history.
Low, B. "Equilibrium and Dynamics of Coronal Magnetic Fields" in *Annual Reviews of Astron. & Astrophys.*, vol. 28, p. 491 (1990).
Moore, R. & Rabin, D. "Sunspots" in *Annual Reviews of Astron. and Astrophys.*, vol. 23, p. 239. (1985)

Newkirk, G. & Frazier, K. "The Solar Cycle" in *Physics Today*, Apr. 1982, p. 25.
Radick, R., et al. "Stellar Activity and Brightness Variations: The Sun's History" in *Science*, Jan. 5, 1990, p. 39.
Snodgrass, H. & Howard, R. "Torsional Oscillations of the Sun" in *Science*, May 24, 1985, p. 945.
Spruit, H., et al. "Solar Convection" in *Annual Reviews of Astron. & Astrophys.*, vol. 28, p. 263 (1990).
Sturrock, P., et al., eds. *Physics of the Sun*. (3 volumes) 1986, Reidel.
von Hippel, F., et al. "Helioseismology" in *Scientific American*, Sep. 1985.
Weneser, J. & Friedlander, G. "Solar Neutrinos: Questions and Hypotheses" in *Science*, Feb. 13, 1987, p. 755.
Wolfson, R. "The Active Solar Corona" in *Scientific American*, Feb. 1983.
Zirker, J. *Total Eclipses of the Sun*. 1984, Van Nostrand Reinhold. Reviews much modern solar physics.
Zwaan, C. "Elements and Patterns in the Solar Magnetic Field" in *Annual Reviews of Astronomy and Astrophysics*, vol. 25, p. 83. (1987)

Recent Audio-Visual Aids

Astronomers of the Past (50 slides and biographies, Astronomical Society of the Pacific)
Day Star (1984, 16 min video on the work at the San Fernando Solar Observatory; University of California Extension Media Ctr., 2223 Fulton St., Berkeley, CA 94720.)
The Sunspot Mystery (1977, episode of *NOVA* TV series, Time-Life Films)

Topics for Discussion and Papers

1. Observing the Sun

With proper precautions, a solar observing session can be the highlight of a discussion of the Sun and an excellent way to give students a feel for the energies in the solar photosphere. Here are some articles to help instructors (and students who have safe filters and know how to use them) to get the most out of a solar observing session:

Burnham, R. "Observing the Sun" in *Astronomy*, Aug. 1984, p. 51.
Chou, R. "Safe Solar Filters" in *Sky & Telescope*, Aug. 1981, p. 119.
Dilsizian, R. "Photographing Our Nearest Star" in *Astronomy*, May 1987, p. 38.
McIntosh, P. & Leinbach, H. "Watching the Premier Star" in *Sky & Telescope*, Nov. 1988, p. 468.

Morel, P. "Drawing the Sun and the Moon" in *Sky & Telescope*, Apr. 1988, p. 404.
Taylor, P. "Watching the Sun" in *Sky & Telescope*, Feb. 1989, p. 220.

2. Fusion on Earth

Much research has gone into trying to duplicate the fusion process that powers the Sun safely here on Earth. This is a complex, difficult problem, which students majoring in the physical sciences may enjoy exploring further. See:

Craxton, R. "Progress in Laser Fusion" in *Scientific American*, Aug. 1986.
Heppenheimer, T. *The Man-made Sun: The Quest for Fusion Power*. 1984, Little Brown.
Kider, T. "Taming a Star: A Report on Fusion" in *Science '82*, Mar. 1982, p. 54.

3. The Neutrino Problem

Kaufmann discusses the dilemma about solar neutrinos and some of the proposed solutions in section 18-3, citing some of the recent hypotheses that astronomers and physicists have come up with. Students who want to delve further into this puzzling, controversial area can consult:

Bahcall, J. "Where are the Solar Neutrinos" in *Astronomy*, Mar. 1990, p. 40.
Bahcall, J. "Neutrinos from the Sun: An Astronomical Puzzle" in *Mercury*, Mar/Apr. 1990, p. 53.
Field, G. "Is the Theory of Stellar Evolution Wrong?" in *Mercury*, Mar/Apr. 1982, p. 42.
Fischer, D. "Closing In on the Solar Neutrino Problem" in *Sky & Telescope*, Oct. 1992, p. 378.
Norman, E. "Neutrino Astronomy: A New Window to the Universe" in *Sky & Telescope*, Aug. 1985, p. 101.

4. The Solar Max Mission

The malfunction and eventual "space rescue" of the Solar Max satellite are a dramatic story in the short history of manned space exploration. And the results from the Solar Max mission also make for an interesting paper topic. See:

Burnham, R. "Solar Max Returns" in *Astronomy*, Sep. 1984, p. 6.
Maran, S. & Woodgate, B. "A Second Chance for Solar Max" in *Sky & Telescope*, June 1984, p. 16.
Morrison, N. & D. "Early Results from the Solar Maximum Mission" in *Mercury*, Sep/Oct. 1981, p. 136.
Nichols, R. "Solar Max: 1980-1989" in *Sky & Telescope*, Dec. 1989, p. 601.

Overbye, D. "Putting the Arm on Solar Max" in *Discover*, June 1984, p. 16.
Ryan, J. "The Solar Max Mission" in *Astronomy*, May 1981, p. 6.
Verschuur, G. "Will Solar Max Be Saved?" in *Astronomy*, Oct. 1988, p. 34.

5. The Ulysses Probe

After the text was in press, the Ulysses probe was successfully launched onto the complex path that will ultimately take it (via Jupiter) above and below the ecliptic plane to study the polar regions of the Sun. For more on the mission, students can see:

Bennett, G. "Ulysses: Voyage into the Third Dimension" in *Astronomy*, May 1987, p. 15.
Talcott, R. "Seeing the Unseen Sun" in *Astronomy*, Jan. 1990, p. 30.

6. Science Fiction About the Sun

Benford, G. & Eklund, G. *If the Stars Are Gods*. 1977, Berkley. What if the Sun had intelligence within?
Clarke, A. "The Wind from the Sun" in *The Wind from the Sun*. 1973, Signet. About sailing using the solar wind.
Clayton, D. *The Joshua Factor*. 1986, Texas Monthly Press. A thriller by an astronomer involving solar neutrinos.
Clement, H. "Proof" in Asimov, I., ed. *Where Do We Go From Here?* 1971, Fawcett. About life in the Sun.
Niven, L. "Inconstant Moon" in *All the Myriad Ways*. 1971, Ballantine. A giant solar flare wreaks havoc with civilization.

CHAPTER 19
The Nature of Stars

Reading Material for the Student

Bruning, D. "Seeing a Star's Surface" in *Astronomy*, Oct. 1993, p. 35.
Cooke, D. *The Life and Death of Stars*. 1985, Crown.
Cooper, W. & Walker, E. *Getting the Measure of the Stars*. 1989. Adam Hilger. A manual with background and observing projects for advanced students and amateurs.
Croswell, K. "Dance of the Double Sun" in *Astronomy*, July 1993, p. 26.
Davis, J. "Measuring the Stars" in *Sky & Telescope*, Oct. 1991, p. 361. On direct measurements of stellar diameters.

Evans, D., et al. "Measuring Diameters of Stars" in *Sky & Telescope*, Aug. 1979, p. 130.
Fortier, E. "Inside the Hyades" in *Astronomy*, Feb. 1990, p. 78.
Griffin, R. "The Radial Velocity Revolution" in *Sky & Telescope*, Sep. 1989, p. 263.
Hodge, P. "How Far Away are the Hyades" in *Sky & Telescope*, Feb. 1988, p. 138.
Kaler, J. "Origins of the Spectral Sequence" in *Sky & Telescope*, Feb. 1986, p. 129. A fine history.
Kaler, J. *Stars*. 1992. Scientific American/W. H. Freeman. Good modern introduction.
Kaler, J. *Stars and Their Spectra*. 1989, Cambridge U. Press. The definitive book on the spectral classification of stars and what their spectra can teach us. Parts of the book appeared in *Sky & Telescope* magazine: see "O Stars" in Nov. 1987, p. 464; "B Stars" in Aug. 1987, p. 147; "A Stars" in May 1987, p. 491; "F Stars" in Feb. 1987, p. 131; "G Stars" in Nov. 1986, p. 450; "K Stars" in Aug. 1986, p. 130; and "M Stars" in May 1986, p. 450.
Kopal, Z. "Eclipsing Binary Stars: Algol and Its Celestial Relations" in *Mercury*, May/June 1990, p. 88.
Mellillo, F. "Getting Started in Photoelectric Photometry" in *Astronomy*, June 1986, p. 65.
Moore, P. *Astronomers' Stars*. 1987, Norton. Focuses on 15 specific stars that have been important in history, including 61 Cygni.
Morrison, N. & D. "The Eclipsing Binary DI Herculis" in *Mercury*, Nov/Dec. 1985, p. 186.
Reddy, F. "How Far the Stars" in *Astronomy*, June 1983, p. 6.
Sneden, D. "Reading the Colors of the Stars" in *Astronomy*, Apr. 1989, p. 36.
Soderblom, D. "The Alpha Centauri System" in *Mercury*, Sep/Oct. 1987, p. 138.
Trimble, V. "A Field Guide to Close Binary Stars" in *Sky & Telescope*, Oct. 1984, p. 306.
Upgren, A. "New Parallaxes for Old" in *Mercury*, Nov/Dec. 1980, p. 143.

Reading Material for the Instructor

Ashbrook, J. *The Astronomical Scrapbook*. 1985, Cambridge U. Press. Has several vignettes from the history of astronomy relevant to this chapter.
Collins, G. *Fundamentals of Stellar Astrophysics*. 1989, Freeman.
DeVorkin, D. "Steps Toward the Hertzsprung-Russell Diagram" in *Physics Today*, Mar. 1978, p. 32.
Evans. A. "The Orbit of a Visual Binary" in *Sky & Telescope*, Sep. 1980, p. 195. A lab exercise.

Hearnshaw, J. *The Analysis of Starlight*. 1986, Cambridge U. Press. A history of spectroscopy.
Jaschek, C. *The Classification of Stars*. 1987, Cambridge U. Press.
Morgan, W. & Keenan, P. "Spectral Classification" in *Annual Reviews of Astron. & Astrophys.*, vol. 11 (1973).
Paczynksi, B. "Binary Stars" in *Science*, July 20, 1984, p. 275.
Phillip, A., ed. *The H-R Diagram: The 100th Anniversary of Henry Norris Russell*. 1978, Reidel.
Popper, D. "Stellar Masses" in *Annual Reviews of Astron. & Astrophys.*, vol. 18, p. 115 (1980).

Recent Audio-Visual Aids

Astronomers of the Past (slide set with photos of Russell, Hertzsprung, and a number of others mentioned in this chapter; Astronomical Society of the Pacific)
Stars and Their Spectra (36 slides and detailed captions by James Kaler, Astronomical Society of the Pacific)

Topics for Discussion and Papers

1. History of Parallax Measurements

As the text discusses, measuring distances even just to those stars whose parallax can be detected was quite a struggle. Students with an interest in the history of science or technology may want to investigate what was required before good parallax measurements could be made. See:

Fernie, J. "The Historical Search for Stellar Parallax" in *Journal of the Royal Astronomical Society of Canada*, Aug. and Oct. 1975.
Ferris, T. *Coming of Age in the Milky Way*. 1988, Morrow.
Ferris, T. "A Plumb Line to the Sun" in *Mercury*, May/June 1989, p. 66.
Hetherington, N. "The First Measurements of Stellar Parallax" in *Annals of Science*, vol. 28, p. 319 (1972).
Jackson, J. "The Distances of the Stars: A Historical Review" in *Vistas in Astronomy*, vol. 2, p. 1018 (1956).
Strand, K. "Determination of Stellar Distances" in *Science*, vol. 144, p. 1299 (1964).

2. History of the H-R Diagram

Few tools have been as useful for astronomers as the H-R diagram. Students may enjoy delving into the origins of the diagram and the thinking of the scientists who came up with it. See:

DeVorkin, D. "Stellar Evolution and the Origin of the H-R Diagram" in O. Gingerich, ed. *Astrophysics and 20th Century Astronomy to 1950*. 1984, Cambridge U. Press.
DeVorkin, D. "Henry Norris Russell" in *Scientific American*, May 1989.
Gingerich, O. "A Search for Russell's Original Diagram" in *Sky & Telescope*, July 1982, p. 36.
Nielsen, A. "E. Hertzsprung—Measurer of Stars" in *Sky & Telescope*, Jan. 1968, p. 4.
Phillip, A. & Green, L. "Henry N. Russell and the H-R Diagram" in *Sky & Telescope*, Apr. 1978, p. 306; "The H-R Diagram as an Astronomical Tool" in May 1978, p. 395.

3. Spectral Types Mnemonic Competition

As Kaufmann mentions, the traditional mnemonic for remembering the spectral types has been "O Be A Fine Girl Kiss Me" (or for the longer version, including the rarer cool star types, "O Be A Fine Girl, Kiss Me Right Now....Smack"). These obviously date from a time when most astronomy students were men. Although one can make the mnemonic less sexist by changing Girl to "Girl or Guy," many astronomy teachers feel it is time for a new mnemonic. For many years, Owen Gingerich has run a competition among his introductory astronomy course students at Harvard for an improved slogan. You can do the same in your classes.

In the meantime, here are a few "gems" from his collection:

"Oh Boy An F Grade Kills Me!"
"Oh Brutal and Fearsome Gorilla, Kill My Roommate Next Saturday"
"Oh Bring Another Full Grown Kangaroo, My Recipe Needs Some".

4. Observing Binary Stars

The observation of binary stars (using the naked eye, binoculars, or small telescopes) is a favorite activity of amateur astronomers and students first getting to know the night sky. As the text discusses, not every star that appears double on the sky is a true binary, and students may want to keep careful track of how many of each they have observed. Some guides:

Couteau, P. *Observing Visual Double Stars*. 1981, MIT Press. An occasionally technical manual.
Harrington, P. "The Ten Best Double Stars" in *Astronomy*, July 1989, p. 78.
MacRobert, A. "Observing Double Stars" in *Sky & Telescope*, Nov. 1984, p.417.
Meeus, J. "Some Bright Visual Binary Stars' in *Sky & Telescope*, Jan. 1971, p. 21; Feb. 1971, p. 88.
Williams, D. "Binoculars and Binaries" in *Astronomy*, Sep. 1983, p. 75.

5. Women in Astronomy

Section 19-4 of the text introduces the pioneering of work of Annie Cannon (and the other women working for Pickering at the Harvard College Observatory) and of Cecilia Payne around the beginning of the 20th century. The discussion of their particular contributions can lead to a more general examination of the role of women in astronomy and the challenges they have faced in a male dominated world. There are many new articles and books being written on this subject; the representative sampling below can get interested students started.

References on women mentioned in this chapter:

Haramundanis, K., ed. *Cecilia Payne Gaposchkin: An Autobiography and Notes.* 1984, Cambridge U. Press.

Hoffleit, D. "Antonia Maury" in *Sky & Telescope*, Mar. 1952, p. 106.

Kidwell, P. "Cecilia Payne-Gaposchkin: Astronomy in the Family" in P. Abir-Am & Outram, D., eds. *Uneasy Careers and Intimate Lives: Women in Science 1789-1979.* 1987, Rutgers U. Press.

Payne-Gaposchkin, C. "Miss Cannon and Stellar Spectroscopy" in *The Telescope* (predecessor to *Sky & Telescope*), vol. 8, p. 63 (1941).

Reed, H. "Women's Work at the Harvard Observatory" in *New England Magazine*, vol. 6, p. 174 (1892).

Spradley, J. "The Industrious Mrs. Fleming" in *Astronomy*, July 1990, p. 48.

Welther, B. "Annie J. Cannon: Classifier of the Stars" in *Mercury*, Jan/Feb. 1984, p. 28.

Whitney, C. "Cecilia Payne-Gaposchkin: An Astronomer's Astronomer" in *Sky & Telescope*, Mar. 1980, p. 212.

References on women in astronomy in general:

Kohlstedt, S. "Maria Mitchell and the Advancement of Women in Science" in P. Abir-Am & Outram, D., eds. *Uneasy Careers and Intimate Lives: Women in Science 1789-1979.* 1987, Rutgers U. Press.

Mack, P. "Straying from Their Orbits: Women in Astronomy in America" in G. Kass-Simon & P. Farnes, eds. *Women of Science.* 1990, Indiana U. Press.

Rossiter, M. *Women Scientists in America.* 1982, Johns Hopkins U. Press.

Rubin, V. "Women's Work: Women in Modern Astronomy" in *Science 86*, Jul/Aug. 1986, p. 58.

Spradley, J. "Women and the Stars" in *Physics Teacher*, Sep. 1990, p. 372.

Warner, D. "Women Astronomers" in *Natural History*, May 1979, p. 12.

Chapter 20
The Birth of Stars

Reading Material for the Student

Arny, T. & Gordon, D. "The Pleiades" in *Mercury*, Sep/Oct. 1980, p. 113.
Bally, J. "Bipolar Gas Jets in Star Forming Regions" in *Sky & Telescope*, Aug. 1983, p. 94.
Bally, J. & Reipurth, B. "Object 50 in Orion: The Birth of a Nebula" in *Mercury*, Mar/Apr. 1987, p. 46.
Beatty, J. "Hails From the Heliopause" in *Sky & Telescope*, Oct. 1993, p. 30.
Bok, B. "Early Phases of Star Formation" in *Sky & Telescope*, Apr. 1981, p. 284.
Brophy, T. "Motes in the Solar System's Eye" in *Astronomy*, Mar. 1993, p. 34.
Churchwell, E. & Anderson, K. "The Anatomy of a Nebula" in *Astronomy*, June 1985, p. 66. On the Omega Nebula.
Cohen, M. *In Darkness Born: The Story of Star Formation*. 1988, Cambridge U. Press. A fine introduction.
Dame, R. "The Molecular Milky Way" in *Sky & Telescope*, July 1988, p. 22.
Davies, J. "The Herbig Stars" in *Astronomy*, Dec. 1985, p. 90.
Goldsmith, D. & Cohen, N. "The GReat Molecule Factory in Orion" in *Mercury*, Sep/Oct. 1991, p. 148.
Hartley, K. "A New Window on Star Birth" in *Astronomy*, Mar. 1989, p. 32.
Gingerich, O. "Robert Trumpler and the Dustiness of Space" in *Sky & Telescope*, Sep. 1985, p. 213.
Hartmann, W., et al. *Cycles of Fire*. 1987, Workman. Lavishly illustrated introduction to stellar evolution.
Higgins, D. & Eicher, D. "The Secret World of Dark Nebulae" in *Astronomy*, Sep. 1987, p. 46. A guide for amateur astronomers.
Kanipe, J. "Inside Orion's Stellar Nursery" in *Astronomy*, Aug. 1989, p. 40.
Lada, C. "Stars in the Making" in *Sky & Telescope*, Oct. 1986, p. 334.
Marschall, L. "The Secrets of Interstellar Clouds" in *Astronomy*, Mar. 1982, p. 6.
Mumford, G. "The Legacy of E. E. Barnard" in *Sky & Telescope*, July 1987, p. 30.
Norris, R. "Cosmic Masers" in *Sky & Telescope*, Mar. 1986, p. 248.
Peterson, I. "The Winds of Starbirth" in *Science News*, June 30, 1990, p. 408.
Reipurth, B. "Bok Globules" in *Mercury*, Mar/Apr. 1984, p. 50.
Schild, R. "A Star is Born" in *Sky & Telescope*, Dec. 1990, p. 600.
Shore, L. & S. "The Chaotic Material Between the Stars" in *Astronomy*, June 1988, p. 6.

Spitzer, L. "Interstellar Matter and the Birth and Death of Stars" in *Mercury*, Sep/Oct. 1983, p. 142.
Stahler, S. & Comin, N. "The Difficult Births of Sunlike Stars" in *Astronomy*, Sep. 1988, p. 22.
van Buren, D. "Bubbles in the Sky" in *Astronomy*, Jan. 1993, p. 46.
Verschuur, G. *Interstellar Matters*. 1989, Springer Verlag. Essays on dust and the interstellar medium.
Verschuur, G. "Interstellar Molecules" in *Sky & Telescope*, Apr. 1992, p. 379.
Verschuur, G. *The Invisible Universe Revealed: The Story of Radio Astronomy*. 1987, Springer Verlag.
Waldrop, M. "Stellar Nurseries" in *Science '83*, May 1983, p. 50.

Reading Material for the Instructor

Bally, J. "Interstellar Molecular Clouds" in *Science*, Apr. 11, 1986, p. 185.
Bertout, C. "T Tauri Stars: Wild as Dust" in *Annual Reviews of Astron. & Astrophys.*, vol. 27, p. 351 (1989).
Blitz, L. "Giant Molecular Cloud Complexes in the Galaxy" in *Scientific American*, Apr. 1982.
Boss, A. "Collapse and Formation of Stars" in *Scientific American*, Jan. 1985.
Cox, D. & Reynolds, R. "The Local Interstellar Medium" in *Annual Reviews of Astron. and Astrophys.*, vol. 25, p. 303. (1987)
Gehrz, R., et al. "The Formation of Stellar Systems from Interstellar Molecular Clouds" in *Science*, May 25, 1984, p. 823.
Genzel, R. & Stutzki, J. "The Orion Molecular Cloud and Star-Forming Region" in *Annual Reviews of Astron. & Astrophys.*, vol. 27, p. 41 (1989).
Greenberg, J. "The Structure and Evolution of Interstellar Grains" in *Scientific American*, June 1984.
Lada, C. "Cold Outflows, Energetic Winds, and Enigmatic Jets Around Young Stellar Objects" in *Annual Reviews of Astron. and Astrophys.*, vol. 23, p. 267. (1985)
Mathis, J. "Interstellar Dust and Extinction" in *Annual Reviews of Astron. & Astrophys.*, vol. 28, p. 37 (1990).
Morfill, G., et al., eds. *Physical Processes in Interstellar Clouds*. 1987, Reidel.
Neugebauer, G., et al. "Early Results from the IRAS Mission" in *Science*, Apr. 6, 1984, p. 13.
Peimbert, M., et al., eds. *Star Forming Regions*. 1986, Reidel.
Schwartz, R. "Herbig Haro Objects" in *Annual Reviews of Astron. & Astrophys.*, vol. 21, p. 209 (1983).
Scoville, N. & Young, J. "Molecular Clouds, Star Formation, and Galactic Structure" in *Scientific American*, Apr. 1984.
Shu, F., et al. "Star Formation in Molecular Clouds: Observation and Theory" in *Annual Reviews of Astron. and Astrophys.*, vol. 25, p. 23. (1987)

Spitzer, L. *Searching Between the Stars*. 1982, Yale U. Press.
Welch, W., et al. "Gas Jets Associated with Star Formation" in *Science*, June 21, 1985, p. 138.
Wynn-Williams, G. *The Fullness of Space: Nebulae, Stardust, and the Interstellar Medium*. 1992, Cambridge U. Press.
Yorke, H. "The Dynamical Evolution of H II Regions" in *Annual Reviews of Astron. and Astrophys.*, vol. 24, p. 49. (1986)

Recent Audio-Visual Aids

Cosmic Clouds (5 slides of nebulae taken with the UK Schmidt Telescope, Hansen Planetarium)
Countdown to the Invisible Universe (58 min film from the *Nova* TV series on IRAS and its results; Coronet Films)
The Infrared Universe: An IRAS Gallery (25 slides, with detailed captions and background information by IRAS Project Scientist Charles Beichman; Astronomical Society of the Pacific)
Southern Wonders (5 slides, including 3 of nebulae, taken with the UK Schmidt Telescope; Hansen Planetarium)
Splendors of the Universe: Nebulae and Galaxies (3 sets of 15 slides each taken by David Malin with the AAT and Uk Schmidt, with detailed booklets of captions and background information; Astronomical Society of the Pacific)
"Stardust" (an episode of *The Astronomers* TV series on the birth and death of stars; from Pacific Arts or the Astronomical Society of the Pacific)

Topics for Discussion and Papers

1. E. E. Barnard

Section 20-2 briefly mentions the pioneering work of Edward Barnard (1857-1923) on the dark clouds of dust in our galaxy. Barnard was a mostly self-educated man, who became one of the leading visual observers and photographers among American astronomers. Among other accomplishments he discovered the fifth satellite of Jupiter, and found the star with the largest known proper motion, now named after him. Students who want to investigate the life of this interesting astronomer further can begin with:

Frost, E. "E. E. Barnard" in *Astrophysical Journal*, vol. 58, p. 1 (1923).
Hardie, R. "The Early Life of E. E. Barnard" in *Leaflets of the Astronomical Society of the Pacific*, no. 415 (1964).
Mumford, G. "The Legacy of E. E. Barnard" in *Sky & Telescope*, July 1987, p. 30.
Verschuur, G. *Interstellar Matters*. 1989, Springer Verlag. See chapters 2-6.
Verschuur, G. "Barnard's Dark Dilemma" in *Astronomy*, Feb. 1989, p. 30.

2. Interstellar Molecules

The presence of complex molecules in interstellar clouds is a relatively recent discovery that surprised many astronomers. In fact, when in the late 1950s, Fred Hoyle wanted to suggest that such molecules could be found in these clouds, the idea was so outside the realm of current astronomical thinking that he wrote it up as a science fiction novel (*The Black Cloud*, Signet 1959). Students may enjoy tracing how this idea became acceptable in astronomy by tracing mentions of it in textbooks over the years and looking at early articles and comparing them to recent ones:

Early references

Buhl, D. "Light Molecules and Dark Clouds" in *Mercury*, Sep/Oct & Nov/Dec. 1972.
Tuner, B. "Interstellar Molecules" in *Scientific American*, Mar. 1973.

Recent references

Darling, D. "Molecules Between the Stars" in *Astronomy*, Mar. 1982, p. 82.
Norris, R. "Cosmic Masers" in *Sky & Telescope*, Mar. 1986, p. 248.
Verschuur, G. "Molecules Between the Stars" in *Mercury*, May/June 1987, p. 66.
Verschuur, G. *Interstellar Matters*. 1989, Springer Verlag. See chapters 19 and 20.

3. Supernovae and Star Birth

Recently, the idea that supernovae might help trigger star formation has generated some intriguing investigations of possible evidence that a supernova blew up "in the neighborhood" just before our Sun formed. Students who want to look into this might consult:

Cohen, M. "Do Supernovae Trigger Star Formation" in *Astronomy*, Apr. 1982, p. 16.
Falk, S. & Schramm, D. "Did the Solar System Start with a Bang?" in *Sky & Telescope*, July 1979, p. 18.

Note: For more on the life and work of Bart Bok, see topic 25.1.

Chapter 21
Stellar Maturity and Old Age

Reading Material for the Student

Clark, G. "Stellar Populations: Key to the Clusters" in *Astronomy*, Oct. 1986, p. 106; "Ancients of the Universe: The Globular Clusters" in May 1985, p. 6.
Cohen, M. "The Vanishing Act of Carbon Giants" in *Astronomy*, June 1982, p. 66.
Cooke, D. *The Life and Death of Stars.* 1985, Crown. A nontechnical introduction.
Croswell, K. "When Stars Coalesce" in *Astronomy*, May 1985, p. 67.
Darling, D. "Breezes, Bangs, and Blowouts: Stellar Evolution Through Mass Loss" in *Astronomy*, Sep. 1985, p. 78; Nov. 1985, p. 94.
Friel, E. "A Symposium on Stellar Populations" in *Mercury*, Nov/Dec. 1984, p. 165.
Fortier, E. "Touring the Stellar Cycle" in *Astronomy*, Mar. 1987, p. 49. Observing objects that show the stages of stellar evolution.
Glasby, J. *Variable Stars.* 1968, Harvard U. Press.
Harris, W., et al. "Probing 47 Tucanae" in *Sky & Telescope*, Sep. 1984, p. 219.
Hartmann, W., et al. *Cycles of Fire.* 1987, Workman. Illustrated book on stellar evolution.
Hawkins, M. "R15: The Most Distant RR Lyrae Variable" in *Mercury*, Sep/Oct. 1985, p. 145.
Johnson, B. "Red Giant Stars" in *Astronomy*, Dec. 1976, p. 26.
Kaler, J. "The Coolest Stars" in *Astronomy*, May 1990, p. 20.
Kaler, J. *Stars.* 1992, Scientific American/ W. H. Freeman.
Kippenhahn, R. *100 Billion Suns: The Birth, Life and Death of Stars.* 1983, Basic Books.
Kopal, Z. "Eclipsing Binary Stars: Algol and Its Celestial Relations" in *Mercury*, May/June 1990, p. 88.
Levy, D. Jedicke, P. "Betelgeuse" in *Astronomy*, Apr. 1987, p. 6.
Moore, P. *Astronomers' Stars.* 1987, Norton.
Percy, J. "Cepheids: Cosmic Yardsticks, Celestial Mysteries?" in *Sky & Telescope*, Dec. 1984, p. 517.
Percy, J., ed. *The Study of Variable Stars Using Small Telescopes.* 1986, Cambridge U. Press. A manual for observers with good background information.
Plavec, M. "IUE Looks at the Algol Paradox" in *Sky & Telescope*, May 1983, p. 413.

Tomkin, J. & Lambert, D. "The Strange Case of Beta Lyrae" in *Sky & Telescope*, Oct. 1987, p. 354.
Trimble, V. "A Field Guide to Close Binary Stars" in *Sky & Telescope*, Oct. 1984, p. 306.
Trimble, V. "How to Survive Cataclysmic Binaries" in *Mercury*, Jan/Feb. 1980, p. 8.
van den Bergh, S. "Star Clusters: Enigmas in Our Backyard" in *Sky & Telescope*, May 1992, p. 508.
White, R. "Globular Clusters: Fads and Fallacies" in *Sky & Telescope*, Jan. 1991, p. 24.
Wyckoff, S. "Red Giants: The Inside Scoop" in *Mercury*, Jan/Feb. 1979, p. 7.

Reading Material for the Instructor

Abbott, D. & Conti, P. "Wolf-Rayet Stars" in *Annual Reviews of Astron. and Astrophys.*, vol. 25, p. 113 (1987).
Bohm-Vitense, E. *Introduction to Stellar Astrophysics*, vol. 1. 1989, Cambridge U. Press.
Chiosi, C. & Maeder, A. "The Evolution of Massive Stars with Mass Loss" in *Annual Reviews of Astron. and Astrophys.*, vol. 24, p. 329 (1986).
Dupree, A. "Mass Loss from Cool Stars" in *Annual Reviews of Astron. and Astrophys.*, vol. 24, p. 377 (1986).
Eggleton, P., et al., eds. *Interacting Binaries*. 1985, Reidel.
Feast, M. & Walker, A. "Cepheids as Distance Indicators" in *Annual Reviews of Astron. and Astrophys.*, vol. 25, p. 345 (1987).
Hack, M. "Epsilon Aurigae" in *Scientific American*, Oct. 1984.
Hoffmeister, C., et al. *Variable Stars*. 1985, Springer-Verlag.
Humphreys, R. & Davidson, K. "The Most Luminous Stars" in *Science*, Jan. 20, 1984, p. 243.
Iben, I. & Renzini, A. "Asymptotic Giant Branch Evolution and Beyond" in *Annual Reviews of Astron. & Astrophys.*, vol. 21, p. 271 (1983).
Kafatos, M. & Michalitsianos, A. "Symbiotic Stars" in *Scientific American*, July 1984.
Kaler, J. *Stars and Their Spectra*. 1989, Cambridge U. Press.
Kenyon, S. *The Symbiotic Stars*. 1986, Cambridge U. Press.
Madore, B., ed. *Cepheids: Theory and Observations*. 1985, Cambridge U. Press.
Morris, M., et al., eds. *Mass Loss from Red Giants*. 1985, Reidel.
Paczyinksi, B. "Binary Stars" in *Science*, July 20, 1984, p. 275.
Payne-Gaposchkin, C. "The development of Our Knowledge of Variable Stars" in *Annual Reviews of Astron. & Astrophys.*, vol. 16, p. 1 (1978).

Pringel, J. & Wade, R., eds. *Interacting Binary Stars*. 1985, Cambridge U. Press.

Wallerstein, G. "Mixing in Stars" in *Science*, June 24, 1988, p. 174.

Recent Audio-Visual Aids

"Lives of the Stars" (Episode of the *Cosmos* TV series, available in home video format from the Astronomical Society of the Pacific)

The Perfect Stargazer (A video on Robotic Observatories and variable star observing, soon to be available from the Astronomical Society of the Pacific)

Topics for Discussion and Papers

1. When the Sun Becomes A Red Giant

Students may find it especially interesting to consider what will happen to our solar system (and to the Earth in particular) when our Sun becomes a red giant. Before they get depressed, it is important to remind them of the enormous time scale of stellar evolution (and how much more likely we are to destroy our civilization through our own follies long before we need to worry about the destructive effects of stellar evolution.) See:

Whitmire, D. & Reynolds, R. "The Fiery Fate of the Solar System" in *Astronomy*, Apr. 1990, p. 20.

(In the science fiction novel *Ossian's Ride* by astronomer Fred Hoyle, aliens come to Earth fleeing the disaster of their star having become a red giant.)

2. Observing Variable Stars

Many amateur astronomers enjoy monitoring and measuring variable stars and contributing their measurements to the growing file of data kept by the American Association of Variable Star Observers (see below). There are many guides and articles for variable star observing for beginners; here are a few to start with:

Hall, D. & Genet, R. *Photoelectric Photometry of Variable Stars*. 1988, Willmann-Bell.

Levy, D. "How to Enjoy Variables" in *Astronomy*, Jan. 1985, p. 74.

Levy, D. *Observing Variable Stars*. 1989, Cambridge U. Press.

MacRoberts, A. "The Lure of the Variables" in *Sky & Telescope*, Feb. 1985, p. 124.

Mattei, J. "Observing Variable Stars" in *Sky & Telescope*, Sep. 1980, p. 180; Oct. 1980, p. 285.
Mellilo, F. "Photometry of Variable Stars and Asteroids" in *Astronomy*, July 1986, p. 58.
Percy, J. "Observing Variable Stars for Fun and Profit" in *Mercury*, May/June 1979, p. 45.
Percy, J., ed. *The Study of Variable Stars Using Small Telescopes*. 1986, Cambridge U. Press. A manual for observers with good background information.

For students with an interest in amateur astronomy projects, a variety of excellent observing aids for variable stars are available from the American Association of Variable Star Observers, 25 Birch St., Cambridge, MA 02138.

3. A History of Finding and Understanding Cepheids

The realization that stars varied in brightness and the eventual discovery of the enormous utility of some variable stars makes for interesting historical reading. Students can consult:

Gilman, C. "John Goodricke and His Variable Stars" in *Sky & Telescope*, Nov. 1978, p. 400.
Fernie, J. "The Period-Luminosity Relation: A Historical Review" in *Publications of the A.S.P.*, vol. 81, p. 707 (1969).
Hoskin, M. "Goodricke, Pigott, and the Quest for Variable Stars" in *Stellar Astronomy*. 1982, Science History Publications.
Kopal, Z. "Eclipsing Binary Stars: Algol and Its Celestial Relations" in *Mercury*, May/June 1990, p. 88. Includes historical material on variables and Cepheids.

4. Observing Star Clusters

For beginning observers, star clusters present an excellent target for binoculars and telescopes. Here are some guides:

Fortier, E. "Inside the Hyades" in *Astronomy*, Feb. 1990, p. 78.
Haas, J. "Galactic Clusters for Binoculars" in *Astronomy*, Feb. 1986, p. 62.
Harrington, P. "An Observer's Guide to Globular Clusters" in *Sky & Telescope*, Aug. 1986, p. 198.
Ling, A. "Standout Winter Star Clusters" in *Astronomy*, Jan. 1989, p. 98.
Schur, C. "Dramatically Diverse Globulars" in *Astronomy*, May 1989, p. 90.

Chapter 22
The Deaths of Stars

Reading Material for the Student

Asimov, I. *The Exploding Suns.* 1985, Dutton. (A well-written primer on supernovae for beginners.)
Balick, B., et al. "The Shaping of Planetary Nebulae" in *Sky & Telescope*, Feb. 1987, p. 125.
Blair, W., et al. "A Powerful Young Supernova Remnant in Another Galaxy" in *Mercury*, May/Jun 1985, p. 80.
Bruning, D. "Neon Nova" in *Astronomy*, July 1993, p. 37.
Clark, D. *Superstars.* 1984, McGraw Hill.
Cooke, D. *The Life and Death of Stars.* 1985, Crown.
deVaucouleurs, G. "The Supernova of 1885 in Messier 31" in *Sky & Telescope*, Aug. 1985, p. 115.
Dopita, M. & Tuohy, I. "The Young Supernova Remnant 1E0102.2-7219" in *Mercury*, Jan/Feb. 1984, p. 25.
Evans, R. & Reeves, R. "Supernova Hunter" in *Astronomy*, Nov. 1989, p. 94.
Fried, E. "The Ungentle Death of a Giant Star" in *Science '86*, Jan/Feb 1986, p. 60.
Graham, J. "The Iron Sun" in *Sky & Telescope*, May 1986, p. 437.
Helfand, D. "Supernovae: Creative Cataclysms in the Galaxy" in *The Universe*, B. Preiss & A. Fraknoi, eds. 1987, Bantam.
Hesser, J. & Fraknoi, A. "The 1985 Amateur Achievement Award to Robert Evans and Gregg Thompson" in *Mercury*, Mar/Apr 1985, p. 44. (On the contributions of amateurs to supernova hunting.)
Hjellming, R., et al. "Mapping Planetary Nebulae with the VLA" in *Sky & Telescope*, Sep. 1978, p. 199.
Kahn, R. "Desperately Seeking Supernovae" in *Sky & Telescope*, Jun. 1987, p. 594.
Kaler, J. "Planetary Nebulae and Stellar Evolution" in *Mercury*, Jul/Aug. 1981, p. 114.
Kaler, J. "Bubbles from Dying Stars" in *Sky & Telescope*, Feb. 1982, p. 129.
Kaler, J. "The Smallest Stars in the Universe" in *Astronomy*, Nov. 1991, p. 50. On white dwarfs, neutron stars, pulsars.
Kawaler, S. & Winget, D. "White Dwarfs: Fossil Stars" in *Sky & Telescope*, Aug. 1987, p. 132.
Marschall, L. *The Supernova Story.* 1988, Plenum. The introduction of choice to supernovae and Supernova 1987A.
Morrison, N. & Gregory, S. "What Makes Massive Stars Explode?" in *Mercury*, May/Jun 1986, p. 77.

Murdin, P. & L. *Supernovae*. 1985, Cambridge U. Press.
Naeye, R. "Supernova 1987A Revisited" in *Sky & Telescope*, Feb. 1993, p. 39.
Nather, R. & Winget, D. "Taking the Pulse of White Dwarfs" in *Sky & Telescope*, Apr. 1992, p. 374.
Nicastro, A. "White Dwarfs" in *Astronomy*, July 1984, p. 6.
Norman, E. "Neutrino Astronomy: A New Window to the Universe" in *Sky & Telescope*, Aug. 1985, p. 101.
Pennypacker, C. "Searching for Supernovae: The Discovery in M99" in *Astronomy*, Jul. 1987, p. 75.
Reddy, F. "Supernovae: Still a Challenge" in *Sky & Telescope*, Dec. 1983, p. 485.
Schmidt, G. "A Supermagnetic Star" in *Mercury*, Jan/Feb. 1987, p. 24. (On a magnetic white dwarf.)
Straka, W. "The Cygnus Loop: An Older Supernova Remnant" in *Mercury*, Sep/Oct 1987, p. 150.
Thompson, G. & Bryan, J. *The Supernova Search Charts and Handbook*. 1989, Cambridge U. Press. A guide for discovering supernovae in nearby galaxies.
Tierney, J. "The Quest for Order: Profile of S. Chandrasekhar" in *Science '82*, sep. 1982, p. 68.
Trimble, V. "White Dwarfs: The Once and Future Suns" in *Sky & Telescope*, Oct. 1986, p. 348.
Tucker, W. "Exploding Stars, Superbubbles, and HEAO Observations" in *Mercury*, Sep/Oct. 1984, p. 130.
Wallerstein, G. & Wolff, S. "The Next Supernova" in *Mercury*, Mar/Apr. 1981, p. 44.
Weiler, K. "A New Look at Supernova Remnants" in *Sky & Telescope*, Nov. 1979, p. 414.
Weiler, K. "3C58: Only the Second Known Plerion" in *Mercury*, mar/Apr. 1981, p. 42.

Reading Material for the Instructor

Bethe, H. & Brown, G. "How a Supernova Explodes" in *Scientific American*, May 1985.
Chandrasekhar, S. "On Stars, Their Evolution, and Their Stability" in *Science*, Nov. 2, 1984, p. 497. His Nobel Prize lecture.
d'Antona, F. & Mazzitelli, I. "Cooling of White Dwarfs" in *Annual Reviews of Astron. & Astrophys.*, vol. 28, p. 139 (1990).
Jeffries, A., et al. "Gravitational Wave Observatories" in *Scientific American*, June 1987.
Kaler, J. "Planetary Nebulae and Their Central Stars" in *Annual Reviews of Astron. & Astrophys.*, vol. 23, p. 89 (1985).

Koshiba, M. "Observational Neutrino Astrophysics" in *Physics Today*, Dec. 1987, p. 38.
Kwok, S., et al., eds. *Late Stages of Stellar Evolution*. 1987, Reidel.
Pottasch, S. *Planetary Nebulae*. 1984, Reidel.
Raymond, J. "Observations of Supernova Remnants" in *Annual Reviews of Astron. & Astrophys.*, vol. 22, p. 75 (1984).
Rees, M., et al. eds. *Supernovae: A Survey of Current Research*. 1989, Reidel/Kluwer.
Seward, F., et al. "Young Supernova Remnants" in *Scientific American*, May 1985.
Wali, K. "Chandrasekhar Versus Eddington: An Unexpected Confrontation" in *Physics Today*, Oct. 1982, p. 33.
Wheeler, J. & Harkness, R. "Helium-rich Supernovae" in *Scientific American*, Nov. 1987, p. 50.
Weiler, K., et al. "Radio Studies of Extragalactic Supernovae" in *Science*, Mar. 14, 1986, p. 125.
Weiler, K. & Sramek, R. "Supernovae and Supernova Remnants" in *Annual Reviews of Astron. & Astrophys.*, vol. 26, p. 295 (1988).
Weideman, V. "Masses and Evolutionary Status of White Dwarfs and Their Progenitors" in *Annual Reviews of Astron. & Astrophys.*, vol. 28, p. 103 (1990).
Woosley, S. & Weaver, T. "The Physics of Supernova Explosions" in *Annual Reviews of Astron. & Astrophys.*, vol. 24, p. 205 (1986).

Reading Material for the Student on Supernova 1987A

Goldsmith, D. *Supernova! The Exploding Star of 1987*. 1989, St. Martin's.
Kirschner, R. "Supernova: The Death of A Star" in *National Geographic*, May 1988, p. 618.
Lattimer, J. & Burrows, A. "Neutrinos from Supernova 1987A" in *Sky & Telescope* Oct. 1988, p. 348.
Lemonick, M. "Supernova!" in *Time*, March 23, 1987. (A nicely done cover story with background and early results.)
Malin, D. & Allen, D. "Echoes of the Supernova" in *Sky & Telescope*, Jan. 1990, p. 22.
Marschall, L. *The Supernova Story*. 1988, Plenum. The introduction of choice to supernovae and Supernova 1987A.
Murdin, P. *End In Fire: The Supernova in the Large Magellanic Cloud*. 1990, Cambridge U. Press.
Schorn, R. "Supernova Shines On" and "Neutrinos From Hell" in *Sky & Telescope*, May 1987, p. 470. (Has full story of discovery, but premature

information on which star exploded; good details on spectra and excellent separate story on neutrinos.)

Schorn, R. "Supernova 1987A: Watching and Waiting" in *Sky & Telescope*, Jul. 1987, p. 14.

Schorn, R. "Supernova 1987A After 200 Days" in *Sky & Telescope*, Nov. 1987, p. 477.

Schorn, R. "Supernova 1987A's Changing Face" in *Sky & Telescope*, July 1988, p. 32.

Talcott, R. "A Burst of Discovery: The First Days of Supernova 1987A" in *Astronomy*, June 1987, p. 90.

Talcott, R. "Insight Into Star Death" in *Astronomy*, Feb. 1988, p. 6. Excellent review.

Thomsen, D. "Neutrino Astronomy Born in a Supernova" in *Science News*, Vol. 131, p. 180.

Thomsen, D. "Supernova: High On Understanding" in *Science News*, May 2, 1987, p. 279. (Nice one-page summary of APS meeting discussion, late April 1987.)

Tierney, J. "Exploding Star Contains Atoms of Elvis Presley's Brain" in *Discover*, July 1987, p. 46. (Despite its title, this a good anecdotal report on the supernova in nontechnical terms.)

Waldrop, M. "Supernova 1987A Shows a Mind of its Own — and a Burst of Neutrinos" in *Science*, Vol. 235, p. 1322 (13 March 1987) (2 pages, mostly on neutrino observations.)

Waldrop, M. "Supernova 1987A: Notes from All Over" in *Science*, Vol. 236, p. 522 (1 May 1987) (2 pp. on recent developments.)

Waldrop, M. "Supernova 1987A on Center Stage" in *Science*, 20 Nov. 1987, p. 1038.

Reading Material for the Instructor on Supernova 1987A

Arnett, D., et al. "Supernova 1987A" in *Annual Reviews of Astron. & Astrophys.*, vol. 27, p. 629 (1989).

Burrows, A. "Neutrinos from Supernova Explosions" in *Annual Reviews of Nuclear and Particle Science*, vol. 40 (1990).

Danziger, I., ed. *Workshop on SN1987A*. 1987, European Southern Observatory, Karl-Schwarzschild Str. 2, D 8046, Garching bei Munchen, Fed. Rep. of Germany.

Helfand, D. "Bang: The Supernova of 1987" in *Physics Today*, Aug. 1987, p. 25.

Woosley, S. & Phillips, M. "Supernova 1987A!" in *Science*, May 6, 1988, p. 750. Nice review.

Woosley, S. & Weaver, T. "The Great Supernova of 1987" in *Scientific American*, Aug. 1989.

Recent Audio-Visual Aids

Death of A Star (1988 Episode of NOVA TV series; 60 min., Coronet Films) On Supernova 1987A.
Supernova 1987A (6 slides, with 24-page booklet of captions, background information, and references; Astronomical Society of the Pacific)

Images of planetary nebulae and supernova remnants can be found in the following slide sets:

Radio Universe (Astronomical Society of the Pacific)
Splendors of the Southern Skies (Hansen Planetarium)
Splendors of the Universe (Astronomical Society of the Pacific)
"Stardust" (an episode of *The Astronomers* TV series on the birth and death of stars: from Pacific Arts or the Astronomical Society of the Pacific)

Topics for Discussion and Papers

1. The Discovery of Sirius B

For more on the discovery of Sirius' faint companion and how it led to a better understanding of white dwarfs, students can consult:

Irwin, J. "The Case of the Degenerate Dwarf" in *Mercury*, Nov/Dec. 1978, p. 125.
Hetherington, N. "Sirius B and the Gravitational Redshift: An Historical Review" in *Quarterly Journal of the Royal Astronomical Society of Canada*, vol. 21, p. 246 (1980).
Parker, B. "Those Amazing White Dwarfs" in *Astronomy*, July 1984, p. 15.

2. Observing Planetary Nebulae and Supernova Remnants

Kaufmann discusses some famous planetary nebulae in Box 22-1 and some of the best-known supernovae remnants in section 22-5. Students who want to observe these "last gasps" of stellar evolution can refer to the following guides:

Eicher, D. "The Art of Observing Planetaries" in *Astronomy*, Apr. 1989, p. 68.
Juhnke, C. "A Delightful Dozen of Planetary Nebulae" in *Astronomy*, Oct. 1986, p. 39.
Marling, J. "In Pursuit of Planetaries" in *Sky & Telescope,* June 1986, p. 631.
Schur, C. "Find a Supernova Remnant" in *Astronomy*, Feb. 1990, p. 72.

3. Historical Supernovae

Section 22-5 lists some of the supernovae that were seen throughout history; students may want to follow up by learning more about them or about older supernova suspects in:

Clark, D. *Superstars.* 1984, McGraw Hill.
Kamper, K. "Tycho's Supernova" in *Mercury*, Jul/Aug. 1980, p. 97.
Maran, S. "The Gum Nebula" in *Scientific American*, Dec. 1971.
Murdin, P. & L. *Supernovae.* 1985, Cambridge U. Press.
Stephenson, F. & Clark, D. "Historical Supernovas" in *Scientific American*, Dec. 1974.
Stephenson, F. & Clark, D. "Ancient Astronomical Records from the Orient" in *Sky & Telescope*, Feb. 1977, p. 84.
Tucker, W. "Tycho's Supernovae" in *Astronomy*, Oct. 1982, p. 74.
Van den Bergh, S. "Optical Remnant of the Supernova of 1181" in *Sky & Telescope*, Mar. 1978, p. 196.
Verschuur, G. "A New Look at the Crab Nebula" in *Astronomy*, May 1990, p. 30.

4. Supernovae and Life

Not only do supernovae serve to inject new elements into the raw material of the cosmos and thus to provide the elemental complexity which we believe to be necessary for life, they can also — it appears — bring life in a star system to an end should they explode too close. Students with a necrological interest can research the devastating effects of a nearby supernova in:

Chapman, C. & Morrison, D. *Cosmic Catastrophes.* 1989, Plenum Press. Chapter 18.
Clark, D. "Extraterrestrial Climate Threats" in *Astronomy*, Aug. 1980, p. 66.
Ruderman, M. "Possible Consequences of Nearby Supernova Explosions" in *Science*, vol. 184, p. 1079 (June 7, 1974). Somewhat technical.
Tucker, W. "Supernovae, Dinosaurs, and Us" in *Mercury*, July/Aug. 1980, p. 95.

5. Supernovae in Science Fiction

A supernova can make quite a splash in a science fiction story and a number of authors have used one or more to good effect. Some examples:

Anderson, P. "Kyrie" in Pournelle, J., ed. *Black Holes.* 1978, Fawcett.
Anderson, P. "Day of Burning" in *Beyond the Beyond.* 1969, Signet.
Clarke, A. "The Star" in *The Nine Billion Names of God.* 1967, Signet.
Cowper, R. *The Twilight of Briarius.* 1974, John Day.
Niven, L. "At the Core" in *Neutron Star.* 1968, Ballantine.

Silverberg, R. "The Iron Star" in Preiss, B. & Fraknoi, A., eds. *The Universe*. 1987, Bantam.

6. Gravity Waves

For more on the interesting subject of gravity waves, introduced in section 22-6, students can see:

Bartusiak, M. "Sensing the Ripples in Space-time" in *Science '85*, Apr. 1985, p. 58. On experiments to detect gravitational radiation.
Bartusiak, M. "Einstein's Unfinished Symphony" in *Discover*, Aug. 1989, p. 62.
Davies, P. *The Search for Gravity Waves*. 1980, Cambridge U. Press.
Kaufmann, W. "The Future of Gravitational Wave Astronomy: An Interview with Kip Thorne" in *Mercury*, May/June 1978, p. 58.
Kaufmann, W. "Listening for the Whisper of Gravity Waves" in *Science '80*, May/June 1980, p. 64.
Thomsen, D. "Trying to Rock with Gravity's Waves" in *Science News*, Aug. 4, 1984, p. 76.
Trimble, V. "Gravity Waves: A Progress Report" in *Sky & Telescope*, Oct. 1987, p. 364.
Will, C. "The Binary Pulsar: Gravity Waves Exist" in *Mercury*, Nov/Dec. 1987, p. 162.

7. Sirius B and the Dogon Tribe

One of the more bizarre little sidelights in astronomy is that a primitive African tribe called the Dogon apparently knew about the white dwarf Sirius B and incorporated it into their tribal mythology. Several sensationalist authors have discussed how this could only be the result of interstellar visitors sharing cosmic secrets with the Dogon, but further investigation has revealed that the most likely source of the information was European visitors, not interstellar ones. Students who want to see how such anthropological discoveries can be exploited—and then set right again—can consult:

Brecher, K. "Sirius Enigmas" in *Astronomy of the Ancients*. 1979, MIT Press.
Ridpath, I. "Investigating the Sirius Mystery" in *Skeptical Inquirer*, Fall 1978, p. 56.
Ridpath, I. "The Amphibians of Sirius" in *Messages from the Stars*. 1978, Harper and Row.
Sagan, C. "White Dwarfs and Little Green Men" in *Broca's Brain*. 1979, Random House.
Steffey, P. "Some Serious Astronomy in the Sirius Mystery" in *Griffith Observer*, Sep. 1980; Feb., Mar., Apr., May, Aug., Oct., Nov. 1982.

CHAPTER 23
Neutron Stars

Reading Material for the Student

Bailyn, C. "Problems with Pulsars" in *Mercury*, MAr/Apr. 1991, p. 55.
Charles, P. The Mysterious SU Ursa Majoris Stars" in *Sky & Telescope*, June 1990, p. 607. A class of cataclysmic variables.
Graham-Smith, F. "Pulsars Today" in *Sky & Telescope*, Sep. 1990, p. 240.
Greenstein, G. *Frozen Star: Of Pulsars, Black Holes and the Fate of Stars.* 1984, Freundlich Books.
Greenstein, G. "Neutron Stars and the Discovery of Pulsars" in *Mercury*, Mar/Apr. 1985, p. 34; May/June 1985, p. 66.
Grindlay, J. "New Bursts in Astronomy" in *Mercury*, Sep/Oct. 1977, p. 6.
Helfand, D. "Pulsars" in *Mercury*, May/June 1977, p. 2.
Hewish, A. "Pulsars After 20 Years" in *Mercury*, Jan/Feb. 1989, p. 12. About the discovery and our current knowledge.
Hurley, K. "What Are Gamma-Ray Bursters" in *Sky & Telescope*, Aug. 1990, p. 143.
Kaler, J. "The Smallest Stars in the Universe" in *Astronomy*, Nov. 1991, p. 50. On white dwarfs, neutron stars, pulsars.
Kwok, S. "A Star That Has Risen from the Dead" in *Astronomy*, Aug. 1984, p. 66. On HM Sge.
Lewin, W. "The Mystery of X-Ray Burst Sources" in Cornell, J. & Gorenstein, P., eds. *Revealing the Universe*. 1982, MIT Press.
Lovell, B. *The Jodrell Bank Telescopes*. 1985, Oxford U. Press.
Mason, K. & Cordova, F. "Satellite Observations of Cataclysmic Variables" in *Sky & Telescope*, July 1982, p. 25.
Preston, R. "The Eclipsing Death Star" in *Discover*, Aug. 1988, p. 40. A pulsar with a white dwarf companion.
Seward, F. "Neutron Stars in Supernova Remnants" in *Sky & Telescope*, Jan. 1986, p. 6.
Seward, F. "The Discovery of a New Pulsar in a Supernova Remnant" in *Mercury*, Mar/Apr. 1983, p. 56.
Sexl, R. & H. "Curved Spacetime Near a Neutron Star" in *Mercury*, Mar/Apr. 1980, p. 38.
Smith, D. "The Melodious Pulsar" in *Sky & Telescope*, Oct. 1983, p. 311. On the first millisecond pulsar.
Taubes, G. "Looking at Matter in a New Light: Synchrotron Radiation" in *Discover*, May 1983, p. 76.
Trimble, V. "How To Survive Cataclysmic Variables" in *Mercury*, Jan/Feb. 1980, p. 8.

Trimble, V. " A Field Guide to Close Binary Stars" in *Sky & Telescope*, Oct. 1984, p. 306.
Verschuur, G. *The Invisible Universe Revealed*. 1987, Springer Verlag. See chapter 11 on pulsars and 13 on SS433.
Verschuur, G. "On the Trail of Exotic Pulsars" in *Astronomy*, Dec. 1988, p. 22.
White, N. "New Wave Pulsars" in *Sky & Telescope*, Jan. 1987, p. 22.
Will, C. "The Binary Pulsar" in *Mercury*, Nov/Dec. 1987, p. 162.

Reading Material for the Instructor

Canal, R., et al. "The Origin of Neutron Stars in Binary Systems" in *Annual Reviews of Astron. & Astrophys.*, vol. 28, p. 183 (1990).
Davidson, K. & Fesen, R. "Recent Developments Concerning the Crab Nebula" in *Annual Reviews of Astronomy and Astrophysics*, vol. 23, p. 119 (1985).
Drechsel, H., et al., eds. *Cataclysmic Variables*. 1987, Reidel.
Gingerich, O. "The Crab Nebula" in *Sky & Telescope*, Nov. 1977, p. 378. A lab exercise.
Gordon, K. "Pulsars" in *Sky & Telescope*, Mar. 1977, p. 178. A lab exercise.
Helfand, D., et al., eds. *The Origin and Evolution of Neutron Stars*. 1987, Reidel.
Hewish, A. "Pulsars and High Density Physics" in *Science*, June 13, 1975, p. 107. His Nobel prize lecture.
Higdon, J. & Lingenfelter, R. "Gamma-Ray Bursts" in *Annual Reviews of Astron. & Astrophys.*, vol. 28, p. 401 (1990).
Joss, P. & Rappaport, S. "Neutron Stars in Interacting Binary Systems" in *Annual Reviews of Astron. & Astrophys.*, vol. 22, p. 537 (1984).
Kafatos, M., et al., eds. *The Crab Nebula and Related Supernova Remnants*. 1986, Cambridge U. Press.
Lewin, W. & Van den Heuvel, E., eds. *Accretion Driven Stellar X- ray Sources*. 1984, Cambridge U. Press.
Margon, B. "Observations of SS 433" in *Annual Reviews of Astron. & Astrophys.*, vol. 22, p. 507 (1984).
Paczynski, B. "Binary Stars" in *Science*, July 20, 1984, p. 275.
Pines, D. "Accreting Neutron Stars, Black Holes, and Degenerate Dwarf Stars" in *Science*, Feb. 8, 1980, p. 597. Review article on interpreting x-ray observations.
Pringle, J. & Wade, R., eds. *Interacting Binary Stars*. 1985, Cambridge U. Press.
Shaham, J. "The Oldest Pulsars in the Universe" in *Scientific American*, Feb. 1987.
Shapiro, S. & Teukolsky, S. *Black Holes, White Dwarfs, and Neutron Stars*. 1983, Wiley.

Taylor, J. & Stinebring, D. "Recent Progress in the Understanding of Pulsars" in *Annual Reviews of Astronomy and Astrophysics*, vol. 24, p. 285 (1986).
Truemper, J., et al., eds. *The Evolution of Galactic X-ray Binaries*. 1986, Reidel.
Van der Klis, M. "Quasi-Periodic Oscillations and Noise in Low-Mass X-ray Binaries" in *Annual Reviews of Astron. & Astrophys.*, vol. 27, p. 517 (1989).
Weisberg, J., et al. "Gravitational Waves from an Orbiting Pulsar" in *Scientific American*, Oct. 1981.

Recent Audio-Visual Aids

Discovery of Pulsars (A BBC Audiotape, featuring interviews with Antony Hewish and Jocelyn Bell, Astronomical Society of the Pacific)
"The Lives of the Stars" (an episode of the *Cosmos* TV series, available on home video from the Astronomical Society of the Pacific)
A Pulsar Discovery (Slides and background material on the first optical identification of a pulsar; American Institute of Physics, 335 E. 45th St., New York, NY 10017.)
"Stardust" (an episode of *The Astronomers* TV series on the birth and death of stars: from Pacific Arts or the Astronomical Society of the Pacific)
Telescopes of the World (50 slides, Astronomical Society of the Pacific)

Topics for Discussion and Papers

1. The Discovery of Pulsars

The work of Antony Hewish and Jocelyn Bell (now Jocelyn Bell Burnell) in discovering pulsars — and thus including neutron stars in the roster of observable objects— makes for interesting reading and discussion. When Hewish received the Nobel Prize, there was a flurry of concern about why Bell had not been included, although she herself later said—what many in the scientific community know—that graduate students do not receive prizes for work done as they earn their PhD. See:

Bell Burnell, J. "Little Green Men, White Dwarfs, or What?" in *Sky & Telescope*, Mar. 1978, p. 218.
Greenstein, G. "Neutron Stars and the Discovery of Pulsars" in *Mercury*, Mar/Apr. 1985, p. 34; May/June 1985, p. 66.
Hewish, A. "Pulsars" in *Scientific American*, Oct. 1968.
Wade, N. "Discovery of Pulsars: A Graduate Student's Story" in *Science*, Aug. 1, 1975; vol. 189, p. 359.

(See also the tape entitled *The Discovery of Pulsars*, under audio-visual aids, above.)

2. Science Fiction about Neutron Stars

Physicist Robert Forward has written two science fiction novels about a lifeform that can evolve and thrive on a neutron stars:

Dragon's Egg (1980, Ballantine) and *Starquake* (1985, Ballantine).

Larry Niven's short story "Neturon Star" has won awards and is reprinted in a collection by the same name (1968, Ballantine). Two of his novels—*The Integral Trees* (1984, Ballantine) and *The Smoke Ring* (1988, Ballantine) take place in a thick ring of gas, stripped from a Jovian planet, in orbit around a neutron star in a double star system.

3. SS 433

The discovery and ultimate interpretation of the bizarre behavior of SS 433 was one of the more interesting astronomical detective stories of the last decade. Students who want to do a report can consult:

Clark, D. *The Quest for SS 433*. 1985, Penguin Books.
Margon, B. "The Bizarre Spectrum of SS 433" in *Scientific American*, Oct. 1980.
Margon, B. "SS 433: One of a Kind?" in *Mercury*, Sep/Oct. 1979, p. 108.
Moore, P. *Astronomers' Stars*. 1987, Norton. Has a chapter on SS 433.
Overbye, D. "Does Anyone Understand SS 433?" in *Sky & Telescope*, Dec. 1979, p. 510.
Schorn, R. "SS 433: Enigma of the Century" in *Sky & Telescope*, Aug. 1981, p. 100.

4. The Development of X-Ray Astronomy

Section 23-3 briefly describes the early development of x-ray astronomy, a science which had to await the coming of the space age and requires technology quite a bit different from optical or radio astronomy. For students who want to investigate how x-ray astronomy progressed from early rocket flights through the HEAO satellites, and now ROSAT, here are some useful references:

Beatty, J. "Rosat and the X-ray Universe" in *Sky & Telescope*, Aug. 1990, p. 128.
Friedman, H. "Discovering the Invisible Universe" in *Mercury*, Jan/Feb. 1991, p. 2.
Friedman, H. "Rocket Astronomy: An Overview" in Hanle, P. & Chamberlain, V., eds. *Space Science Comes of Age*. 1981, Smithsonian Instit. Press.

Giacconi, R. "The Einstein X-ray Observatory" in *Scientific American*, Feb. 1980.
Hirsch, R. *Glimpsing an Invisible Universe: The Emergence of X- ray Astronomy*. 1983, Cambridge U. Press.
Margon, B. "Exploring the High-Energy Universe" in *Sky & Telescope*, Dec. 1991, p. 607.
Overbye, D. "Riccardo Giacconi: The Man with X-Ray Vision" in *Discover*, Dec. 1980, p. 66.
Overbye, D. "The X-ray Eyes of Einstein" in *Sky & Telescope*, June 1979, p. 527.
Tucker, W. *The Star Splitters*. 1984, NASA SP-466. On the Einstein x-ray observatory.
Tucker, W. & Giacconi, R. *The X-Ray Universe*. 1985, Harvard U. Press.
Tucker, W. & Giacconi, R. "The Birth of X-ray Astronomy" in *Mercury*, Nov/Dec. 1985, p. 178; Jan/Feb. 1986, p. 13.
Tucker, W. & K. *The Cosmic Inquirers: Modern Telescopes and Their Makers*. 1986, Harvard U. Press. Chapter 3 is about the Einstein x-ray observatory.
van den Heuvel, E. & van Paradijs, J. "X-Ray Binaries" in *Scientific American*, Nov. 1993, p. 64.

CHAPTER 24
Black Holes

Reading Material for the Student

Block, D. "Black Holes and Their Astrophysical Implications" in *Sky & Telescope*, July 1975, p. 20; Aug. 1975, p. 87.
Darling, D. "The Quest for Black Holes" in *Astronomy*, July 1983, p. 6.
Davies, P. *The Edge of Infinity*. 1981, Simon & Schuster.
Davies, P. "Wormholes and Time Machines" in *Sky & Telescope*, Jan. 1992, p. 20.
Folger, T. "The Ultimate Vanishing" in *Discover*, Oct. 1993, p. 98.
Greenstein, G. *Frozen Star*. 1984, Freundlich Books.
Hawking, S. *A Brief History of Time*. 1988, Bantam. Has sections on black holes.
Hutchings, J., et al. "LMC X-3: A Black Hole in a Neighbor Galaxy" in *Mercury*, July/Aug. 1984, p. 106.
Kaufmann, W. *Black Holes and Warped Spacetime*. 1979, W. H. Freeman.
Kaufmann, W. *Cosmic Frontiers of General Relavity*. 1977, Little Brown.
Kaufmann, W. "The Black Hole" in *The Universe*, B. Preiss & A. Fraknoi, eds. 1987, Bantam.

LoPresto, J. "The Geometry of Space and Time" in *Astronomy*, Oct. 1987, p. 6.
McClintock, J. "Stalking the Black Hole in the Star Garden of the Unicorn" in *Mercury*, Jul/Aug. 1987, p. 108. On another black hole candidate.
McClintock, J. "Do Black Holes Exist?" in *Sky & Telescope*, Jan. 1988, p. 28.
Parker, B. "In and Around Black Holes" in *Astronomy*, Oct. 1986, p. 6.
Sheldon, E. "Opinautics: A Matter of Much Gravity" in *Sky & Telescope*, Aug. 1982, p. 118. About traveling through rotating black holes.
Stokes, G. & Michalsky, J. "Cygnus X-1" in *Mercury*, May/June 1979, p. 60.
Sullivan, W. *Black Holes*. 1979, Doubleday.
Thorne, K. "The Search for Black Holes" in *Scientific American*, Dec. 1974.
Tucker, W. *The Star Splitters: The High Energy Astronomy Observatories*. 1984, NASA SP-466.
Will, C. *Was Einstein Right?—Putting General Relativity to the Test*. 1986, Basic Books.
Will, C. "The Binary Pulsar: Gravity Waves Exist" in *Mercury*, Nov/Dec. 1987, p. 162.
Will, C. "Testing General Relativity: 20 Years of Progress" in *Sky & Telescope*, Oct. 1983, p. 295.
Zee, A. *An Old Man's Toy: Gravity in Einstein's Universe*. 1989, Macmillan.
Zirker, J. "Testing Einstein's General Relativity During Eclipses" in *Mercury*, Jul/Aug. 1985, p. 98.

For articles about gravity waves, see activity 22.6.

Reading Material for the Instructor

Backer, D. & Hellings, R. "Pulsar Timing and General Relativity" in *Annual Reviews of Astronomy and Astrophysics*, vol. 24, p. 537 (1986).
Bekenstein, J. "Black Hole Thermodynamics" in *Physics Today*, Jan. 1980, p.24.
Blandford, R., et al. "Gravitational Lens Optics" in *Science*, Aug. 25, 1989, p. 824.
Eardley, D. & Press, W. "Astrophysical Processes Near Black Holes" in *Annual Reviews of Astron. & Astrophys.*, vol. 13 (1975).
Lewin, W. & Van den Heuvel, E., eds. *Accretion Driven Stellar X- ray Sources*. 1984, Cambridge U. Press.
Pacini, F., ed. *High-Energy Phenomena Around Collapsed Stars*. 1987, Reidel.
Powell, C. "Inconstant Cosmos" in *Scientific American*, May 1993, p. 110.
Price, R. & Thorne, K. "The Membrane Paradigm for Black Holes" in *Scientific American*, Apr. 1988.
Pringle, J. "Accretion Disks in Astrophysics" in *Annual Reviews of Astron. & Astrophys.*, vol. 19 (1981).

Schastok, J., et al. "Stellar Sky As Seen from the Vicinity of a Black Hole" in *American Journal of Physics*, Apr. 1987, p. 336.
Schutz, B. *A First Course in General Relativity*. 1985, Cambridge U. Press.
Shapiro, S. & Teukolsky, S. *Black Holes, White Dwarfs and Neutron Stars*. 1983, John Wiley.
Turner, E. "Quasars and Gravitational Lenses" in *Science*, Mar. 23, 1984, p. 125.

Recent Audio-Visual Aids

Black Holes and Warped Spacetime (a lecture by William Kaufmann taped in 1992, available to adoptors of this textbook from W. H. Freeman)
A Brief History of Time (1992, 84 min., a documentary look at the life and theories of Stephen Hawking, available from Facets Video)
A Brief History of Time (audio-tapes of the complete text of Stephen Hawking's best-selling book; 6 hours; Astronomical Society of the Pacific)
Curved Space and Black Holes (a video-tape lesson, part of the "Mechanical Universe" series; Calif. Institute of Technology, Mail Code 1-70, Pasadena, CA 91125.)
"The Lives of the Stars" (part of the *Cosmos* TV series; available in home video format from the Astronomical Society of the Pacific)
"Searching for Black Holes" (episode 2 of *The Astronomers* TV series, from the Pacific Arts Video or the Astronomical Society of the Pacific)
"Shades of Black" (1979, 28 min., part of the *Understanding Space and Time* series by George Abell, University of California Extension Media)

Topics for Discussion and Papers

1. Gravitational Lenses

For more on gravitational lenses and how we have observed the images they form, students can consult:

Chaffee, F. "The Discovery of a Gravitational Lens" in *Scientific American*, Nov. 1980.
Cohen, M. *Gravity's Lens: Views of the New Cosmology*. 1988, John Wiley. Chapter 7 is a good review of gravitational lenses.
Gorenstein, M. "Charting Paths Through Gravity's Lens" in *Sky & Telescope*, Nov. 1983, p. 390.
Gott, R. "Gravitational Lenses" in *American Scientist*, Mar/Apr. 1983, p. 150.
Lawrence, J. "Gravitational Lenses and the Double Quasar" in *Mercury*, May/June 1980, p. 66.
Morrison, N. & D. "Gravitational Lenses and QSO's" in *Mercury*, Jan/Feb. 1982, p. 28.

Overbye, D. "Lenses in the Sky" in *Discover*, Mar. 1984, p. 30.
Robinson, L. "Early Thoughts About Lenses in the Sky" in *Sky & Telescope*, Nov. 1983, p. 387.
Schild, R. "Gravity is my Telescope" in *Sky & Telescope*, Apr. 1991, p. 375.
Smith, D. "Arcs Galore" in *Sky & Telescope*, Oct. 1988, p. 358.
Turner, E. "Gravitational Lenses" in *Scientific American*, July 1988.
Verschuur, G. "A New Yardstick for the Universe" in *Astronomy*, Nov. 1988, p. 60.

2. John Wheeler

The term black hole was coined by physicist John Wheeler, who has one of the most interesting and far-ranging intellects in science today. His articles speculating about the nature of the universe, ultimate reality, the implications of physics, etc. are fascinating for both students and instructors. To begin with, you might look up:

Overbye, D. "God's Turnstile: The Work of John Wheeler and Stephen Hawking" in *Mercury*, Jul/Aug. 1991, p. 98.
Overbye, D. "John A. Wheeler: Messenger at the Gate of Time" in *Science '81*, June 1981, p. 60.
Wheeler, J. "Our Universe" in *American Scholar*, Spring 1968.
Wheeler, J. "The Universe as Home for Man" in *American Scientist*, Nov/Dec. 1974.

3. Black Holes in Fiction

Black Holes have made their way into serious fiction as well as science fiction, and poetry. Students with an interest in literature may be able to find other examples, but they can begin with these:

Anderson, P. "Kyrie" in Pournelle, J., ed. *Black Holes*. 1978, Fawcett. Explores the distortion of time near a black hole.
Benford, G. "Mandikini" in Preiss, B. & Fraknoi, A., eds. *The Universe*. 1987, Bantam. Action takes place near the black hole at the center of the Milky Way. Continued in a novel entitled *The Great Sky River*.
Brin, D. "The Crystal Spheres" in *The River of Time*. 1987, Bantam. Using black holes to bear the loneliness of an uninhabited universe.
Niven, L. *World Out of Time*. 1976, Ballantine. Time travel using a black hole.
Oates, Joyce Carol: "Passions and Meditations" in *The Seduction and Other Stories*. 1975, Black Sparrow Press. Story about a disturbed person using black hole metaphors.
Pohl, F. *Gateway*. 1977, Ballantine. A marvelous novel of black hole guilt and many other subjects.

348 Resource Guide

Sagan, C. *Contact*. 1985, Simon & Schuster. Features an interstellar "subway" system, using black holes.
Watson, R. "Swan X One (Or, the Coal Bag)"—a poem in *Selected Poems*. 1974, Atheneum.
Willis, C. "Schwarzschield Radius" in Preiss, B. & Fraknoi, A., eds. *The Universe*. 1987, Bantam. Haunting, surrealistic story combining the life of Schwarzschield and black hole images.

We might just add one more reference: In Isaac Asimov's short story "The Billiard Ball" (in *Asimov's Mysteries*, 1968, Dell) murder is committed using general relativity.

CHAPTER 25
Our Galaxy

Reading Material for the Student

Bok, B. "The Bigger and Better Milky Way" in *Astronomy*, Jan. 1984, p. 6.
Bok, B. & P. *The Milky Way*, 5th ed. 1981, Harvard U. Press.
Chaisson, E. "Journey to the Center of the Galaxy" in *Astronomy*, Aug. 1980, p. 6.
Dame, T. "The Molecular Milky Way" in *Sky & Telescope*, July 1988, p. 22.
Davis, J. *Journey to the Center of Our Galaxy*. 1991. Contemporary Books.
Ferris, T. *Coming of Age in the Milky Way*. 1988, Morrow.
Gingerich, O. "Robert Trumpler and the Dustiness of Space" in *Sky & Telescope*, Sep. 1985, p. 213.
Goldsmith, D. & Cohen, N. *Mysteries of the Milky Way*. 1991, Contemporary Books.
Kraus, J. "The Center of our Galaxy" in *Sky & Telescope*, Jan. 1983, p. 30.
Morrison, N. & D. "The Nucleus and Nuclear Bulge of our Galaxy" in *Mercury*, Jul/Aug. 1985, p. 119.
Mulholland, D. "The Beast at the Center of the Galaxy" in *Science '85*, Sep. 1985, p. 50.
Palmer, S. "Unveiling the Hidden Milky Way" in *Astronomy*, Nov. 1989, p. 32.
Smith, D. & Robinson, L. "Dissecting the Hub of our Galaxy" in *Sky & Telescope*, Dec. 1984, p. 494.
Twarog, B. "Chemical Evolution of the Galaxy" in *Mercury*, Jul/Aug. 1985, p. 107.
Verschuur, G. "In the Beginning" in *Astronomy*, Jan. 1993, 40.
Verschuur, G. *The Invisible Universe Revealed*. 1987, Springer Verlag. Chap-

ters on galactic structure and the galactic center as revealed through radio astronomy.
Verschuur, G. *Interstellar Matters*. 1989, Springer
Verlag. Verschuur, G. "The Magnetic Milky Way" in *Astronomy*, June 1990, p. 32.
Verschuur, G. "Journey Into the Galaxy" in *Astronomy*, Jan. 1993, p. 33.
Waldrop, M. "The Core of the Milky Way" in *Science*, Oct. 11, 1985, p. 158.

Reading Material for the Instructor

Bahcall, J. "Star Counts and Galactic Structure" in *Annual Reviews of Astronomy and Astrophysics*, vol 24, p. 577 (1986).
Brown, R. & Liszt, H. "Sagittarius A and Its Environment" in *Annual Reviews of Astron. & Astrophys.*, vol. 22, p. 223 (1984).
Dickey, J. & Lockman, F. "H I in the Galaxy" in *Annual Reviews of Astron. & Astrophys.*, vol. 28, p. 215 (1990).
Freeman, K. "The Galactic Spheroid and Old Disk" in *Annual Reviews of Astronomy and Astrophysics*, vol.25, p. 603 (1987).
Genzel, R. & Townes, C. "Physical Conditions, Dynamics, and Mass Distribution in the Center of the Galaxy" in *Annual Reviews of Astronomy and Astrophysics*, vol. 25, p. 377 (1987).Gilmore, G. & Carswell, B., eds. *The Galaxy*. 1987, Reidel.
Gilmore, G., et al. "Kinematics, Chemistry, and Structure of the Galaxy" in *Annual Reviews of Astron. & Astrophys.*, vol. 27, p. 555 (1989).
Hirschfeld, A. "How Far is the Galactic Center?" in *Sky & Telescope*, Dec. 1984, p. 498. A lab exercise on Shapley's work.
Kuhn, L. *The Milky Way*. 1982, Wiley.
Lo, K. "The Galactic Center: Is It a Massive Black Hole?" in *Science*, Sep. 26, 1986, p. 139.
Madore, B., ed. *Cepheids: Theory and Observations*. 1985, Cambridge U. Press.
Sandage, A. "The Population Concept, Globular Clusters, Subdwarfs, Ages, and the Collapse of the Galaxy" in *Annual Reviews of Astronomy and Astrophysics*, vol. 24, p. 421 (1986).
Scoville, N. & Young, J. "Molecular Clouds, Star Formation, and Galactic Structure" in *Scientific American*, Apr. 1984.
Struve, O. & Zebergs, V. *Astronomy in the 20th Century*. 1962, Macmillan. A good history.
van den Bergh, S., & Hesser, J., "How the Milky Way Formed" in *Scientific American*, Jan. 1993.
van Woerden, H., et al., eds. *The Milky Way Galaxy*. 1985, Reidel.
van Woerden, H., et al. *Oort and the Universe*. 1980, Reidel.

Recent Audio-Visual Aids

The Countdown to the Invisible Universe (1987, 58-min film, part of the *Nova* TV series, distributed by Coronet Films) Planning for and results from the IRAS mission.

Slide sets from the Astronomical Society of the Pacific that include images of the galactic center:

The Radio Universe
The Infrared Universe: An IRAS Gallery
The Sky at Many Wavelength (A set of 11 slides from the Astronomical Society of the Pacific) Shows the sky in galactic coordinates in all the wavelengths bands of the electromagnetic spectrum; includes an excellent recent map of the 21-cm radiation.

Topics for Discussion and Papers

1. Bart Bok

Shapley's student and colleague Bart Bok (whose work on star formation was discussed in chapter 20) also made important contributions to the understanding of the Milky Way. He was an astronomer whose knowledge and enthusiasm were a great inspiration to his many students, other astronomers, and to the public. For more on his life and work, see:

The special Bart Bok issue of *Mercury* magazine, Mar/Apr. 1984.
White, R. "Bart J. Bok (1906-1983)" in *Sky & Telescope*, Oct. 1983, p. 303.

2. The History of Our Understanding of the Milky Way

The history of our quest to understand the existence, extent, and shape of our galaxy is considerably more complex than introductory textbooks can cover. But the twists and turns of trying to piece together what the galaxy looks like (while inside it) makes for fascinating discussion or research. Students can see:

Berendzen, R., et al. *Man Discovers the Galaxies*. 1976, Neale Watson.
Gingerich, O. "The Discovery of the Milky Way's Spiral Arms" in *Sky & Telescope*, July 1984, p. 10.
Gingerich, O. "Robert Trumpler and the Dustiness of Space" in *Sky & Telescope*, Sep. 1985, p. 213.
Jaki, S. *The Milky Way: An Elusive Road for Science*. 1972, Neale Watson.
Kopal, Z. *Widening Horizons: Man's Quest to Understand the Structure of the Universe*. 1970, Taplinger.
Oort, J. "The Development of Our Insight into the Structure of the Galaxy

Between 1920 and 1940" in *Annals of the New York Academy of Science*, Aug. 25, 1972.

Pfeiffer, J. *The Changing Universe*. 1956, Random House. A history of the early days of radio astronomy.

Weaver, H. "Steps Toward Understanding the Large-Scale Structure of the Milky Way" in *Mercury*, Sep/Oct. 1975, p. 18; Nov/Dec. 1975, p. 18; Jan/Feb. 1976, p. 19.

Whitney, C. *The Discovery of the Galaxy*. 1971, A. Knopf.

See also the references under the Shapley-Curtis debate in the next chapter.

3. William Herschel

The pioneering studies of the structure of the Galaxy were not the only area where Herschel made great contributions to astronomy. We have already seen that he discovered the planet Uranus and was the first to detect infrared rays from the Sun. Students who want to find out more about his life and work can consult:

Chapman, A. "William Herschel and the Measurement of Space" in *Quarterly Journal of the Royal Astronomical Society*, vol. 30, p. 399 (Dec. 1989).

Gingerich, O. "Herschel's 1784 Autobiography" in *Sky & Telescope*, Oct. 1984, p. 317.

Hoskin, M. "Interview with William Herschel" in *The Mind of the Scientist*. 1971, Taplinger.

Hoskin, M. "William Herschel and the Making of Modern Astronomy" in *Scientific American*, Feb. 1986.

Hoskin, M. *William Herschel and the Construction of the Heavens*. 1963, Neale Watson.

Hoskin, M. *Stellar Astronomy: Historical Studies*. 1982, Science History Publications.

Jones, B. "William Herschel: Pioneer of the Stars" in *Astronomy*, Nov. 1988, p. 40.

Ronan, C. "William Herschel and His Music" in *Sky & Telescope*, Mar. 1981, p. 195.

4. Harlow Shapley

Shapley's work establishing our position in the Galaxy was in many ways as seminal a contribution to humanity's view of itself as Copernicus'. Shapley himself was an interesting and sometimes controversial scientist as well as an enthusiastic popularizer of astronomy, active in politics, and never afraid to speak his mind. He would make a excellent subject for a historical report or a discussion of the role of the scientist in society:

Bok, B. "Harlow Shapley: Cosmographer and Humanitarian" in *Sky & Telescope*, Dec. 1972, p. 354.
Bok, B. "Harlow Shapley and the Discovery of the Center of Our Galaxy" in Neyman, J., ed. *The Heritage of Copernicus*. 1974, MIT Press.
Hoagland, H. "Harlow Shapley—Some Recollections" in *Publications of the A.S.P.*, vol. 77, p. 422 (Dec. 1965).
Jeffers, H. "The Awards of the Bruce Medal to Professor Harlow Shapley" in *Publications of the A.S.P.*, vol. 51, p. 77 (1939).
Shapley, H. *Through Rugged Ways to the Stars*. 1969,. Scribners. An autobiography.
Wright, F. "Harlow Shapley: A Tribute to a Great Man" in *Mercury*, Mar/Apr. 1973, p. 3.

There are several historical articles in the technical volume *The Harlow Shapley Symposium on Globular Cluster Systems*, ed. J. Grindlay & A. Philip (1988, Kluwer).

5. Dark Matter in the Galaxy

Section 25-3 briefly introduces the mystery of the dark matter in our galaxy, which is an area of intense current research on both the theoretical and observational front. It has also recently generated a whole slew of books and articles for the layperson:

Bartusiak, M. "Wanted: Dark Matter" in *Discover*, Dec. 1988, p. 62.
Krauss, L. *The Fifth Essence: Dark Matter in the Universe*. 1989, Basic Books.Overbye, D. "The Shadow Universe" in *Discover*, May 1985, p. 12.
Parker, B. *Invisible Matter and the Fate of the Universe*. 1989, Plenum Press.
Riordan, M. & Schramm, D. *The Shadows of Creation: dark Matter and the Structure of the Universe*. 1991, W. H. Freeman.
Stephens, S. "Vera Rubin: An Unconventional Career" in *Mercury*, Jan/Feb. 1992, p. 38.
Trefil, J. *The Dark Side of the Universe*. 1988, Scribners.
Trimble, V. "The Search for Dark Matter" in *Astronomy*, Mar. 1988, p. 18.
Tucker, W. & K. *The Dark Matter*. 1988, Morrow.
Tucker, W. & K. "Dark Matter in Our Galaxy" in *Mercury*, Jan/Feb. 1989, p. 2; Mar/Apr. 1989, p. 51.

CHAPTER 26
Galaxies

Historical Reading Material for the Student

Berendzen, R., et al. *Man Discovers the Galaxies*. 1976, Neale Watson.
Burstein, D. & Manly, P. "Cosmic Tug of War" in *Astronomy*, July 1993, p. 40.
Corwin, M. & Wachowiak, D. "Discovering the Expanding Universe" in *Astronomy*, Feb. 1985, p. 18.
Croswell, K. "Intruder Galaxies" in *Astronomy*, Nov. 1993, p. 28.
Dressler, A. "Galaxies Far Away and Long Ago" in *Sky & Telescope*, Apr. 1993, p. 22
Dressler, A. "Observing Galaxies Through Time" in *Sky & Telescope*, Aug. 1991, p. 126.
Elmegreen, D., & Elmegreen, B. "What Puts the Spiral in Spiral Galaxies?" in *Astronomy*, Sep. 1993, p. 34.
Ferris, T. *The Red Limit*. 1983, Morrow.
Harris, W. "Globular Clusters in Distant Galaxies" in *Sky & Telescope*, Feb. 1991, p. 148.
Hodge, P. "The Extragalactic Distance Scale: Agreement at last?" in *Sky & Telescope*, Oct. 1993, p. 16.
Odenwald, S. & Fienberg, R. "Galaxy Redshifts Reconsidered" in *Sky & Telescope*, Feb. 1993, p. 31.
Parker, B. "The Discovery of the Expanding Universe" in *Sky & Telescope*, Sep. 1986, p. 227.
Phillips, S. "Counting to the Edge of the Universe" in *Astronomy*, Apr. 1993, p. 38.
Sandage, A. "Inventing the Beginning" in *Science '84*, Nov. 1984, p. 111. A brief biography of E. Hubble.
Schramm, D. "The Origin of Cosmic Structure" in *Sky & Telescope*, Aug. 1991, p. 140.
Smith, R. *The Expanding Universe: Astronomy's Great Debate*. 1982, Cambridge U. Press.

Reading Material for the Student on Current Research

Allen, D. "Star Formation and IRAS Galaxies" in *Sky & Telescope*, Apr. 1987, p. 372.
Comins, N. & Marschall, L. "How Do Spiral Galaxies Spiral" in *Astronomy*, Dec. 1987, p. 6.

Davies, J., et al. "Are Spiral Galaxies Heavy Smokers?" in *Sky & Telescope*, July 1990, p. 37.

Geller, M. "Mapping the Universe: Slices and Bubbles" in *Mercury*, May/June 1990, p. 66.

Gregory, S. & Morrison, N. "New Observations of Three Nearby Galaxies" in *Mercury*, May/June 1987, p. 84.

Gregory, S. & Morrison, N. "The Largest Supercluster Filament" in *Mercury*, Mar/Apr. 1986, p. 54.

Hartley, K. "Elliptical Galaxies Forged by Collision" in *Astronomy*, May 1989, p. 42.

Hodge, P. *Galaxies*. 1986, Harvard U. Press.

Hodge, P. "The Local Group: Our Galactic Neighborhood" in *Mercury*, Jan/Feb. 1987, p. 2.

Hodge, P. "M-31: The Andromeda Galaxy" in *Mercury*, Jul/Aug. 1982, p. 118.

Hunter, D. & Wolff, S. "Star Formation in Irregular Galaxies" in *Mercury*, May/June 1985, p. 76.

Kaufman, M. "Tracing M-81's Spiral Arms" in *Sky & Telescope*, Feb. 1987, p. 135.

Keel, W. "Crashing Galaxies, Cosmic Fireworks" in *Sky & Telescope*, Jan. 1989, p. 18.

Marschall, L. "Superclusters: Giants of the Cosmos" in *Astronomy*, Apr. 1984, p. 6.

Parker, B. "Celestial Pinwheels: The Spiral Galaxies" in *Astronomy*, May 1985, p. 14.

Silk, J. "Formation of the Galaxies" in *Sky & Telescope*, Dec. 1986, p. 582.

Smith, D. "Spirals from Order and Chaos" in *Sky & Telescope*, Aug. 1987, p. 136.

Spinrad, H. "Galaxies and Clusters" in *The Universe*, B. Preiss & A. Fraknoi, eds. 1987, Bantam.

Struble, M. & Rood, H. "Diversity Among Galaxy Clusters" in *Sky & Telescope*, Jan. 1988, p. 16.

Trefil, J. "Galaxies" in *Smithsonian*, Jan. 1989, p. 36. Nice long review article.

Tully, R. "Unscrambling the Local Supercluster" in *Sky & Telescope*, June 1982, p. 550.

van den Bergh, S. "The Golden Anniversary of Hubble's Classification System" in *Sky & Telescope*, Dec. 1976, p. 410.

Vogel, S. "Star Attractor" in *Discover*, Nov. 1989, p. 20. (On the Great Attractor.)

Waldrop, M. "Why Do Galaxies Exist?" in *Science*, May 24, 1985, p. 978.

Reading Material for the Instructor

Athanassoula, E. & Bosma, A. "Shells and Rings Around Galaxies" in *Annual Reviews of Astron. and Astrophys.*, vol. 23, p. 147 (1985).

Bahcall, N. "Large-scale Structure in the Universe Indicated by Galaxy Clusters" in *Annual Reviews of Astron. & Astrophys.*, vol. 26, p. 631 (1988).

Burns, J. "Very Large Structures in the Universe" in *Scientific American*, July 1986.

Chiosi, C., et al., eds. *Spectral Evolution of Galaxies*. 1987, Reidel.

DeZeeuw, T., ed. *Structure and Dynamics of Elliptical Galaxies*.

Dressler, A. "The Large-scale Streaming of Galaxies" in *Scientific American*, Sep. 1987.

Dressler, A. "The Evolution of Galaxies in Clusters" in *Annual Reviews of Astron. & Astrophys.*, vol. 22, p. 185 (1984).

Gallagher, J. & Hunter, D. "Structure and Evolution of Irregular Galaxies" in *Annual Reviews of Astron. & Astrophys.*, vol. 22, p. 37 (1984).

Geller, M. & Huchra, J. "Mapping the Universe" in *Science*, Nov. 17, 1989, p. 897.

Gorenstein, P. & Tucker, W. "Rich Clusters of Galaxies" in *Scientific American*, Nov. 1978.

Helfand, D. "Superclusters and the Large-scale Structure of the Universe" in *Physics Today*, Oct. 1983, p. 17.

Hunter, D. & Gallagher, J. "Star Formation in Irregular Galaxies" in *Science*, Mar. 24, 1989, p. 155.

Kormendy, J., et al, eds. *Dark Matter in the Universe*. 1987, Reidel.

Kormendy, J. & Djorgovsky, S. "Surface Photometry and Structure of Elliptical Galaxies" in *Annual Reviews of Astron. & Astrophys.*, vol. 27, p. 235 (1989).

Madore, B., et al., eds. *Galaxy Distances and Deviations from Universal Expansion*. 1986, Reidel.

Mathewson, D. "The Clouds of Magellan" in *Scientific American*, Apr. 1985.

Miley, G. & Chambers, K. "The Most Distant Radio Galaxies" in *Scientific American*, June 1993, p. 54.

Peebles, P. "The Origin of Galaxies and Clusters of Galaxies" in *Science*, June 29, 1984, p. 138.

Rood, H. "Voids" in *Annual Reviews of Astron. & Astrophys.*, vol. 26, p. 245 (1988).

Rowan-Robinson, M. *The Cosmological Distance Ladder*. 1985, W. H. Freeman.

Schulman, L. & Seiden, P. "Perculation and Galaxies" in *Science*, July 25, 1986, p. 425.

Schweizer, F. "Colliding and Merging Galaxies" in *Science*, Jan. 17, 1986, p. 227.
Soifer, B., et al. "The IRAS View of the Extragalactic Sky" in *Annual Reviews of Astronomy and Astrophysics*, vol. 25, p. 187 (1987).
Tully, R. & Fisher, J. *Nearby Galaxies Atlas*. 1987, Cambridge U. Press.
van den Bergh, S. & Pritchett, C. *The Extra-galactic Distance Scale*. 1988, Astronomical Society of the Pacific Conference Series, vol. 4.

Recent Audio-Visual Aids

Images of galaxies can be found in the following new slide sets:

Galaxies (Hansen Planetarium)
Hubble Space Telescope (2 slides sets from the A.S.P.)
The Infrared Universe (Astronomical Society of the Pacific)
Mauna Kea Slide Set (Hansen)
The Radio Universe (A.S.P.)
So Many Galaxies . . . So Little Time (1992, 40 min. Available from the Astronomical Society of the Pacific)
Splendors of the Southern Sky (Hansen)
Splendors of the Universe (3 slide sets from A.S.P.)
"Where is the Rest of the Universe" (episode 1 of *The Astronomers*, from Pacific Arts Video or the Atsronomical Society of the Pacific)
Where the Galaxiees Are (1991, Astronomical Society of the Pacific) 8-minute award-winning video on the work of Geller and Huchra mapping the large-scalle distribution of galaxies.

Topics for Discussion and Papers

1. Lord Rosse and His Giant Reflectors

The days of Lord Rosse's lonely efforts to build huge telescopes seem to belong to an era of 19th century science that has been completely replaced by the large scientific institutions of our time, operated by governments or university consortia. Students who would like to read more about William Parsons and his mammoth telescopes, can consult:

Ashbrook, J. "The Discovery of Spiral Structure in Galaxies" in *Sky & Telescope*, June 1968, p. 366.
Hirschfeld, A. "The Leviathan of Parsonstown" in *Griffith Observer*, Oct. 1974, p. 2.
King, H. *The History of the Telescope*. 1955/1979, Dover.
Learner, R. *Astronomy Through the Telescope*. 1981, Van Nostrand Reinhold.

2. The Shapley-Curtis Debate

The 1920 debate at the National Academy of Sciences was really two debates: one on the nature of the spiral nebulae and one on the extent of the Milky Way Galaxy. Students often enjoy researching the debate and its antecedents, and even—if time permits—recreating a bit of it. For more information, see:

Hoskin, M. "The Great Debate: What Really Happened" in *Stellar Astronomy*. 1982, Science History Publications.
Hoskin, M. "Shapley's Debate" in Grindlay, J. & Philip, A., eds. *The Harlow Shapley Symposium on Globular Cluster Systems*. 1988, Kluwer.
Seeley, D. & Berendzen, R. "Astronomy's Great Debate" in *Mercury*, July/Aug. 1978, p. 67.
Smith, R. "The Great Debate Revisited" in *Sky & Telescope*, Jan. 1983, p. 28.
Smith, R. *The Expanding Universe: Astronomy's Great Debate*. 1982, Cambridge U. Press.

3. Edwin Hubble

Hubble is a towering figure in 20th century astronomy and a very appropriate person for the naming of the Space Telescope. For more on his life and work, students can consult:

Babcock, H. "The Award of the Bruce Medal to Dr. Edwin Hubble" in *Publications of the A.S.P.*, vol. 50, p. 87 (Apr. 1938.)
Goldsmith, D. "Edwin Hubble and the Universe Outside Our Galaxy" in Neyman, J., ed. *The Heritage of Copernicus*. 1974, MIT Press.
Jones, B. "The Legacy of Edwin Hubble" in *Astronomy*, Dec. 1989, p. 38.
Lake, G. "Understanding the Hubble Sequence [of Galaxies]" in *Sky & Telescope*, May 1992, p. 515.
Osterbrock, D., et al, "Edwin Hubble and the Expanding Universe" in *Scientific American*, July 1993, p. 84.
Osterbrock, D., et al. "Young Edwin Hubble" in *Mercury*, Jan/Feb. 1990, p. 2.
Robertson, H. "Edwin Powell Hubble" in *Publications of the A.S.P.*, vol. 66, p. 120 (June 1954).
Sandage, A. "Inventing the Beginning" in *Science '84*, Nov. 1984, p. 111. A brief biography.
Smith, R. "The Origin of the Velocity-Distance Relation" in *Journal for the History of Astronomy*, vol. 10, p. 133 (Oct. 1979).

A poem, "In Memory of Dr. Edwin Hubble," by Lord Dunsany, appeared in *Publications of the A.S.P.*, Aug. 1954, p. 184. A strange novel entitled *Hubble Time* by Tom Bezzi (1987, Mercury House) is set as the diary of Hubble's (fictional) granddaughter.

4. Dark Matter

See topic 25.5 for good readings on the problem of dark matter in the universe. (Many of the books and articles that begin by discussing dark matter in our galaxy, then expand the discussion to the topic of dark matter between galaxies as well.)

5. Measuring Cosmic Distances

This chapter introduces the main ideas in establishing the cosmic distance scale—standard candles, Hubble's Law, the Tully-Fisher relation, etc. This is a complex, challenging area that lies at the heart of much current research. More advanced students may find aspects of the distance scale a "meaty" topic for further investigation. See:

DeVaucouleurs, G. "The Distance Scale of the Universe" in *Sky & Telescope*, Dec. 1983, p. 511.
Kiernan, V. "How Far to the Galaxies" in *Astronomy*, June 1989, p. 48.
McCarthy, P. "Measuring Distances to Remote Galaxies and Quasars" in *Mercury*, Jan/Feb. 1988, p. 19.
Morrison, N. "The Extragalactic Distance Scale" in *Mercury*, Nov/Dec. 1988, p. 171.
Rowan-Robinson, M. *The Cosmological Distance Ladder*. 1985, W. H. Freeman. Technical, but very clear.
Shipman, H. *Black Holes, Quasars, and the Universe*, 2nd ed. 1982, Houghton Mifflin. Chapter 14 has a good discussion of the pyramid of distance methods astronomers use.
Smith, D. "Supernovae: Mileposts of the Universe" in *Sky & Telescope*, Jan. 1985, p. 18.
Tenn, J. "Cosmic Distances and QSO's" in *Mercury*, July/Aug. 1979, p. 67.

See also topics 19.1 and 21.3

6. Observing Galaxies

Observing galaxies through binoculars or a small telescope can be a very rewarding and satisfying project. Students can consult back issues of *Astronomy* and *Sky & Telescope* magazine and any good manual on amateur astronomy for suggestions and observing lists. Some specific guides they might consult include:

Crossen, D. "Observing Spring Galaxies with Binoculars" in *Astronomy*, May 1987, p. 62.
Eischer, D. "Observing the Local Group of Galaxies" in *Astronomy*, Nov. 1984, p. 35.

Goldstein, A. "Galaxy Hunting Around the Big Dipper" in *Astronomy*, Mar. 1989, p. 78.
Goldstein, A. "Observing the Andromeda Galaxy" in *Astronomy*, Nov. 1991, p. 84.
Harrington, P. "A Bowl Full of Galaxies" in *Sky & Telescope*, Apr. 1989, p. 444.
Pommier, R. "Observing the Sculptor Group of Galaxies" in *Astronomy*, Dec. 1991, p. 84.

Chapter 27
Quasars, Blazars, and Active Galaxies

Reading Material for the Student

Note: For more on supermassive black holes, see also the readings in chapter 24.

Balick, B. "Quasars with Fuzz" in *Mercury*, May/June 1983, p. 81.
Burns, J. "Chasing the Monster's Tail: New Views of Cosmic Jets" in *Astronomy*, Aug. 1990, p. 28.
Croswell, K. "Have Astronomers Solved the Quasar Enigma?" in *Astronomy*, Feb. 1993, p. 28.
Downes, A. "Radio Galaxies" in *Mercury*, Mar/Apr. 1986, p. 34.
Finkbeiner, A. "Active Galactic Nuclei: Sorting Out the Mess" in *Sky & Telescope*, Aug. 1992, p. 138.
Gregory, S. & Morrison, N. "Visible Synchrotron Emission from the Lobes of a Radio Galaxy" in *Mercury*, Jul/Aug. 1986, p. 120.
Gregory, S. "Active Galaxies and Quasars: A Unified View" in *Mercury*, Jul/Aug. 1988, p. 111.
Harrington, S., et al. *Learning About Quasars*. 1989, Astronomical Society of the Pacific. An information packet.
Hodge, P. *Galaxies*. 1986, Harvard U. Press.
Kanipe, J. "M87: Describing the Indescribable" in *Astronomy*, May 1987, p. 6; "Anatomy of a Cosmic Jet" in July 1988, p. 30.
Kanipe, J. "Quest for the Most Distant Objects in the Universe" in *Astronomy*, June 1988, p. 20.
Mood, J. "Star Hopping to a Quasar" in *Astronomy*, Apr. 1987, p. 49. On finding 3C273 in the sky.
Morrison, N. & Gregory, S. "A Remarkable Radio Image of Cygnus A" in *Mercury*, Mar/Apr. 1985, p. 55.
Parker, B. "In and Around Black Holes" in *Astronomy*, Oct. 1986, p. 6.

Posey, D. "Three Recent Snapshots on How Quasar Activity Can be Triggered" in *Mercury*, Jan/Feb. 1988, p. 22.

Preston, R. *First Light*. 1987, Atlantic Monthly Books. An intriguing book on doing astronomy with the 200-inch telescope at Palomar, with a number of sections on quasar research.

Schendel, J. "Looking Inside Quasars" in *Astronomy*, Nov. 1982, p. 6.

Schorn, R. "The Extragalactic Zoo" [3 articles on various active galaxies and how they got their names] in *Sky & Telescope*, Jan. 1988, p. 23; Apr. 1988, p. 376; July 1988, p. 36.

Shipman, H. *Black Holes, Quasars, and the Universe*, 2nd ed. 1980, Houghton Mifflin.

Smith, D. "Mysteries of Cosmic Jets" in *Sky & Telescope*, Mar. 1985, p. 213.

Smith, D. "Black Hole Weighs in at One Billion Suns" in *Sky & Telescope*, Aug. 1984, p. 127.

Smith, H. "Quasars and Active Galaxies" in *The Universe*, B. Preiss & A. Fraknoi, eds. 1987, Bantam.

Verschuur, G. *The Invisible Universe Revealed: The Story of Radio Astronomy*. 1987, Springer Verlag. Part II is a good introduction to quasars and active galaxies.

Weedman, D. "Quasars: A Progress Report" in *Mercury*, Jan/Feb. 1988, p. 12.

Wilkes, B. "The Emerging Picture of Quasars" in *Astronomy*, Dec. 1991, p. 35.

Wright, A. & H. *At the Edge of the Universe*. 1989, Ellis Horwood/John Wiley. Best introductory book.

Reading Material for the Instructor

Arp, H. *Quasars, Redshifts, and Controversies*. 1987, Interstellar Media. [See review in *Sky & Telescope*, Jan. 1988.]

DeYoung, D. "Jets in Extragalactic Sources" in *Science*, Aug. 17, 1984, p. 677.

Guiricin, G., et al., eds. *Structure and Evolution of Active Galactic Nuclei*. 1986, Reidel.

Hartwick, F. & Schade, D. "The Space Distribution of Quasars" in *Annual Reviews of Astron. & Astrophys.*, vol. 28, p. 437 (1990).

Kellermann, K. & Thompson, A. "The Very Long Baseline Array" in *Science*, July 12, 1985, p. 123.

Kundt, W., ed. *Astrophysical Jets and Their Engines*. 1987, Reidel.

Miller, J., ed. *Astrophysics of Active Galaxies and Quasi-Stellar Objects*. 1985, University Science

Osmer, P. "Quasars as Probes of the Distant and Early Universe" in *Scientific American*, Feb. 1982.

Osterbrock, D. & Matthews, W. "Emission-Line Regions of Active Galaxies and QSO's" in *Annual Reviews of Astron. & Astrophys.*, vol. 24, p. 171 (1986).

Saikia, D. & Salter, C. "Polarization Properties of Extragalactic Radio Sources" in *Annual Reviews of Astron. & Astrophys.*, vol. 26, p. 93 (1988).
Schwarzschild, B. "Probing the Early Universe with Quasar Light" in *Physics Today*, Nov. 1987, p. 11.
Soifer, B., et al. "The IRAS View of the Extragalactic Sky" in *Annual Reviews of Astron. & Astrophys.*, vol. 25, p. 187 (1987).
Swarup, G., et al., eds. *Quasars*. 1986, Reidel.
Trimble, V. & Woltjer, L. "Quasars at 25" in *Science*, Oct. 10, 1986, p. 155.
Weedman, D. *Quasar Astronomy*. 1986, Cambridge U. Press.

Recent Audio-Visual Aids

Images of galaxies can be found in the following new slide sets:

Hubble Space Telescope (2 slide sets from the A.S.P.).
"Searching for Black Holes" (episode 2 of *The Astronomers* TV series, from Pacific Arts Video or the Astronomical Society of the Pacific).
The Infrared Universe: An IRAS Gallery (25 slides and booklet by Charles Beichman; Astronomical Society of the Pacific) Includes one image of a peculiar galaxy, Arp 220.
The Radio Universe (40 slides with detailed captions booklet; Astronomical Society of the Pacific) Features a significant number of radio images of active galaxies, jets, and quasars.
Splendors of the Universe (3 sets of 15 slides each by David Malin; Astronomical Society of the Pacific) Include some images of active galaxies.

Topics for Discussion and Papers

1. The Discovery of Quasars

As the text recounts, the "discovery" of quasars took place in several steps, starting with the identification of optical counterparts for certain radio sources and proceeding through the identification of the spectral lines by Maarten Schmidt. Students who want to read about the details of the story can consult:

Ferris, T. *The Red Limit*, 2nd ed. 1983, Morrow Quill.
Greenstein, J. "Quasi-stellar Radio Sources" in *Scientific American*, Dec. 1963.
Preston, R. *First Light*. 1987, Atlantic Monthly Press.
Preston, R. "Beacons in Time: Maarten Schmidt and the Discovery of Quasars" in *Mercury*, Jan/Feb. 1988, p. 2.

2. The Work of Halton Arp

The efforts of Halton Arp and his colleagues to show that some extragalactic objects may not obey Hubble's Law led to one of the most interesting controversies of modern astronomy. Although the preponderance of evidence and opinion in the astronomical community is settling on the side of the conventional view, the questions Arp raised have been good for encouraging research and inspiring new techniques for investigation.

For more on Arp's work, see:

Arp, H. "Related Galaxies with Different Redshifts" in *Sky & Telescope*, Apr. 1983, p. 307.
Burbidge, G. "Quasars, Redshifts, and Controversies" in *Sky & Telescope*, Jan. 1988, p. 38. A somewhat one-sided discussion of the Arp controversy in the form of a long review of Arp's book.
Burbidge, G. "Quasars in the Balance" in *Mercury*, Sep/Oct. 1988, p. 136.
Ferris, T. "The Spectral Messenger: The Redshift Controversy" in *Science '81*, Oct. 1981, p. 66.
Field, G., et al. *The Redshift Controversy*. 1973, Benjamin-Cummings. Technical and somewhat dated, but a superb introduction to both sides at the time it was published.
Sulentic, J. "Are Quasars Far Away?" in *Astronomy*, Oct. 1984, p. 66. By a student and coworker of H. Arp's.
Tenn, J. "Quasars and Active Galaxies" in *Mercury*, Jul/Aug. 1979, p. 67; Sep/Oct. 1979, p. 101.

3. Distances Become Model Dependent

Kaufmann's text is one of the few introductory books to give such clear emphasis to the idea that as you get to high red-shifts, the distances we derive for objects like quasars are highly model-dependent, requiring assumptions about the curvature of the universe. This idea can make for interesting discussions about what it means to know a distance in the universe, about the construction and evaluation of scientific models, and about stories in the media which blithely talk about the most distant object in the universe being x light-years away.

For more on quasar distances, see:

McCarthy, M. "Measuring Distances to Remote Galaxies and Quasars" in *Mercury*, Jan/Feb. 1988, p. 19.

See also topic 3 for chapter 8 for references on measuring continental drift using quasars.

Chapter 28
Cosmology: The Creation and Fate of the Universe

Reading Material for the Student

Note: Many of the reading materials for Chapter 28 include topics covered in Chapter 29 and vica versa. Each reference is only listed once, and it is probably best to consider these two chapters together when recommending reading material or a term paper topic.

Albers, D. "The Meaning of Curved Space" in *Mercury*, Jul/Aug. 1975, p. 16.
Barrow, J. & Silk, J. *The Left Hand of Creation: Origin and Evolution of the Universe*. 1983, Basic Books.
Bartusiak, M. *Thursday's Universe*. 1986, Times Books. A well-written introduction to a number of topics in cosmology, by a science journalist.
Callahan, J. "The Curvature of Space in a Finite Universe" in *Scientific American*, Aug. 1976.
Chaisson, E. *Relatively Speaking: Relativity, Black Holes, and the Fate of the Universe*. 1988, W. W. Norton.
Cohen, N. *Gravity's Lens: Views of the New Cosmology*. 1988, Wiley.
Davies, P. "The Arrow of Time" in *Sky & Telescope*, Sep. 1986, p. 239.
Davies, P. "Everyone's Guide to Cosmology" in *Sky & Telescope*, March 1991, p. 250. Good introductory article.
Davies, P. "The First One Second of the Universe" in *Mercury*, May/June 1992, p. 82.
Ferris, T. *Coming of Age in the Milky Way*. 1988, Morrow. Excellent history of cosmology.
Ferris, T. *The Red Limit*, 2nd ed. 1983, Morrow. A history.
Ferris, T. "Where Are We Going?" in *Sky & Telescope*, May 1987, p. 486.
Gribbin, J. *In Search of the Big Bang*. 1986, Bantam.
Harrison, E. *Cosmology*. 1981, Cambridge U. Press. An excellent textbook.
Harrison, E. *Masks of the Universe*. 1985, Macmillan. Philosophical and historical views of cosmology.
Harrison, E. *Darkness at Night*. 1987, Harvard U. Press. On Olbers' Paradox.
Jayawardhana, R. "The Age Paradox" in *Astronomy*, June 1993, p. 38.
Kippenhahn, R. *Light From the Depths of Time*. 1987, Springer-Verlag.
Kippenhahn, R. "Light from the Depths of Time" in *Sky & Telescope*, Feb. 1987, p. 140. On the cosmic background.
Lemonick, M. *The Light at the Edge of the Universe: Leading Cosmologists on the Brink of a Scientific Revolution*. 1993, Villard Books.
Lightman, A. *Ancient Light*. 1991, Harvard U. Press. Nice concise, modern introduction.

Lightman, A. & Brawer, R. *Origins: The Lives and Worlds of Modern Cosmologists*. 1990, Harvard U. Press. Interviews with active researchers in this field.
LoPresto, J. "The Geometry of Space and Time" in *Astronomy*, Oct. 1987, p. 6.
Odenwald, S. "Einstein's Fudge Factor" in *Sky & Telescope*, Apr. 1991, p. 362. On the cosmological constant.
Overbye, D. *Lonely Hearts of the Cosmos*. 1991, Harper Collins. Wonderful biographical introduction to the way cosmology is being done today, with a focus on Allan Sandage.
Pagels, H. *The Cosmic Code*. 1982, Bantam.
Peratt, A. "Plasma Cosmology" in *Sky & Telescope*, Feb. 1992, p. 136.
Rothman, T. "Is Cosmology a Sometime Thing" in *Astronomy*, Jul. 1991, p. 38.
Rothman, T. "This is the Way the World Ends" in *Discover*, July 1987, p. 82. On the future of the universe in the different models.
Rothman, T. "The Seven Arrows of Time" in *Discover*, Feb. 1987, p. 62.
Sandage, A. "Cosmology: The Quest to Understand the Creation and Expansion of the Universe" in *The Universe*, B. Preiss & A. Fraknoi, eds. 1987, Bantam.
Scherrer, R. "From the Cradle of Creation" in *Astronomy*, Feb. 1988, p. 40.
Schramm, D. "The Origin of Cosmic Structure" in *Sky & Telescopes*, Aug. 1991, p. 140.
Shipman, H. *Black Holes, Quasars, and the Universe*, 2nd ed. 1980, Houghton-Mifflin.
Shu, F. "The Expanding Universe and the Large-scale Geometry of Space-time" in *Mercury*, Nov/Dec. 1983, p. 162.
Silk, J. *The Big Bang*, 2nd ed. 1989, Freeman. Slightly advanced, but an excellent resource.
Trefil, J. *The Dark Side of the Universe*. 1988, Scribners. A quick guide to cosmology, with emphasis on dark matter.
Wagoner, R. & Goldsmith, D. *Cosmic Horizons: Understanding the Universe*. 1983, W. H. Freeman.
Will, C. *Was Einstein Right? Putting General Relativity to the Test*. 1986, Basic Books.

Reading Material for the Instructor

Note: Many of the reading materials for chapter 28 include topics covered in chapter 29 and vica versa. Each reference is only listed once, and it is probably best to consider these two chapters together when recommending reading material or a term paper topic.

Berger, A., ed. *The Big Bang and Georges Lemaitre*. 1984, Reidel.
Bernstein, J., et al., eds. *Cosmological Constants*. 1986, Columbia U. Press. A reprint of the major papers in cosmology.

Boesgaard, A. "Big Bang Nucleosynthesis: Theories and Observations" in *Annual Reviews of Astron. and Astrophys.*, vol 23, p. 319 (1985).
Boslaugh, J. *Masters of Time: Cosmology at the End of Innocence.* 1992. Addison-Wesley.
Contopoulos, G., et al., eds. *Cosmology: The Structure and Evolution of the Universe.* 1987, Springer-Verlag.
Freedman, W. "The Expansion Rate of the Universe" in *Scientific American*, Nov. 1992, p. 54.
Gott, J., et al. "Will the Universe Expand Forever?" in *Scientific American*, Mar. 1976.
Harrison, E. "The Dark Night Sky Riddle" in *Science*, Nov. 23, 1984, p. 941.
Hewitt, A., et al., eds. *Observational Cosmology.* 1987, Reidel.
Muller, R. "The Cosmic Background Radiation and the New Aether Drift" in *Scientific American*, May 1978.
Sandage, A. "Observational Tests of World Models" in *Annual Reviews of Astron. & Astrophys.*, vol. 26, p. 561 (1988).
Webster, A. "The Cosmic Background Radiation" in *Scientific American*, Aug. 1974.
Wilkinson, D. "Anisotropy of the Cosmic Blackbody Radiation" in *Science*, June 20, 1986, p. 151.

Recent Audio-Visual Aids

A Brief History of Time (1992, 84 min., a documentary look at the life and theories of Stephen Hawking, available from Facets video).
The Creation of the Universe (90 min; PBS television special on the Big Bang, narrated by Timothy Ferris; Astronomical Society of the Pacific)
Coming of Age in the Milky Way (audiotape from the book by Timothy Ferris; 2 hrs. 48 min., Astronomical Society of the Pacific) — includes a section on modern cosmology
Cosmic Background Explorer (slide set from the Astronomical Society of the Pacific) Includes images of and from the COBE spacecraft.

Topics for Discussion and Papers

1. Olbers' Paradox

Olbers' Paradox (which, as the book mentions, has a long history) is a fascinating topic for further analysis in class discussion or a paper. Beginning students can look at the history of the problem, while more advanced students can explore (especially through the writings of Edward Harrison) the physics of how the paradox can be resolved. See:

Harrison, E. "The Paradox of Olbers' Paradox" in *Mercury*, Jul/Aug. 1980, p. 83.
Harrison, E. *Darkness at Night*. 1987, Harvard U. Press. The definitive book on the paradox.
Jaki, S. *The Paradox of Olbers' Paradox: A Case History of Scientific Thought*. 1969, Herder and Herder.
Wesson, P. "Olbers Paradox Solved At Last" in *Sky & Telescope*, June 1989, p. 594.

2. The Discovery of the Background Radiation

The detection of the cosmic microwave background by two Bell Labs scientists working on something completely different is one of the most interesting examples of serendipity in science. Students who would like to investigate the story in more detail can consult:

Bernstein, J. *Three Degrees Above Zero*. 1984, Scribner's. A history of Bell Labs, including a section on Penzias & Wilson's work.
Ferris, T. "The Radio Sky and the Echo of Creation" in *Mercury*, Jan/Feb. 1984, p. 2 (see also his book *The Red Limit*.)
Gribbin, J. *In Search of the Big Bang*. 1986, Bantam. See chapter 5.
Peebles, P. & Wilkinson, D. "The Primeval Fireball" in *Scientific American*, Jan. 1967.

3. The COBE Satellite

As this book goes to press, the Cosmic Background Explorer satellite continues to send back information about the background radiation and the microwave sky. For more on COBE, see:

Bauer, D. "COBE's Quest" in *Astronomy*, Aug. 1986, p. 74.
Fienberg, R. "COBE Confronts the Big Bang" in *Sky & Telescope*, July 1992, p. 34. Good summary of the temperature fluctuations discovery.
Kanipe, J. "Too Smooth: COBE's Perfect Universe" in *Astronomy*, June 1990, p. 20.
Silk, J. "Probing the Primeval Fireball" in *Sky & Telescope*, June 1990, p. 600.

Compare COBE's results to terrestrial observations, summarized in:

Smith, D. "Cosmic Fire, Terrestrial Ice" in *Sky & Telescope*, Nov. 1989, p. 471.

4. The Ultimate Fate of the Universe

Few discussions in astronomy draw out students more than considering the ultimate fate of the universe in the various models. Whether it is the Big Crunch or the cold dark universe that continues to expand forever, students are upset to find how little comfort the far future gives humanity and how little our fate seems to matter on the cosmic stage. These ideas can lead to good class discussion and can inspire heartfelt papers by students with a philosophical bent. Some references:

Darling, D. "Deep Time: The Fate of the Universe" in *Astronomy*, Jan. 1986, p. 6.
Dicus, D., et al. "The Future of the Universe" in *Scientific American*, March 1983.
Islam, J. "The Ultimate Fate of the Universe" in *Sky & Telescope*, Jan. 1979, p. 13.
Islam, J. *The Ultimate Fate of the Universe*. 1983, Cambridge U. Press.
Lawrence, J. "The Future History of the Universe" in *Mercury*, Nov/Dec. 1978, p. 132.
Rothman, T. "This is the Way the World Ends" in *Discover*, July 1987, p. 82. On the future of the universe in the different models.

5. The Evaporation of Black Holes

The work of Stephen Hawking and others on the evaporation of black holes and on primordial black holes that could be evaporating now has received quite a bit of media attention, especially after Hawking's popular book became a surprise best seller. Director Stephen Spielberg is even making a feature film about Hawking and his work. For students who want to follow up on the scientific ideas in this area, the following references are good places to start:

Boslaugh, S. *Stephen Hawking's Universe*. 1984, Morrow.
Boslaugh, J. "The Unfettered Mind: Stephen Hawking" in *Science '81*, Nov. 1981, p. 66.
Hawking, S. *A Brief History of Time*. 1988, Bantam.
Hawking, S. "The Quantum Mechanics of Black Holes" in *Scientific American*, Jan. 1977.
Kaufmann, W. "An Interview with Stephen Hawking" in *Mercury*, Nov/Dec. 1975, p. 13.
Kaufmann, W. *Black Holes and Warped Spacetime*. 1979, Freeman.
Kaufmann, W. "Primordial Black Holes" in *Mercury*, Jan/Feb. 1980, p. 1.
Porter, N. & Weekes, T. "A Search for Exploding Black Holes" in *Sky & Telescope*, Feb. 1978, p. 113.

See topic 24.3 for black holes in science fiction stories.

Chapter 29
Exploring the Early Universe

Reading Material for the Student

Note: Many of the reading materials for Chapter 29 include topics covered in Chapter 28 and vica versa. Each reference is only listed once, and it is probably best to consider these two chapters together when recommending reading material or a term paper topic.

Barrow, J. *Theories of Everything*. 1991. Oxford U. Press.
Bartusiak, M. "The Cosmic Burp: Genesis of the Inflationary Universe Hypothesis" in *Mercury*, Mar/Apr. 1987, p. 34.
Bartusiak, M. "Before the Big Bang: The Big Foam" in *Discover*, Sep. 1987, p. 76.
Bartusiak, M. "Wanted: Dark Matter" in *Discover*, Dec. 1988, p. 62.
Burns, J. "Dark Matter in the Universe" in *Sky & Telescope*, Nov. 1984, p. 396.
Cornell, J., ed. *Bubbles, Voids, and Bumps in Time*. 1988, Cambridge U. Press.
Crease, R. & Mann, C. *The Second Creation: Makers of the Revolution in 20th Century Physics*. 1986, Macmillan. A widely praised history.
Davies, P. *The Forces of Nature*, 2nd ed. 1986, Cambridge U. Press.
Davies, P. "Particle Physics for Everybody" in *Sky & Telescope*, Dec. 1987, p. 582.
Davies, P. "The New Physics and the Big Bang" in *Sky & Telescope*, Nov. 1985, p. 406.
Davies, P. "Relics of Creation" in *Sky & Telescope*, Feb. 1985, p. 112. On cosmic strings and more.
Davies, P. & Gribbin, J. *The Matter Myth*. 1992. Simon & Schuster.
Disney, M. *The Hidden Universe*. 1984, Macmillan. On the "missing" mass.
Finkbeiner, A. "Fossils of Something Interesting" in *Astronomy*, Nov. 1984, p. 18.
Freedman, D. "Maker of Worlds: Profile of [Cosmologist] Sidney Coleman" in *Discover, July 1990, p. 46.*
Geller, M. "Mapping the Universe: Slices and Bubbles" in *Mercury*, May/June 1990, p. 66.
Greenstein, G. "Through the Looking Glass" in *Astronomy*, Oct. 1989, p. 20. On inflation, GUT's, other universes.
Gregory, S. "The Structure of the Visible Universe" in *Astronomy*, Apr. 1988, p. 42.
Krauss, L. *The Fifth Essence: Dark Matter in the Universe*. 1989, Basic Books.
Lederman, L. & Schramm, D. *From Quarks to the Cosmos*. 1989, W. H. Freeman.

Lightman, A. & Brawer, R. *Origins: The Lives and Worlds of Modern Cosmologists.* 1990, Harvard U. Press. Includes interviews with many of the people mentioned in this chapter.
Odenwald, S. "The Planck Era" in *Astronomy*, Mar. 1984, p. 66.
Odenwald, S. "To the Big Bang and Beyond" in *Astronomy*, May 1987, p. 90.
Odenwald, S. "Does Space Have More than 3 Dimensions?" in *Astronomy*, Nov. 1984, p. 66.
Overbye, D. "The Shadow Universe" in *Discover*, May 1985, p. 12.
Parker, B. "The Mystery of the Missing Mass" in *Astronomy*, Nov. 1984, p. 9.
Parker, B. *Search for a Supertheory: From Atoms to Superstrings.* 1987, Plenum. Particle physics and a bit on cosmology.
Riordan, M. & Schramm, D. *The Shadows of Creation: Dark Matter and the Structure of the Universe.* 1991, W. H. Freeman.
Rothman, T. & Ellis, G. "Has Cosmology Become Metaphysical?" in *Astronomy*, Feb. 1987, p. 6.
Sadoulet, B. & Cronin, J. "Subatomic Astronomy" in *Sky & Telescope*, Jan. 1992, p. 25.
Taubes, G. "Everything's Now Tied into Strings" in *Discover*, Nov. 1986, p. 34.
Taubes, G. "The Ultimate Theory of Everything" in *Discover*, Apr. 1985, p. 52.
Thomsen, D. "In the Beginning was Quantum Mechanics" in *Science News*, May 30, 1987, p. 346.
Thomsen, D. "The Quantum Universe: A Zero-Point Fluctuation" in *Science News*, Aug. 3, 1985, p. 72.
Trefil, J. *The Moment of Creation.* 1983, Macmillan.
Trefil, J. "The New Physics and the Universe" in *The Universe*, B. Preiss & A. Fraknoi, ed. 1987, Bantam.
Wagoner, R. & Goldsmith, D. "Quarks, Leptons, and Bosons: A Particle Physics Primer" in *Mercury*, Jul/Aug. 1983, p. 98.
Waldrop, M. "New Light on Dark Matter" in *Science*, June 1, 1984, p. 971; "Why Do Galaxies Exist?" in *Science*, May 24, 1985, p. 978.
Waldrop, M. "The Quantum Wave Function of the Universe" in *Science*, Dec. 2, 1988, p. 124. On Hawking's work and quantum cosmology.
Weinberg, S. *The First Three Minutes*, 2nd ed. 1988, Basic Books. Warning: Only an afterword makes this book different from the 1977 edition.

Reading Material for the Instructor

Note: Many of the reading materials for chapter 29 include topics covered in chapter 28 and vica versa. Each reference is only listed once, and it is probably best to consider these two chapters together when recommending reading material or a term paper topic.

Abbott, L., et al., eds. *Inflationary Cosmology.* 1986, World Scientific Publishing.
Burns, J. "Very Large Structures in the Universe" in *Scientific American*, July 1986.
Carrigan, R. & Trower, W. *Particle Physics in the Cosmos* and *Particles and Forces: At the Heart of Matter.* 1989, Freeman. Collections of articles from *Scientific American.*
Dressler, A. "The Large-scale Streaming of Galaxies" in *Scientific American*, Sep. 1987.
Freedman, D. & van Hieuwenhuizen, P. "The Hidden Dimensions of Spacetime" in *Scientific American*, Mar. 1985.
Gibbons, G., et al., eds. *Supersymmetry and Its Applications.* 1986, Cambridge U. Press.
Haber, H. & Kane, G. "Is Nature Supersymmetric?" in *Scientific American*, June 1986.
Kolb, E., et al., eds. *Inner Space Outer Space.* 1986, U. of Chicago Press. Proceedings of a symposium on particle physics and cosmology.
Kormendy, J., et al., eds. *Dark Matter in the Universe.* 1987, Reidel.
Krauss, L. "Dark Matter in the Universe" in *Scientific American*, Dec. 1986.
Lake, G. "Windows on a New Cosmology" in *Science*, May 18, 1984, p. 675.
Narlikar, J., et al., eds. *Gravity, Gauge Theories, and Quantum Cosmology.* 1986, Reidel.
Ponman, T. "Hot Gas and Dark Matter in a Compact Galaxy Group" in *Nature*, May 6, 1993, p. 51.
Primack, J., et al. "Detection of Cosmic Dark Matter" in *Annual Reviews of Nuclear and Particle Science*, vol. 38 (1988).
Schramm, D. & Steigman, G. "Particle Accelerators Test Cosmological Theory" in *Scientific American*, June 1988.
Schwarzschield, B. "Gigantic Structures Challenge Standard Views of Cosmic Evolution" in *Physics Today*, June 1990, p. 20.
Setti, G., et al., eds. *Cosmology, Astronomy, and Fundamental Physics.* 1986, European Southern Observatory.
Silk, J., et al. "The Large-scale Structure of the Universe" in *Scientific American*, Oct. 1983.
Trimble, V. "Existence and Nature of Dark Matter in the Universe" in *Annual Review of Astronomy and Astrophysics*, vol. 25, p. 425 (1987).

Recent Audio-Visual Aids

What Einstein Never Knew (58 min. episode of the *Nova* TV series on GUT's, supersymmetry, Kaluza-Klein theory, etc.; Coronet Films)

Topics for Discussion and Papers

1. Antimatter and the Cosmos

Students are fascinated by the role of antimatter in the universe—both in the early moments of the big bang and now. The text discusses symmetry breaking and the nonequivalence of matter and antimatter that led to the universe we see now. Students who want to follow up on the ideas relating to antimatter (and even investigate proposals for antimatter technology—in the works of Robert Forward) can see:

Adair, R. "A Flaw in the Universal Mirror" in *Scientific American*, Feb. 1988.
Davies, P. "Matter-Antimatter" in *Sky & Telescope*, Mar. 1990, p. 257.
Forward, R. & Davis, J. *Mirror Matter: Pioneering Antimatter Physics*. 1988, J. Wiley.
Goldman, T., et al. "Gravity and Antimatter" in *Scientific American*, Mar. 1988.
Guillen, M. "The Paradox of Antimatter" in *Science Digest*, Feb. 1985, p. 32.
Trefil, J. "Matter Versus Antimatter" in *Science '81*, Mar. 1981, p. 66.
Wilczek, F. "The Cosmic Asymmetry of Matter and Antimatter" in *Scientific American*, Dec. 1980.

2. Cosmic Strings

Section 29-6 introduces cosmic strings, places of unbroken symmetry that could serve as "seeds" for structure in the universe. It is important to note that these cosmic strings are quite a different concept from "superstrings," a currently popular concept in particle physics in which fundamental particles may be made of tiny one-dimensional loops.

For more on cosmic strings, students can see:

Bartusiak, M. "Cosmic Strings" in *Discover*, Apr. 1988, p. 60.
Davies, P. "Relics of Creation" in *Sky & Telescope*, Feb. 1985, p. 112.
Gribbin, J. & Rees, M. "Cosmic Strings" in *Cosmic Coincidences*. 1989, Bantam.
Vilenkin, A. "Cosmic Strings" in *Scientific American*, Dec. 1987.

3. The Anthropic Principle

Students with an interest in philosophy and its connections with science may want to investigate a fascinating idea which has become known as the *anthropic principle*. It asks whethers there is a connection between the fact that we are here thinking about the universe and the laws of physics which lie at the heart of the cosmos. The fact that we are here means that the universe had to have a set of rules that make life-forms like ours possible. Could there have been other universes with different laws that would not have given rise to intelligent creatures? Is there more we can conclude about the universe from our presence?

For more on the various versions and implications of the principle, students can see:

Davies, P. "The Anthropic Principle and the Early Universe" in *Mercury*, May/June 1981, p. 66.
Finkbeiner, A. "A Universe in Our Own Image" in *Sky & Telescope*, Aug. 1984, p. 107.
Gale, G. "The Antrhopic Principle" in *Scientific American*, Dec. 1981.
Greenstein, G. *The Symbiotic Universe*. 1988, Morrow.
Gribbin, J. & Rees, M. *Cosmic Coincidences: Dark Matter, Mankind, and Anthropic Cosmology*. 1989, Bantam.
Robson, J., ed. *Origin and Evolution of the Universe: Evidence for Design*. 1987, McGill-Queen's Univ. Press. Proceedings of a conference on science, religion, and philosophy.

A technical but comprehensive book about this subject is *The Anthropic Cosmological Principle* by J. Barrow and F. Tipler (1986, Oxford U. Press.)

4. The Inflationary Universe

Section 29-4 introduces the idea of an inflationary epoch in the history of the universe and its connection to some of the large-scale properties we observe in the cosmos. The inflationary model has taken the world of cosmologists by storm in the last decade; students could investigate both the science of the inflationary epoch and the sociology of the accepatance of a new idea which solves a long-standing problem or issue in science. See:

Bartusiak, M. "The Cosmic Burp: The Genesis of the Inflationary Universe Hypothesis" in *Mercury*, Mar/Apr. 1987, p. 34.
Guth, A. "The Inflationary Universe " in Cornell, J., ed. *Bubbles, Voids, and Bumps in Time*. 1988, Cambridge U. Press.
Guth, A. & Steinhardt, P. "The Inflationary Universe" in *Scientific American*, May 1984.
Mallove, E. "The Self-Reproducing Universe" in *Sky & Telescope*, Sep. 1988, p. 253.
Overbye, D. "The Universe According to Guth" in *Discover*, June 1983, p. 92.
Smith, D. "The Inflationary Universe Lives?" in *Sky & Telescope*, Mar. 1983, p. 207.
Waldrop, M. "Before the Beginning: The Inflationary Universe" in *Science '84*, Jan/Feb. 1984, p. 44.

See also the chapter by Alan Guth in Cornell, J., ed. *Bubbles, Voids, and Bumps in Time*. 1988, Cambridge U. Press.

5. Cosmology in the Humanities

Cosmological ideas have influenced both literature and music in our day, as artists try to come to grips with the human meaning of some of the ideas discussed in the last chapters of the text. Here are a few suggestions for students who'd like to investigate the interdisciplinary effects of cosmology:

a. *Literature*

The American novelist John Updike has written a series of books about an "everyman" named Rabbit Angstrom, beginning with *Rabbit Run*. Through these books, he has peppered thoughts about cosmology and the beginning and end of the universe. Some of these ideas are worked out more explicitly in his novel *Roger's Version* in which a computer student and a divinity professor grapple with issues of cosmology and the existence of God.

The California poet Robinson Jeffers wrote eloquent and informed poetry using astronomical images, in part because his brother was an astronomer at the Lick Observatory. Such poems as "Margrave" and "The Great Explosion" incorporate cosmological questions.

Science fiction novels which consider cosmological ideas include:

Anderson, P. *Tau Zero*. 1970, Doubleday.
Asimov, I. *The Gods Themselves*. 1972, Doubleday.
Preuss, P. *Broken Symmetries*. 1983, Pocket Books. Particle physics fiction.

b. *Music*

In a fascinating piece of electronic music called *Ylem* (Deutsche Grammophone 2530442), composer Karlheinz Stockhausen incoporates ideas about the Big Bang and the oscillating universe. The players actually stand in a tight circle and play a "big bang" and then expand outward, moving into the concert hall!

AFTERWORD
The Search for Extraterrestrial Life

Afterword Summary

This brief supplement reviews information relating to the existence of life beyond the Earth. Organic molecules are identified, the Miller-Urey experiment is discussed and radio astronomy searches for evidence of other civilizations beyond the solar system are described. The Drake equation is used to estimate the number of technologically advanced civilizations in the Galaxy. Future plans to search for extraterrestrial life are noted.

Teaching Hints and Strategies

When discussing the role of carbon in forming the vast number of *organic molecules* it can be of interest to refer to the Periodic Table of the Elements (Figure 7-6). Explain how the table is arranged with each column being composed of elements having similar chemical and physical properties. Many science fiction works have suggested the existence of silicon-based life forms. Note that silicon is the element immediately below carbon in the table. You might also review the information about the abundance of silicon in the universe. As shown in Table 7-4 silicon is much less abundant in the universe. This is due to the fact that carbon is more readily generated in nuclear reactions during stellar evolution as noted in sections 21-2 and 22-3 on stellar evolution. Be sure to note also that vast amounts of the minerals in the terrestrial planets contain silicon. Sand is primarily silicon!

Students should be cautioned repeatedly when discussing the *Drake equation* that we have very little firm information about the value of *any* of the terms needed to compute N. This material can provide an opportunity to review our current state of knowledge about a wide variety of topics covered in earlier chapters. It leads naturally to discussions of the nature of the solar system and the differences between terrestrial and Jovian planets, to applications of information about stellar evolution and to analysis of our current state of information about the Galaxy and how it might vary with time due to the evolution of its stars and nebulae.

The *star formation rate* in the Galaxy is determined for our immediate location and can only be assumed to vary with time in some fashion. Stress that we have what amounts to a snapshot of the current stellar population in a small portion of the Galaxy centered on the Sun. We must try to account for the numbers of various stars based on the information we have about the evolution of stars. If we assume that stars of given masses form at some specified rate and evolve as expected, we can predict the relative numbers of stars and compare with our current stellar census. There is no guarantee that there is not more than one way to account for the current stellar population. The very great differences that exist within spiral galaxies should cause us to be very skeptical about an assumption that local star formation rates can be assumed to be the same throughout the Galaxy.

Our direct information about the *number of planets which might be suitable for life* is worthy of some discussion as well. Remind students that planets in this category are terrestrial planets and not Jovian planets. The terrestrial planets are low mass objects which make them much more difficult to detect than Jovian planets. We have not yet detected even Jovian planets around any other star with certainty. It might be wise at this point to talk briefly about the definitions of planets and stars as suggested in Chapters 7 and 19. Although we have not yet detected objects having masses like those of planets around other stars, our ability to measure proper motions and radial velocities has improved dramatically over the past twenty years and we may now be capable of such measurements. Studies designed to detect accelerations of stars produced by planets require several years of observations due to the orbital periods of such systems.

The *remaining terms in the Drake equation* are even less certain than the first two. While one might speculate that the probabilies are nearly one, is may well be that any or all of them has a near zero probability. Note that just because one can produce a logical formula for the calculation of N is no assurance that we have sufficient knowledge about the individual terms to make reliable calculations.

When discussing the funding of the projects which are designed to search for extraterrestrial life such as *Project Cyclops* and *SETI*, be sure to indicate that the instruments used for such activities often are used for other scientific research projects as well.

Reading Material for the Student

The Jul/Aug. 1992 issue of *Mercury*, June 1985 issue of *Science '85*, and the March 1983 issue of *Discover* magazines each had several articles devoted to this subject.

Beatty, J. "The New Improved SETI [Search for Extra-Terrestrial Intelligence]" in *Sky & Telescope*, May 1983, p. 411.

Bova, B. & Preiss, B., eds. *First Contact: The Search for Extraterrestrial Intelligence*. 1990, New American Library.

Cooper, H. *The Search for Life on Mars*. 1980, Holt Rinehart and Winston.

Drake F. & Sobel, D. *Is Anyone Out There: The Scientific Search for Extraterrestrial Intelligence*. 1992. Delcorte Press.

Eberhart, J. "Listening for ET" in *Science News*, May 13, 1989, p. 296.

Feinberg, G. & Shapiro, R. *Life Beyond the Earth*. 1980, Morrow. Informed speculations on what alien life might be like.

Fienberg, R. "Pulsars, Planets and Pathos" in *Sky & Telescope*, May 1992, p. 493.

Goldsmith, D., ed. *The Quest for Extraterrestrial Life*. 1980, University Science Books. An excellent collection of readings.

Goldsmith, D. & Owen, T. *The Search for Life in the Universe*. 1980, Benjamin/Cummings.

Goldsmith, D. "SETI: The Search Heats Up" in *Sky & Telescope*, Feb. 1988, p. 141. A conference report.
Horowitz, N. *To Utopia and Back: The Search for Life in the Solar System.* 1986, Freeman. Mainly on the Viking project.
Klass, P. *UFO's Explained.* 1974, Vintage. *UFO's: The Public Deceived.* 1983, Prometheus. Superb debunking of UFOs as alien visitors by a distinguished aerospace journalist.
Kuiper, T. & Brin, D., eds. *Extraterrestrial Civilizations.* 1989, American Association of Physics Teachers. A collection of articles.
Kutter, G. *The Universe and Life.* 1987, Jones & Bartlett. A textbook on the evolution of the universe and of life on Earth.
McDonough, T. *The Search for Extraterrestrial Intelligence.* 1987, John Wiley.
Olson, E. "Intelligent Life in Space" in *Astronomy*, July 1985, p. 6.
Overbye, D. "Is Anyone Out There?" in *Discover*, Nov. 1983, p. 50.
Overbye, D. "The Big Ear" in *Omni*, Dec. 1990, p. 42.
Papagiannis, M. "Bioastronomy: The Search for Extraterrestrial Life" in *Sky & Telescope*, June 1984, p. 508.
Regis, R., ed. *Extraterrestrials: Science and Alien Intelligence.* 1985, Cambridge U. Press. A series of philosophical and scientific essays.
Rood, R. & Trefil, J. *Are We Alone?* 1981, Scribners.
Sagan, C. *Cosmos.* 1980, Random House. An eloquent introduction to astronomical thought; the "Encyclopedia Galactica" chapter is a fine way to become familiar with SETI ideas.
Sagan, C., et al. *Murmurs of Earth.* 1978, Ballantine. The story of the Voyager audio and video record and an introduction to considerations of extraterrestrial communication.
Sagan, C. and Drake, F. "The Search for Extraterrestrial Intelligence" in *Scientific American*, May 1975.
Shostak, S. "The New Search for Intelligent Life" in *Mercury*, Jul/Aug. 1992, p. 114.
Swift, D. *SETI Pioneers.* 1990, U. of Arizona Press. Interviews with scientists active in the field.
Tarter, J. "Searching for Them: Interstellar Communication" in *Astronomy*, Oct. 1982, p. 10.
Tipler, F. "The Most Advanced Civilization in the Galaxy is Ours" in *Mercury*, Jan/Feb. 1982, p. 5.
Volkomir, R. "Alien Worlds: The Search Heats Up" in *Discover*, Oct. 1987, p. 66.
Whetherill, C. & Sullivan, W. "Eavesdropping on the Earth" in *Mercury*, Mar/Apr. 1979, p. 23.

The June 1985 issue of *Science '85* and the March 1983 issue of *Discover* magazines each had several articles devoted to this subject.

Reading Material for the Instructor

Baugher, J. *On Civilized Stars: The Search for Intelligent Life in Outer Space.* 1985, Prentice Hall.
Billingham, J., ed. *Life in the Universe.* 1982, MIT Press. Proceedings of a 1979 conference held at NASA Ames.
Black, D., ed. *Project Orion: A Design Study of a System for Detecting Extrasolar Planets.* 1980, NASA Special Publication #436.
Brin, D. "The Great Silence: The Controversy Concerning Extraterrestrial Intelligent Life" in *Quarterly Journal of the Royal Astronomical Society*, vol. 24, p. 283 (1980).
Crowe, M. *The Extraterrestrial Life Debate: 1750-1900.* 1986, Cambridge U. Press.
Dick, S. *Plurality of Worlds.* 1982, Cambridge U. Press. Early ideas about SETI (complements the book by Crowe, above.)
Finney, B. & Jones, E., eds. *Interstellar Migration and the Human Experience.* 1985, U. of California Press. Conference on past migrations on Earth and their relevance to space migration.
Hart, M. & Zuckerman, B., eds. *Extraterrestrials—Where Are They?* 1982, Pergamon Press.—Proceedings of a conference on "the implications of our failure to observe extraterrestrials."
Hartman, H., et al. *The Search for Universal Ancestors.* 1985, NASA SP-477.
Iversen, W. "Looking for ET" in *Supercomputing Review*, May 1991, p. 24.
Kellermann, K. & Seielstad, G., eds. *The Search for Extraterrestrial Intelligence.* 1986, National Radio Astronomy Observatory. Conference proceedings.
Marx, G., ed. *Bioastronomy.* 1988, Reidel. IAU conference proceedings.
Papagiannis, M., ed. *The Search for Extraterrestrial Life: Recent Developments.* 1986, Reidel. IAU conference proceedings.
Papagiannis, M., ed. *Strategies for the Search for Life in the Universe.* 1983, Reidel. An international conference on SETI sponsored by the International Astronomical Union.
Ponnamperuma, C., ed.*Cosmochemistry and the Origin of Life.* 1983, Reidel.
Tipler, F. "Extaterrestrial Beings Do Not Exist" in *Physics Today*, Apr. 1981, p. 9. (Responses to this provocative and controversial essay appeared in several subsequent issues, especially Apr. 1982, p. 26 ff.)
White, F. *The SETI Factor.* 1990, Walker.

Recent Audio-Visual Aids

Cosmos (1980 Public Television series, now available on home video from the Astronomical Society of the Pacific) The episodes entitled "Encyclopedia

Galactica", "One Voice in the Cosmic Fugue", and "The Persistence of Memory" all include SETI ideas.

Is Anybody Out There? (58 min. film or videotape, 1986, from the *Nova* TV series; Coronet Films) Written by Donald Goldsmith and narrated by Lili Tomlin.

Murmurs of Earth (The Voyager Spacecraft Record, 1992, Warner New Media or the Astronomical Society of the Pacific)

Quest for Contact (videotape, 1987, 32 min., Astronomical Society of the Pacific) Narrated by Jill Tarter; on the scientific ideas behind the search for life elsewhere.

The Search for Extraterrestrial Intelligence (20 slides, 1988, Astronomical Society of the Pacific) Selected by Frank Drake; comes with extensive captions and background material.

Topics for Class Discussion and Papers

1. What Do UFO's Tell Us

The science fiction writer and futurist Arthur C. Clarke once wrote that UFO's tell us nothing about intelligence in the universe at large, but do tell us how rare it is down here on Earth. Despite the claims of UFO enthusiasts over the years, not a shred of concrete evidence can be found that UFO's are extraterrestrial space ships or probes. Most UFO sightings turn out to be honest mistakes by people not familiar with the sky (Jimmy Carter once reported the planet Venus as a UFO), report of rare terrestrial phenomena (such as ball lightning or migratory swarms of luminous insects), or frauds perpetrated for the sake of publicity or book sales.

For students who want to look into the UFO debate, we recommend starting with the books of Philip Klass, a veteran aerospace writer and editor who has spent decades carefully investigating UFO reports. See:

Klass, P. *UFO's Explained* (1974, Vintage paperback); *UFO's: The Public Deceived* (1983, Prometheus Books); *UFO Abductions: A Dangerous Game* (1988, Prometheus Books).

Menzel, D. & Taves, E. *The UFO Enigma* (1977, Doubleday)—An astronomer and psychoanalyst examine many facets of the UFO phenomenon.

Oberg, J. *UFO's and Outer Space Mysteries* (1982, Donning) — A noted space science writer evaluates UFO claims and other forms of pseudoscience.

Ridpath, I. "Astronomical UFO's" in *Astronomy*, Dec. 1988, p. 114.

Sagan, C. & Page, T. *UFO's: A Scientific Debate* (1972, Norton paperback)—From a symposium held by the American Association for the Advancement of Science, presenting a wide range of views.

Shaeffer, R. *The UFO Verdict: Examining the Evidence* (1981, Prometheus

Books)—Thorough, responsible review by a respected investigator; a good source book for skeptics.

A good magazine for keeping up with skeptical news and information about UFO's is *The Skeptical Inquirer* (CSICOP, Box 229, Central Park Station, Buffalo, NY 14215.)

2. Science Fiction and Alien Life

There are countless science fiction stories involving life elsewhere in the universe. An interesting student project can be to analyze such a story based on what the student has learned in the astronomy course: how plausible are the aliens, their world, their mode of transportation, etc.

Here are a few such stories based at least in part on reasonable science:

Asimov, I. *The Gods Themselves*. 1972, Fawcett. Features imaginative aliens and "solves" the problems of quasars and the big bang.
Benford, G. *In the Ocean of Night*. 1977, Dell. Lyrical story of the confrontation of organic and machine intelligences in the universe, written by a physicist. (Sequel: *Across the Sea of Suns*.)
Brin, D. "The Crystal Spheres" in *The River of Timey*. 1987, Bantam. Intriguing story to explain why we have not yet found life elsewhere in the universe.
Brin, D. *Startide Rising*. 1983, Bantam. Interesting ideas about alien life, genetic uplifting of species & a galactic library.
Clarke, A. *2010*. 1982, Ballantine. Life on Europa and beyond.
Clement, H. "Uncommon Sense" in *Space Lash*. 1969, Dell. Life-forms with liquid metal blood which "see" by smell.
Clement, H. *Mission of Gravity*. 1954, Pyramid. Life on a massive, rapidly rotating planet. (Clement is a high school science teacher.)
Crichton, M. *The Andromeda Strain*. 1969, Dell. Doctors and scientists battle extra-terrestrial micro-organisms; by a doctor.
Forward, R. *Dragon's Egg*. 1980, Ballantine. A first novel by a physicist, featuring life on the surface of a neutron star. (Sequel: *Starquake*.)
Gunn, J. *The Listeners*. 1972, NAL. Fictional depiction of a realistic SETI program and making contact.
Hoyle, F. *The Black Cloud*. 1957, NAL. Postulates intelligence arising in giant interstellar dust clouds.
LeGuin, U. *The Left Hand of Darkness*. 1969, Ace. Thought-provoking novel of aliens who are alternately one sex then the other.
McDevitt, J. *The Hercules Text*. 1986, Ace. A flawed, but interesting new novel about extraterrestrial contact via radio messages.
Moffitt, D. *The Jupiter Theft*. 1977, Ballantine. Exciting novel with unusual alien life that steals hydrogen from gas planets.

Niven, L. *World of Ptaavs*. 1965, Ballantine. Vast realms of space and time with a rich variety of alien life forms.

Niven, L. *Protector*. 1973, Ballantine. Interesting novel about an extraterrestrial (and slightly Freudian) origin for life on Earth.

Pohl, F. *Man Plus*. 1976, Bantam. On converting humans into "aliens" who can survive on Mars unprotected.

Preiss, B., ed. *The Planets*. 1985, Bantam. A collection of science and science fiction about the solar system; several stories feature informed speculation about possible life-forms.

Sagan, C. *Contact*. 1985, Simon & Schuster. Bestselling novel about a successful SETI search and its effects. Good twist at end and careful portrayal of science throughout.

Scheckley, R. "Specialist" in *Contact*, ed. N. Keyes. 1963, Paperback Library. Life in the universe is all specialized, except on Earth.

Sheffield, C. *Between the Strokes of Night*. 1985, Baen Books. Imaginative novel proposing a life-form in inter*galactic* space.

Tiptree, J. "Love is the Plan the Plan is Death" in *The Alien Condition*, ed. S. Golden. 1973, Ballantine. A haunting story with truly alien beings.

Varley, J. *The Ophiuchi Hotline*. 1977, Dell. Rich, fascinating novel with a war between terrestrial and jovian planet life in universe.

3. Space Travel

In science fiction films or the television program *Star Trek* there never seems to be much trouble in making trips among the stars. In the real universe, the vast distances between stars and the cost of traveling at speeds fast enough to cross those distances in reasonable times make it pretty unlikely that quick jaunts to other star systems will become commonplace.

Students with an interest in space or science fiction may enjoy investigating the problems of space travel, and some ingenious (though not always practical) suggestions for overcoming them:

Calder, N. *Spaceships of the Mind*. 1978, Viking.

Chiao, R., et al, "Faster Than Light?" in *Scientific American*, Aug. 1993, p. 52.

Darling, D. "Star Trek: The Adventure Begins" in *Astronomy*, Mar. 1987, p. 94; "Star Trek: Interstellar Space Flight" in *Astronomy*, Apr. 1987, p. 94.

Dyson, F. "Interstellar Transport" in *Physics Today*, Oct. 1968, p. 41. Technical.

Forward, R. & Davis, J. *Mirror Matter*. 1988, John Wiley. On space travel using antimatter.

Friedman, L. *Starsailing: Solar Sails and Interstellar Travel*. 1988, John Wiley.

Mallove, E. & Matloff, G. *The Starflight Handbook*. 1989, John Wiley. Best popular-level introduction.

Martin, A. & Bond, A. *Project Daedelus*. 1978, British Interplanetary Society.

Patton, P. "Daedelus: Design for a Spaceship" in *Astronomy*, Oct. 1983, p. 6.

Appendix 1

Addresses of Organizations and Audio-Visual Material Suppliers

American Association of Physics Teachers, Business Office, 5112 Berwyn Rd., College Park, MD 20740. Has catalog of slides and films strips with a few about astronomy; publishes journals and sponsors conferences; has a committee on astronomy education.

American Astronomical Society, Education Office, c/o Charles Tolbert, U. of Virginia, P.O. Box 3818, University Station, Charlottesville, VA 22903. Has excellent brochure on astronomy as a career, organizes workshops, oversees national education projects.

Astronomical Society of the Pacific, 390 Ashton Ave., San Francisco, CA 94112. Has a full catalog of slides, software, videos, observing aids, etc. for teaching astronomy; sponsors conferences and workshops for teachers, distributes monthly sky calendar, publishes *Mercury* magazine, a technical journal, information packets, and a newsletter for teachers; has an astronomy news hotline. (Despite its name, the A.S.P. is a national and international organization.)

Astronomy Magazine, c/o Kalmbach Publishing, P.O. Box 1612, Waukesha, WI 53187. Publishes several magazines and sells posters, slides, and observing aids by mail.

Committee for the Scientific Investigation of Claims of the Paranormal (CSICOP), Box 229, Central Park Station, Buffalo, NY 14215. Publishes *The Skeptical Inquirer*; superb source for debunking "pseudo-science" such as astrology, UFO's, ancient astronauts, etc.

Coronet Films and Video, 108 Wilmot Rd., Deerfield, IL 60015. Major distributor of science films.

Hansen Planetarium, Publications Dept., 1098 South 200 West, Salt Lake City, UT 84101. Has catalog of slides, posters, gift items in astronomy; produces planetarium shows for national distribution.

Indiana University, Film Library, Audio-visual Center, Bloomington, IN 47405. Film rentals.

Lunar and Planetary Institute, Order Dept., 3303 NASA Road One, Houston, TX 77058. Produces several specialized slide sets on solar system phenomena.

NASA: Educational Programs, Office of Public Affairs, Code FC-9, NASA Headquarters, Washington, DC 20546. Publishes an often boring newsletter for teachers, a series of very interesting "Educational Briefs", and booklets on NASA programs. Can also refer teachers to NASA Centers around the country, each of which has programs for educators and students.

National Science Teachers' Association, 1742 Connecticut Ave., NW,

Washington, DC 20009. Has a number of journals and newsletters, conferences, plus a sometimes active branch called the Association of Astronomy Educators, which organizes sessions at NSTA meetings.

The Planetary Society, 65 N. Catalina Ave., Pasadena, CA 91106. Lobbies for more planetary exploration; publishes a newsletter and has a catalog of slides, videos, gift items.

Sky Publishing, P. O. Box 9111, Belmont MA 02178. Publishes *Sky & Telescope*, and has a catalog of atlases and other observing aids.

University Media, Suite J, 11526 Sorrento Valley Rd., San Diego, CA 92121. Distributes some educational videos and films.

Willmann-Bell Publishers, P.O. Box 35025, Richmond, VA 23235. Publishes an excellent series of books on observing, computers, and amateur astronomy.

APPENDIX 2

References on Teaching Astronomy

Acker, A. & Jaschek, C. *Astronomical Methods and Calculations*. 1986, Wiley. Brings together a wide variety of basic equations and solutions.

Astronomical Society of the Pacific: *Interdisciplinary Approaches to Astronomy*. 1988; available for a donation of $4.00 to the nonprofit Society. Reprints a number of articles from *Mercury* on astronomy and science fiction, poetry, and music, and provides a 4-page reading list on interdisciplinary resources.

Brewer, S. *Do-It-Yourself Astronomy*. 1988, Columbia U. Press. A set of traditional observing activities, using some algebra and geometry, developed in Scotland.

Fraknoi, A. "Scientific Responses to Pseudoscience Related to Astronomy: A Bibliography" in *Mercury*, Sep./Oct. 1990, p. 144.

Friedman, A., et al. *Planetarium Educator's Workshop Guide*, 2nd ed. 1990, Lawrence Hall of Science, Astronomy Education Program, U. of California, Berkeley, CA 94720. Includes both resources and activities for teachers with access to a planetarium.

Gibson, B. *The Astronomer's Sourcebook*, 1992, Woodbine House (5615 Fishers Ln, Rockville, MD 20852.) Excellent compilation of resources and how to obtain them.

Goldman, S. "Prime-time Astronomers" in *Sky & Telescope*, May 1991, p. 472. How the PBS series was made.

Grice, N. *Touch the Stars: An Introduction to Astronomy in Braille*. 1990, Boston Museum of Science. Elementary package; available from the Astronomical Society of the Pacific.

Lusis, A. *Astronomy and Astronautics: An Enthusiast's Guide to Books and Periodicals.* 1986, Facts on File. A librarian's guide to 968 books and magazines in astronomy and space science.

Makower, J., ed. *The Air And Space Catalog: The Complete Sourcebook to Everything in the Universe.* 1989, Tilden/Random House. Lists hundreds of places to get information and teaching aids.

Moeschl, R. *Exploring the Sky: 100 Projects for Beginning Astronomers.* 1989, Chicago Review Press. An activity oriented introduction, concentrating on observations.

Pasachoff, J. & Percy, J. *The Teaching of Astronomy.* 1990 Cambridge U. Press. Proceddings of an IAU Colloquium in 1988. Has an international emphasis; excellent reference.

Sunal, D. *Astronomy Education Materials Resource Guide.* Ongoing listing, includes activities; write: Dennis Sunal, Dept. of Curriculum & Instruction, West Virginia Univ., Morgantown, WV 26506.

Regular book reviews or bibliographies in astronomy appear in *Sky & Telescope, Mercury, Astronomy,* and other astronomy magazines, as well as *The Science Teacher, The Physics Teacher,* and *Science Books & Films.*

APPENDIX 3

Astronomy Lab Manuals

Bruck, M. *Exercise in Practical Astronomy Using Photographs.* 1990, Adam Hilger.

Christiansen, W. & Kaitchuk, R. & M. *Investigations in Observational Astronomy.* 1978, Paladin House.

Culver, R. *An Introduction to Experimental Astronomy.* 1984, Freeman.

Ferguson, D. *Introductory Astronomy Exercises.* 1990, Wadsworth.

Gainer, M. *Astronomy Laboratory and Observation Manual.* 1989, Prentice-Hall.

Gingerich, O., et al. *Laboratory Exercises in Astronomy* (reprints from *Sky & Telescope* magazine) [Available from Sky Publishing, P.O. Box 9111, Belmont, MA 02178.]

Hall, J. & Clouser, K. *Laboratory Astronomy: Observations and Analysis for Undergraduates.* 1992, Kendall/Hunt.

Hoff, D., et al. *Activities in Astronomy,* 3rd ed. 1992, Kendall/Hunt.

Holtzinger, J. & Seeds, M. *Laboratory Exercises in Astronomy.* 1976, Macmillan.

Icke, V. *Astronomical Experiments,* 3rd ed. 1983, Burgess.

Johnson, P. & Canterna, R. *Laboratory Experiments for Astronomy.* 1987, Saunders.

Kafatos, M. *Astronomy Laboratory Manual.* 1984, Kendall/Hunt.
Kelsey, L., et al. *Astronomy: Activities and Experiments,* 2nd ed. 1983, Kendall/Hunt.
Kleczek, J & Minnaert, M. *Exercises in Astronomy.* 1987, Reidel.
Lacy, C. *Astronomy Laboratory Exercises.* 1981, Kendall/Hunt.
Lomaga, G., et al. *Astronomy Through Practical Investigations.* 1973-1980, LSW Publications, P.O. Box 82, Mattituck, NY 11952.
Robbins, R. & Hemenway, M. *Modern Astronomy: An Activities Approach.* 1982, U. of Texas Press.
Safko, J. *Laboratory Exercises in Astronomy,* 2nd ed. 1978, Kendall/Hunt.
Safko, J. *Self-Paced Study Guide and Laboratory Exercises in Astronomy,* 6th ed. 1991, Kendall/Hunt.
Schlosser, W., et al. *Challenges of Astronomy: Hands-on Experiments for the Sky & Laboratory.* 1991, Springer-Verlag.
Waxman, J. *A Workbook for Astronomy.* 1984, Cambridge U. Press.
Wooley, J. *Voyages Through Space and Time: Projects for the "Voyager" Software.* 1992, Wadsworth.

APPENDIX 4

A Selected Bibliography for Sky Observing

1. Basic Guides: Books

Berry, R. *Discover the Stars.* 1987, Harmony/Crown. A fine introductory book by the editor of *Astronomy* magazine with clear maps and text.
Beyer, S. *The Star Guide.* 1986, Little Brown. An introduction to the 100 brightest stars and how to find them in the sky.
Chartrand, M. *Audobon Society Field Guide to the Night Sky.* 1991, Knopf. For birdwatchers who want night activities.
Chartrand, M. *Skyguide.* 1982, Golden Press. Compact, inexpensive handbook for the novice, full of useful information.
Cox, J. & Monkhouse, C. *Philip's Color Star Atlas, Epoch 2000.* 1990, Kalmbach. A good atlas for beginning stargazers.
Dickinson, T. *NightWatch: An Equinox Guide to Viewing the Universe.* 1989, Camden House. Nicely illustrated guide by a Canadian amateur astronomer and science writer.
Dickinson, T. & Dyer, A. *The Backyard Astronomer.* 1991, Camden House. A well thought out, rich guide by two experienced observers and astronomy writers.

Eicher, D. *The Universe from Your Backyard.* 1988, Cambridge U. Press. For deep-sky observers with a small telescope.

Harrington, P. *Touring the Universe Through Binoculars.* 1990, John Wiley.

Harrington, S. *Selecting Your First Telescope.* 1988, Astronomical Society of the Pacific. A pamphlet that explains the basics of telescopes and how to buy one that's right for you. (Available for a donation of $4.00)

Matloff, G. *The Urban Astronomer.* 1991, John Wiley. A manual for city dwellers who want to get to know the sky.

Mayer, B. *Starwatch.* 1984, Putnam. A novel way to learn the stars, using coathangers, cellophane, and the large maps in the book; some people love it, some people hate it. See also his *Astrowatch*, 1988, Putnam.

Menzel, D. & Pasachoff, J. *A Field Guide to the Stars and Planets*, 2nd ed. 1983, Houghton-Mifflin. One of the best pocket observing guides, with much information and detailed sky charts.

Moore, P. *Exploring the Night Sky with Binoculars.* 1986, Cambridge U. Press. A friendly introduction by a prolific British author and astronomy popularizer, with clear instructions.

Pasachoff, J. *Peterson First Guide to Astronomy.* 1988, Houghton Mifflin. A brief pocket-size primer.

Pasachoff, J. & Menzel, D. *A Field Guide to the Stars and Planets.* 1992, Houghton Mifflin Co.

Peltier, L. *Guide to the Stars: Exploring the Sky with Binoculars.* 1986, Kalmbach/Astromedia.

Rey, H. A. *The Stars: A New Way to See Them.* 1967, Houghton Mifflin. A classic guide to the constellations that introduced a simplified way to keep track of them. (First issued in 1952.)

Ridpath, I. & Tirion, W. *The Universe Guide to Stars and Planets.* 1985, Universe Books. A fine handbook for beginners with clear sky maps for both the northern and southern hemispheres.

Schaaf, F. *Wonders of the Sky.* 1983, Dover. An enthusiastic, friendly guide to naked-eye observing by an experienced amateur.

Sky Watcher's Handbook: The Expert Reference Source for the Amateur Astronomer. 1993, W. H. Freeman.

Whitney, C. *Whitney's Star Finder*, 5th ed. 1989, Random House. Excellent paperbound guide for beginners, with clear text and figures, by a Harvard astronomer.

2. Basic Guides: Not in Book Form

Astronomical Society of the Pacific: *Tapes of the Night Sky.* 2 audio cassettes that familiarize the listener with the stars and constellations of each season; comes with star maps and full transcripts.

Chandler, D. *Night Sky Star Dial.* This is perhaps the best of the many cardboard or plastic star wheels that can show you the stars visible on any night at your latitude. Two-sided to minimize distortion and inexpensive, the Chandler star-wheel is available from many science or museum stores and mail-order science suppliers.

Nightstar. A flexible plastic hemisphere that can be adjusted to show the night sky at any time from any place. Available at many science stores and through science mail order catalogs.

Sky Challenger. A series of star wheels and activities for students (and teachers) to help explore the constellations. (Write to: Discovery Corner, Lawrence Hall of Science, U. of California, Berkeley, CA 94720.)

3. More Advanced Observing Aids

Bishop, R., ed. *The Observer's Handbook*. Annual volume; Royal Astronomical Society of Canada. The standard North American reference book for keeping track of sky events; also has many useful tables of astronomical information.

Burnham, R. *Burnham's Celestial Handbook*. 1978, Dover Books. A mammoth 2,138-page (3 volume) guide to objects of all kinds that can be found with the naked eye and telescopes.

Levy, D. *The Sky: A User's Guide*. 1991, Cambridge U. Press.

Newton, J. & Teece, P. *The Guide to Amateur Astronomy*. 1988, Cambridge U. Press. Practical advice for serious observing projects.

Ottewell, G. *Astronomical Calendar*. Annual volume; Astronomy Workshop, Furman University, Greenville, SC 29613. A large-size illustrated guide to sky events and observations.

Price, F. *The Moon Observer's Handbook*. 1989, Cambridge U. Press.

Ridpath, I. *Norton's 2000.0 Star Atlas and Reference Handbook*. 1987, Crown. A guide to equipment and projects.

Schaff, F. *Seeing the Solar System*. 1991, John Wiley. Telescopic projects and observations.

Sherrod, P. *A Complete Manual to Amateur Astronomy*. 1981, Prentice Hall. A good guide for those who want to move from simple stargazing to more extensive projects in amateur astronomy.

Shields, J. *The Amateur Radio Astronomer's Handbook*. 1987, Crown. A guide to equipment and projects.

Tirion, W. *Sky Atlas 2000.0*. 1981, Sky Publ. & Cambridge U. Press. The best modern sky atlas for serious observers, includes 43,000 stars and 2,200 deep-sky objects.

Webb Society: *Deep Sky Observer's Handbooks*. 5 volumes; Enslow. A series of thorough guides to double stars, nebulae, clusters, and galaxies for the more advanced amateur astronomer.

APPENDIX 5

References on Computer Software for Teaching or Learning Astronomy

Berry, R. *Introduction to Astronomical Image Processing.* 1991, Willmann-Bell. Comes with software.

Boulet, D. *Methods of Orbit Determination for the Microcomputer.* 1991, Willmann-Bell. Includes many program listings in BASIC.

Burgess, E. *Celestial Basic: Astronomy on Your Computer.* 1982, Sybex.

Duffett-Smith, P. *Astronomy with Your Personal Computer*, 3rd ed. 1990, Cambridge U. Press.

Dukes, R. "Microcomputers in the Teaching of Astronomy" in *The Teaching of Astronomy.* J. Pasachoff & J. Percy, eds. 1990, Cambridge U. Press.

Kanipe, J., et al. "Personal Planetariums: The Night Sky in Your Computer" in *Astronomy*, Mar. 1988, p. 36.

Lawrence, J. *Introduction to Basic Astronomy with a PC.* 1989, Willmann-Bell.

Meeus, J. *Astronomical Algorithms.* 1991, Willmann-Bell. For anyone who wants to write their own programs.

Meeus, J. *Astronomical Formulae for Calculators*, 2nd ed. 1982, Willmann-Bell. An excellent compendium for those who want to write their own programs.

Mosley, J. "Exploring the Sky on Computer" in *Sky & Telescope*, June 1990, p. 650.

Mosley, J. "The CD-ROM Comes of Age" in *Sky & Telescope*, July 1992, p. 77.

Mosley, J. & Fraknoi, A. "Computer Software for Astronomy" in *Mercury*, Sep/Oct. 1986, p. 152.

Mumford, G. "Science in an Imaginary Sky" in *Sky & Telescope*, Feb. 1992, p. 146.

Vogt, G., et al. *Software for Aerospace Education.* 1989, NASA PED-106. (Published by the Educational Technology Branch, Educational Affairs Div., NASA Headquarters, Washington, DC 20546.)

John Mosley writes an excellent column of reviews of astronomy software in *Sky & Telescope* magazine.

A very interesting series of books on the use of microcomputers in advanced amateur astronomy is available from a small firm called AutoScope, P.O. Box 40488, Mesa AZ 85274. Write for a current catalog.